Geophysical Monograph Series

Geophysical Monograph Series

224 **Hydrodynamics of Time-Periodic Groundwater Flow: Diffusion Waves in Porous Media** *Joe S. Depner and Todd C. Rasmussen (Auth.)*

225 **Active Global Seismology** *Ibrahim Cemen and Yucel Yilmaz (Eds.)*

226 **Climate Extremes** *Simon Wang (Ed.)*

227 **Fault Zone Dynamic Processes** *Marion Thomas (Ed.)*

228 **Flood Damage Survey and Assessment: New Insights from Research and Practice** *Daniela Molinari, Scira Menoni, and Francesco Ballio (Eds.)*

229 **Water-Energy-Food Nexus – Principles and Practices** *P. Abdul Salam, Sangam Shrestha, Vishnu Prasad Pandey, and Anil K Anal (Eds.)*

230 **Dawn–Dusk Asymmetries in Planetary Plasma Environments** *Stein Haaland, Andrei Rounov, and Colin Forsyth (Eds.)*

231 **Bioenergy and Land Use Change** *Zhangcai Qin, Umakant Mishra, and Astley Hastings (Eds.)*

232 **Microstructural Geochronology: Planetary Records Down to Atom Scale** *Desmond Moser, Fernando Corfu, James Darling, Steven Reddy, and Kimberly Tait (Eds.)*

233 **Global Flood Hazard: Applications in Modeling, Mapping and Forecasting** *Guy Schumann, Paul D. Bates, Giuseppe T. Aronica, and Heiko Apel (Eds.)*

234 **Pre-Earthquake Processes: A Multidisciplinary Approach to Earthquake Prediction Studies** *Dimitar Ouzounov, Sergey Pulinets, Katsumi Hattori, and Patrick Taylor (Eds.)*

235 **Electric Currents in Geospace and Beyond** *Andreas Keiling, Octav Marghitu, and Michael Wheatland (Eds.)*

236 **Quantifying Uncertainty in Subsurface Systems** *Celine Scheidt, Lewis Li, and Jef Caers (Eds.)*

237 **Petroleum Engineering** *Moshood Sanni (Ed.)*

238 **Geological Carbon Storage: Subsurface Seals and Caprock Integrity** *Stephanie Vialle, Jonathan Ajo-Franklin, and J. William Carey (Eds.)*

239 **Lithospheric Discontinuities** *Huaiyu Yuan and Barbara Romanowicz (Eds.)*

240 **Chemostratigraphy Across Major Chronological Eras** *Alcides N.Sial, Claudio Gaucher, Muthuvairavasamy Ramkumar, and Valderez Pinto Ferreira (Eds.)*

241 **Mathematical Geoenergy: Discovery, Depletion, and Renewal** *Paul Pukite, Dennis Coyne, and Daniel Challou (Eds.)* 242 **Ore Deposits: Origin, Exploration, and Exploitation** *Sophie Decree and Laurence Robb (Eds.)*

243 **Kuroshio Current: Physical, Biogeochemical and Ecosystem Dynamics** *Takeyoshi Nagai, Hiroaki Saito, Koji Suzuki, and Motomitsu Takahashi (Eds.)*

244 **Geomagnetically Induced Currents from the Sun to the Power Grid** *Jennifer L. Gannon, Andrei Swidinsky, and Zhonghua Xu (Eds.)*

245 **Shale: Subsurface Science and Engineering** *Thomas Dewers, Jason Heath, and Marcelo Sánchez (Eds.)*

246 **Submarine Landslides: Subaqueous Mass Transport Deposits From Outcrops to Seismic Profiles** *Kei Ogata, Andrea Festa, and Gian Andrea Pini (Eds.)*

247 **Iceland: Tectonics, Volcanics, and Glacial Features** *Tamie J. Jovanelly*

248 **Dayside Magnetosphere Interactions** *Qiugang Zong, Philippe Escoubet, David Sibeck, Guan Le, and Hui Zhang (Eds.)*

249 **Carbon in Earth's Interior** *Craig E. Manning, Jung-Fu Lin, and Wendy L. Mao (Eds.)*

250 **Nitrogen Overload: Environmental Degradation, Ramifications, and Economic Costs** *Brian G. Katz*

251 **Biogeochemical Cycles: Ecological Drivers and Environmental Impact** *Katerina Dontsova, Zsuzsanna Balogh-Brunstad, and Gaël Le Roux (Eds.)*

252 **Seismoelectric Exploration: Theory, Experiments, and Applications** *Niels Grobbe, André Revil, Zhenya Zhu, and Evert Slob (Eds.)*

253 **El Niño Southern Oscillation in a Changing Climate** *Michael J. McPhaden, Agus Santoso, and Wenju Cai (Eds.)*

254 **Dynamic Magma Evolution** *Francesco Vetere (Ed.)*

255 **Large Igneous Provinces: A Driver of Global Environmental and Biotic Changes** *Richard. E. Ernst, Alexander J. Dickson, and Andrey Bekker (Eds.)*

256 **Coastal Ecosystems in Transition: A Comparative Analysis of the Northern Adriatic and Chesapeake Bay** *Thomas C. Malone, Alenka Malej, and Jadran Faganeli (Eds.)*

257 **Hydrogeology, Chemical Weathering, and Soil Formation** *Allen Hunt, Markus Egli, and Boris Faybishenko (Eds.)*

258 **Solar Physics and Solar Wind** *Nour E. Raouafi and Angelos Vourlidas (Eds.)*

259 **Magnetospheres in the Solar System** *Romain Maggiolo, Nicolas André, Hiroshi Hasegawa, and Daniel T. Welling (Eds.)*

260 **Ionosphere Dynamics and Applications** *Chaosong Huang and Gang Lu (Eds.)*

261 **Upper Atmosphere Dynamics and Energetics** *Wenbin Wang and Yongliang Zhang (Eds.)*

262 **Space Weather Effects and Applications** *Anthea J. Coster, Philip J. Erickson, and Louis J. Lanzerotti (Eds.)*

263 **Mantle Convection and Surface Expressions** *Hauke Marquardt, Maxim Ballmer, Sanne Cottaar, and Jasper Konter (Eds.)*

264 **Crustal Magmatic System Evolution: Anatomy, Architecture, and Physico-Chemical Processes** *Matteo Masotta, Christoph Beier, and Silvio Mollo (Eds.)*

265 **Global Drought and Flood: Observation, Modeling, and Prediction** *Huan Wu, Dennis P. Lettenmaier, Qiuhong Tang, and Philip J. Ward (Eds.)*

266 **Magma Redox Geochemistry** *Roberto Moretti and Daniel R. Neuville (Eds.)*

267 **Wetland Carbon and Environmental Management** *Ken W. Krauss, Zhiliang Zhu, and Camille L. Stagg (Eds.)*

268 **Distributed Acoustic Sensing in Geophysics: Methods and Applications** *Yingping Li, Martin Karrenbach, and Jonathan B. Ajo-Franklin (Eds.)*

269 **Congo Basin Hydrology, Climate, and Biogeochemistry: A Foundation for the Future (English version)** *Raphael M. Tshimanga, Guy D. Moukandi N'kaya, and Douglas Alsdorf (Eds.)*

269 **Hydrologie, climat et biogéochimie du bassin du Congo: une base pour l'avenir (version française)** *Raphael M. Tshimanga, Guy D. Moukandi N'kaya, et Douglas Alsdorf (Éditeurs)*

270 **Muography: Exploring Earth's Subsurface with Elementary Particles** *László Oláh, Hiroyuki K. M. Tanaka, and Dezső Varga (Eds.)*

271 **Remote Sensing of Water-Related Hazards** *Ke Zhang, Yang Hong, and Amir AghaKouchak (Eds.)*

272 **Geophysical Monitoring for Geologic Carbon Storage** *Lianjie Huang (Ed.)*

273 **Isotopic Constraints on Earth System Processes** *Kenneth W. W. Sims, Kate Maher, and Daniel P. Schrag (Eds.)*

274 **Earth Observation Applications and Global Policy Frameworks** *Argyro Kavvada, Douglas Cripe, and Lawrence Friedl (Eds.)*

275 **Threats to Springs in a Changing World: Science and Policies for Protection** *Matthew J. Currell and Brian G. Katz (Eds.)*

Geophysical Monograph 275

Threats to Springs in a Changing World
Science and Policies for Protection

Matthew J. Currell
Brian G. Katz

Editors

This Work is a co-publication of the American Geophysical Union and John Wiley and Sons, Inc.

WILEY

This edition first published 2023
© 2023 American Geophysical Union

All rights reserved. No part of this publication may be reproduced, stored in a retrieval system, or transmitted, in any form or by any means, electronic, mechanical, photocopying, recording or otherwise, except as permitted by law. Advice on how to obtain permission to reuse material from this title is available at http://www.wiley.com/go/permissions.

Published under the aegis of the AGU Publications Committee

Matthew Giampoala, Vice President, Publications
Carol Frost, Chair, Publications Committee
For details about the American Geophysical Union visit us at www.agu.org.

The right of Matthew J. Currell and Brian G. Katz to be identified as the editors of this work has been asserted in accordance with law.

Registered Office
John Wiley & Sons, Inc., 111 River Street, Hoboken, NJ 07030, USA

Editorial Office
111 River Street, Hoboken, NJ 07030, USA

For details of our global editorial offices, customer services, and more information about Wiley products visit us at www.wiley.com.

Wiley also publishes its books in a variety of electronic formats and by print-on-demand. Some content that appears in standard print versions of this book may not be available in other formats.

Limit of Liability/Disclaimer of Warranty
While the publisher and authors have used their best efforts in preparing this work, they make no representations or warranties with respect to the accuracy or completeness of the contents of this work and specifically disclaim all warranties, including without limitation any implied warranties of merchantability or fitness for a particular purpose. No warranty may be created or extended by sales representatives, written sales materials or promotional statements for this work. The fact that an organization, website, or product is referred to in this work as a citation and/or potential source of further information does not mean that the publisher and authors endorse the information or services the organization, website, or product may provide or recommendations it may make. This work is sold with the understanding that the publisher is not engaged in rendering professional services. The advice and strategies contained herein may not be suitable for your situation. You should consult with a specialist where appropriate. Further, readers should be aware that websites listed in this work may have changed or disappeared between when this work was written and when it is read. Neither the publisher nor authors shall be liable for any loss of profit or any other commercial damages, including but not limited to special, incidental, consequential, or other damages.

Library of Congress Cataloging-in-Publication Data

Names: Currell, Matthew J., editor. | Katz, Brian G., editor. | John Wiley & Sons, publisher. | American Geophysical Union, publisher.
Title: Threats to springs in a changing world : science and policies for protection / Matthew J. Currell, Brian G. Katz, editors. Other titles: Geophysical monograph series
Description: Hoboken, NJ : Wiley-American Geophysical Union, 2023. | Series: Geophysical monograph series | Includes bibliographical references and index.
Identifiers: LCCN 2022034908 (print) | LCCN 2022034909 (ebook) | ISBN 9781119818595 (cloth) | ISBN 9781119818601 (adobe pdf) | ISBN 9781119818618 (epub)
Subjects: LCSH: Springs–Pollution. | Spring–Management. | Water quality management.
Classification: LCC TD420 .T495 2023 (print) | LCC TD420 (ebook) | DDC 628.1/68–dc23/eng/20221011
LC record available at https://lccn.loc.gov/2022034908
LC ebook record available at https://lccn.loc.gov/2022034909

Cover Design: Wiley
Cover Image: Madison Blue Spring in Florida, USA. Courtesy of Brian G. Katz

Set in 10/12pt Times New Roman by Straive, Pondicherry, India

CONTENTS

List of Contributors vii

Preface xi

1 Protecting Springs in a Changing World Through Sound Science and Policy 1
   Matthew J. Currell and Brian G. Katz

Part I Threats to Springs and Their Values

2 Assessing Pollution and Depletion of Large Artesian Springs in Florida's Rapidly Developing
  Water-Rich Landscape 9
   Robert L. Knight and Angeline Meeks

3 Regional Passive Saline Encroachment in Major Springs of the Floridan Aquifer System in
  Florida (1991–2020) 19
   Rick Copeland, Gary Maddox, and Andy Woeber

4 Karst Spring Processes and Storage Implications in High Elevation, Semiarid Southwestern
  United States 35
   Keegan M. Donovan, Abraham E. Springer, Benjamin W. Tobin, and Roderic A. Parnell

5 Nitrogen Contamination and Acidification of Groundwater Due to Excessive Fertilizer Use
  for Tea Plantations 51
   Hiroyuki Ii

6 Springs of the Southwestern Great Artesian Basin, Australia: Balancing Sustainable Use
  and Cultural and Environmental Values 69
   Gavin M. Mudd and Matthew J. Currell

Part II Methods, Tools, and Techniques to Understand Spring Hydrogeology

7 Environmental Tracers to Study the Origin and Timescales of Spring Waters 87
   Axel Suckow and Christoph Gerber

8 Assessment of Water Quality and Quantity of Springs at a Pilot-Scale: Applications in
  Semiarid Mediterranean Areas in Lebanon 111
   *Joanna Doummar, Marwan Fahs, Michel Aoun, Reda Elghawi, Jihad Othman,
   Mohamad Alali, and Assaad H. Kassem*

9 Uncertainties in Understanding Groundwater Flow and Spring Functioning in Karst 131
   *Francesco Fiorillo, Mauro Pagnozzi, Rosangela Addesso, Simona Cafaro, Ilenia M. D'Angeli,
   Libera Esposito, Guido Leone, Isabella S. Liso, and Mario Parise*

10 **The Great Subterranean Spring of Minneapolis, Minnesota, USA, and the Potential Impact of Subsurface Urban Heat Islands** 145
Greg Brick

Part III Policy and Governance Approaches for the Protection of Springs

11 **Community-Based Water Resource Management: Pathway to Rural Water Security in Timor-Leste?** 157
Tanja Rosenqvist, George Goddard, Jack Nugent, Nick Brown, Eugenio Lemos, Elsa Ximenes, and Aleixo Santos

12 **Setting Benthic Algal Abundance Targets to Protect Florida Spring Ecosystems** 171
Robert A. Mattson

13 **Protecting Springs in the Southwest Great Artesian Basin, Australia** 181
Mark Keppel, Anne Jensen, Melissa Horgan, Aaron Smith, and Simone Stewart

14 **Patterns in the Occurrence of Fecal Bacterial Indicators at Public Mineral Springs of Central Victoria, 1986–2013** 199
Andrew Shugg

15 **Towards a Collective Effort to Preserve and Protect Springs** 209
Brian G. Katz and Matthew J. Currell

Index 213

LIST OF CONTRIBUTORS

Rosangela Addesso
MIDA Foundation
Pertosa, Italy

Mohamad Alali
Department of Geology
American University of Beirut
Beirut, Lebanon

Michel Aoun
Department of Geology
American University of Beirut
Beirut, Lebanon

Greg Brick
Lands and Minerals Division
Minnesota Department of Natural Resources
St. Paul, Minnesota, USA

Nick Brown
Royal Melbourne Institute of Technology
Melbourne, Australia

Simona Cafaro
MIDA Foundation
Pertosa, Italy

Rick Copeland
AquiferWatch Inc.
Tallahassee, Florida, USA

Matthew J. Currell
School of Engineering
Royal Melbourne Institute of Technology
Melbourne, Australia

Ilenia M. D'Angeli
Department of Earth and Environmental Sciences
University Aldo Moro
Bari, Italy

Keegan M. Donovan
School of Earth and Sustainability
Northern Arizona University
Flagstaff, Arizona, USA

Joanna Doummar
Department of Geology
American University of Beirut
Beirut, Lebanon

Reda Elghawi
Department of Geology
American University of Beirut
Beirut, Lebanon

Libera Esposito
Department of Science and Technology
University of Sannio
Benevento, Italy

Marwan Fahs
Institut Terre et Environnement de Strasbourg
University of Strasbourg
Strasbourg, France

Francesco Fiorillo
Department of Science and Technology
University of Sannio
Benevento, Italy

Christoph Gerber
CSIRO Land and Water
Adelaide, Australia

George Goddard
Engineers Without Borders Australia
Brisbane, Australia

Melissa Horgan
South Australian Arid Lands Landscape Board
Port Augusta, Australia

Hiroyuki Ii
Faculty of Systems Engineering
Wakayama University
Wakayama, Japan

LIST OF CONTRIBUTORS

Anne Jensen
Environmental Consultant
Adelaide, Australia

Assaad H. Kassem
Department of Geology
American University of Beirut
Beirut, Lebanon

Brian G. Katz
Environmental Consultant
Weaverville, North Carolina, USA

Mark Keppel
Department for Environment and Water
Government of South Australia
Adelaide, Australia

Robert L. Knight
Florida Springs Institute
High Springs, Florida, USA

Eugenio Lemos
Permatil
Dili, Timor-Leste

Guido Leone
Department of Science and Technology
University of Sannio
Benevento, Italy

Isabella S. Liso
Department of Earth and Environmental Sciences
University Aldo Moro
Bari, Italy

Gary Maddox
AquiferWatch Inc.
Tallahassee, Florida, USA

Robert A. Mattson
St. Johns River Water Management District
Palatka, Florida, USA

Angeline Meeks
Florida Springs Institute
High Springs, Florida, USA

Gavin M. Mudd
School of Engineering
Royal Melbourne Institute of Technology
Melbourne, Australia

Jack Nugent
Engineers Without Borders Australia
Melbourne, Australia

Jihad Othman
Department of Geology
American University of Beirut
Beirut, Lebanon

Mauro Pagnozzi
Department of Science and Technology
University of Sannio
Benevento, Italy

Mario Parise
Department of Earth and Environmental Sciences
University Aldo Moro
Bari, Italy

Roderic A. Parnell
School of Earth and Sustainability
Northern Arizona University
Flagstaff, Arizona, USA

Tanja Rosenqvist
Royal Melbourne Institute of Technology
Melbourne, Australia

Aleixo Santos
Plan International
Dili, Timor-Leste

Andrew Shugg
Federation University Australia
Ballarat, Australia

Aaron Smith
South Australian Arid Lands Landscape Board
Port Augusta, Australia

Abraham E. Springer
School of Earth and Sustainability
Northern Arizona University
Flagstaff, Arizona, USA

Simone Stewart
Department for Environment and Water
Government of South Australia
Adelaide, Australia

Axel Suckow
CSIRO Land and Water
Adelaide, Australia

Benjamin W. Tobin
Kentucky Geological Survey
University of Kentucky
Lexington, Kentucky, USA

Andy Woeber
AquiferWatch Inc.
Tallahassee, Florida, USA

Elsa Ximenes
Engineers Without Borders Australia
Dili, Timor-Leste

PREFACE

Natural springs are special places where groundwater reaches the land surface, providing a reliable source of water. Spring waters are critical to sustaining plant, animal, microbial, and human communities throughout the world. Springs hold a special place in our imaginations; in many languages and cultures, they are sites of deep spiritual significance, featuring prominently in religious ceremonies, folklore, songs, and art. Some languages describe springs as "windows" or "eyes" to the underground world and celebrate their powers of healing and regeneration. It is hard to think of a place more central to sustaining life than a permanent spring pool, particularly in the world's drylands.

It is now widely accepted that humanity is having an unprecedented impact on our planet's life-support systems. Climate change, land clearing, chemical pollution, wastewater disposal, and direct human interventions in hydrological processes (such as megadams and large-scale groundwater and surface water extraction schemes) are all dramatically altering Earth systems. As connected, integral parts of the global hydrological cycle, springs are highly vulnerable to the effects of these changes. Springs and their associated ecosystems typically depend (for their existence and healthy function) on the maintenance of aquifer water levels and quality within narrow tolerance thresholds. It can thus be argued that they are some of the most sensitive, high-value sites (systems) in danger from humanity's actions in the epoch many people now call the Anthropocene.

Threats to Springs in a Changing World: Science and Policies for Protection examines case studies from around the world where springs are facing the threat of pollution and depletion as a result of local, regional, or global-scale anthropogenic impacts. In Part I we document these threats; in Part II we provide guidance on scientific methods that can be employed to better understand spring hydrology, water quality, and vulnerability; and in Part III we examine policy and management approaches for protection of springs and the values they sustain. We hope readers find the book and its topics timely and that it inspires them to get involved in spring science, conservation work, and springs protection activities.

We dedicate this book to the countless generations of people who have cared for springs since time immemorial, so that these unique systems may survive and thrive for future generations of people, plants, and animals to enjoy. The springs we cherish today are the legacy of these peoples' selfless work, and their care for the future of the human and nonhuman world.

We would like to acknowledge and thank our peers and colleagues who contributed to the development of the ideas outlined in the book, and who helped us see and learn about springs and their incredible values over the years. These include Rod Fensham, Dongmei Han, Ian Cartwright, Clint Hansen, Angus Campbell, Peter Dahlhaus, Adrian Werner, John Webb, Derec Davies, Jim Stevenson, Wes Skiles, Richard Hicks, Stacie Greco, David Hornsby, Cynthia Barnett, Ken Ringle, Craig Pittman, and Sam Upchurch.

We sincerely thank all the peer reviewers who generously gave their time and effort to help improve the 15 chapters of the book.

A special thanks to Ritu, Keerthana, Noel, Layla, and Lesley at Wiley; Jenny and Lieke at AGU Books; and the rest of their teams who worked hard to make the book happen and keep it on track.

Finally, thank you to our dear families who inspire us and support the work we do, day after day.

Matthew J. Currell
Royal Melbourne Institute of Technology, Australia
Brian G. Katz
Environmental Consultant, USA

1

Protecting Springs in a Changing World Through Sound Science and Policy

Matthew J. Currell[1] and Brian G. Katz[2]

ABSTRACT

For many communities, springs represent, both literally and symbolically, a source of life. Springs have long been sites of spiritual significance for many of the world's Indigenous peoples and remain so today. They figure prominently in stories, songs, fables, and artworks throughout the world's cultures. Sadly, mounting evidence has emerged showing that many of the world's springs are in decline, and numerous springs have either been lost or rendered inactive. With new and emerging pressures created by global climate change and ever-increasing demands to develop water, mineral deposits, and land for economic purposes, the threat of degradation of springs is likely to intensify, leading to significant harm for many people and ecological communities. Carefully developed policies and management plans will thus be needed to safeguard springs and their incalculable values across diverse geological, climatic, and anthropogenic settings. These must be informed by high-quality scientific data collection and analysis programs, extensive community participation, and effective monitoring, reporting, and governance.

1.1. INTRODUCTION

For many communities, springs represent, both literally and symbolically, a source of life. Springs have long been sites of spiritual significance for many of the world's Indigenous peoples and remain so today. They figure prominently in countless stories, songs, fables, and artworks throughout the world's regions and cultures (Ah Chee, 2002; Palmer, 2015; Brake et al., 2019).

Springs are an essential source of water upon which many rare ecological communities depend, including endemic species that would otherwise not exist (e.g., Ponder, 2002; Fensham et al., 2010; Rossini et al., 2020). Their role as refuges for people, plants, and animals in varied and often harsh climates over geologic timescales is well recognized. This is evident in extensive geochemical and archaeological evidence (Hughes & Lampert, 1985; Cuthbert et al., 2017), high levels of endemism in spring biota and associated genetic evidence (e.g., Murphy et al., 2009; Rossini et al., 2018; Fahey et al., 2019), and the stories and songs of many Indigenous people (Ah Chee, 2002; Wangan & Jagalingou Family Council, 2015; Moggridge, 2020). Springs also provide the primary source of drinking water for millions of people worldwide (including many major cities), supply public baths and other tourism sites (e.g., geothermal springs), and provide water for agriculture and industries, such as bottled spring water, throughout the world (Kresic & Stevanovic, 2010). It is hard to overstate the immense value of spring waters to humanity and the global biosphere (Cantonati et al., 2020).

Sadly, mounting evidence has emerged over recent decades showing that many of the world's springs and spring systems are in decline, and numerous springs have either been irreparably lost or rendered inactive (Powell

[1] School of Engineering, Royal Melbourne Institute of Technology, Melbourne, Australia
[2] Environmental Consultant, Weaverville, North, Carolina, USA

et al., 2015; Powell & Fensham, 2016). Some springs that have been sites of great significance for countless generations recorded in written and oral histories are now sustained only artificially, for example, using bore water to maintain flows at the spring outlet, wetland, or pool, to prevent complete loss of their value (e.g., Ponder, 2002; Zhu et al., 2020). Damage to and loss of springs is primarily due to water extraction in the aquifers sustaining them (Knight, 2015; Powell & Fensham, 2016; Fensham et al., 2016), but also occurs due to land degradation, livestock damage, and colonization with invasive species (Brake et al., 2019; Cantonati et al., 2020). Chemical and biological pollution is another major problem, which to date has received relatively little attention in the literature. Pollution has left the waters emanating from many springs degraded or unusable for drinking, irrigation, or recreational purposes and has damaged spring-dependent ecology, including triggering toxic algal blooms (Knight, 2015; Katz, 2020).

Together, the loss of spring flows and degraded water quality result in major negative consequences for water supplies, human health, environment, and culture. Spring quality and quantity degradation most commonly relate to the following anthropogenic activities:

1. Agricultural operations (fertilized cropland and animal farming), which extract large quantities of groundwater and/or release high levels of nutrients, bacteria, and agrichemicals to the land and waterways (Katz, 2020)

2. Mining, oil, gas extraction, which dewater aquifers and discharge pollutants that migrate to springs through surface and subsurface pathways (Martin & Dowling, 2013; Llewellyn et al., 2015; Currell et al., 2017)

3. Inappropriate sanitation and other waste management systems, which contaminate groundwater and associated springs with nutrients and bacteria (Graham & Polizzotto, 2013)

The current and looming impacts of global climate change on rainfall patterns, sea levels, and evapotranspiration rates, which can impact on springs directly and indirectly (e.g., through changing recharge patterns or stimulating increased demand to extract groundwater), add further pressure, as do demands for domestic and municipal water from aquifers sustaining springs in areas of population or water demand growth.

It is impossible here to adequately summarize the cultural, ecological, and economic values that are at stake or lost when springs are damaged and/or threatened by the above processes. The Wangan and Jagalingou People, who petitioned the United Nations to intervene to protect the sacred Doongmabulla Springs (in northeast Australia) from coal mining on their country, summarized the significance of the issue as follows:

> These springs are the starting point of our life, and our dreaming totem, the Mundunjudra (also known as the Rainbow Serpent), travelled through the springs to form the shape of the land. Today, our songlines describe the path of the Mundunjudra and the shape of the land, and tell us how to move through our country …. We perform ceremonies and rituals at the springs and other sacred places, like along the Carmichael River, to obtain access to the Mundunjudra and other ancestral beings and spiritual powers …. The mine is very likely to devastate Doongmabulla Springs, which are the starting point of our life and through which our dreaming totem, the Mundunjudra, travelled to form the shape of the land. If our land and waters are destroyed, our culture will be lost, and we become nothing. Our children and grandchildren will never know their culture or who they are, and will suffer significant social, cultural, economic, environmental and spiritual damage and loss. (Wangan & Jagalingou Family Council, 2015, p. 19)

With new and emerging pressures created by global climate change and ever-increasing demands to develop water, mineral resources, and land for economic purposes, the threat of degradation of springs is likely to intensify, leading to realization of such consequences for many people and ecological communities. In line with recent calls for improved global stewardship of springs (Cantonati et al., 2020; Rossini et al., 2020), carefully developed policies and management plans will be needed to safeguard springs and their incalculable values across diverse geological, climatic, and anthropogenic settings. These policies must be informed by (1) high-quality scientific data collection and analysis programs; (2) extensive community consultation and participation; and (3) effective monitoring, reporting, and governance mechanisms.

The aim of this volume is to provide case studies and guidance toward these goals, helping practitioners, policy makers, scientists, and the public to work together (and advocate) to better preserve, protect, and/or enhance springs and the many unique values associated with them. The volume is structured into three major parts, designed to give readers overviews of key topics and examples from around the world. The major contributions in each section are briefly summarized below.

1.2. THREATS TO SPRINGS AND THEIR VALUES

Part I, "Threats to Springs and Their Values," explores and examines causes of their degradation in a variety of contexts. In Chapter 2, Robert L. Knight and Angeline Meeks document the causes of declining water quality in the springs of Florida, the world's largest concentration of artesian karst springs. They show how integrated analysis of land and water-use data in a GIS model can help identify the dominant source(s) of pollution to springs and uncover links between groundwater extraction and water quality degradation. Their Blue Water Audit tool and associated resources have spread awareness of threats to the springs, educating many people about these. In Chapter 3, also focused on Florida, Rick Copeland and coauthors describe how climatic drivers have contributed to the salinization of Florida's spring waters, through passive encroachment of saltwater under conditions of reduced rainfall and rising sea level. Their study serves as a warning of what may come with further unabated climate change in coastal aquifer systems, which sustain springs and water supplies around the world. Moving to the southwest United States in Chapter 4, Keegan Donovan and coauthors use coupled hydrograph and stable isotopic analysis to demonstrate the threat posed by climate change to ecologically significant springs of the Colorado Plateau in Arizona. They show how the timing and duration of seasonal snowpack melt (under threat due to climate change) is crucial for buffering spring flows against seasonal precipitation fluctuations. In the process, they illustrate the value of stable isotopes in hydrograph analysis, and the importance of high-resolution climate and spring-flow monitoring in settings containing vulnerable spring ecosystems.

In Chapter 5, Hiroyuki Ii reviews the impacts of tea plantation fertilization on spring, groundwater, and surface water quality in Japan, showing how reducing fertilizer application rates can reduce spring nutrient pollution, up to a point. In Chapter 6, Gavin M. Mudd and Matthew J. Currell present an analysis of the effects of artesian groundwater extraction for Australia's largest mining operation on the culturally and ecologically significant Mound Springs of the southwestern Great Artesian Basin (Australia's largest interconnected aquifer system). They combine spring-flow and artesian bore water level measurements to illustrate the effects of mine water extraction on different springs in the unique Kati-Thanda complex.

1.3. METHODS, TOOLS, AND TECHNIQUES TO UNDERSTAND SPRING HYDROGEOLOGY

Part II, "Methods, Tools, and Techniques to Understand Spring Hydrogeology," explores methods for understanding spring hydrogeology (including the sources and timescales of water flows) and determining the causes of pollution and associated dynamics. In Chapter 7, Axel Suckow and Christoph Gerber provide an in-depth review of the use of environmental tracers to study the origins and timescales of spring water and solute flows, with a particular focus on radioactive isotopes. They explain the complexities involved in the analysis and interpretation of such tracer data and show the value of multitracer "snapshot" sampling and time-series isotope records for understanding the age profile of spring waters. They utilize an unusually long time-series record of tritium measurements at the Fischa-Dagnitz spring to demonstrate these concepts.

In Chapter 8, Joanna Doummar and coauthors demonstrate the integrated use of high-resolution climatic and spring monitoring data, along with artificial tracer tests and targeted analysis of micropollutants, to provide insights into recharge, flow, and pollution sources in critically important water-supply springs of Lebanon. Their study shows how these techniques can be combined to improve conceptual hydrogeological models (e.g., defining fast and slow flow pathways) and understand pollution sources and transport, including estimation of key solute transport parameters, for use in modeling studies.

In Chapter 9, Francesco Fiorillo and coauthors demonstrate the value of direct geological investigation for springs within karst terrains focusing on the Alburni Massif in Italy. Their chapter shows how cave mapping and exploration (in conjunction with other hydrogeological investigation techniques) are vitally important for constraining drainage and flow patterns, which in turn can be applied in the estimation of recharge to springs across karst terrains (and specific zones therein). Further illustrating the importance and value of cave exploration in spring vulnerability studies, in Chapter 10 Greg Brick provides an example of how direct observations of geology, and measurement of water temperatures and flow rates, can identify the causes and extent of anthropogenically driven thermal anomalies focusing on a subterranean spring in the midwestern United States.

1.4. POLICY AND GOVERNANCE APPROACHES FOR THE PROTECTION OF SPRINGS

Part III, "Policy and Governance Approaches for the Protection of Springs," examines different approaches to management of springs, their water quality, and associated values. In Chapter 11, Tanja Rosenqvist and coauthors investigate the strengths and limitations of community-based water resources management (CBWRM) for springs that are vital drinking water sources in rural communities of Timor-Leste. Maintaining water security in the face of natural and anthropogenic pressures on these springs presents significant challenges for these communities. Through field research and interviews, they show that there is a high willingness among some community members to adopt CBWRM. However, the resilience of water supply systems was not guaranteed under this model, as revealed when analyzed through a socioecological-technical (SETS) lens. The authors provide recommendations to strengthen institutional and governance arrangements to better define and support community roles and responsibilities.

Returning to Florida in Chapter 12, Robert A. Mattson explains the use of benthic algal targets as a management strategy for the protection of spring-fed streams, which have been experiencing ecological degradation due to increased algal abundance. Based on a comparison of targets proposed in the literature with data on algal cover, dry weight, and chlorophyll *a* from springs in north and central Florida, he finds that such targets have promise as management tools for preserving values associated with the springs, albeit with further detailed research required to determine the most appropriate targets for specific areas and springs. In Chapter 13, Mark Keppel and colleagues present an overview of the scientific evidence-based policies that are being used to manage the springs of the southwest Great Artesian Basin in northern South Australia (including those at Kati-Thanda examined by Mudd and Currell earlier) and the outstanding ecological and cultural values associated with these. The primary policy tools include the licensing and capping of groundwater extractions, with the goal of maintaining artesian pressures near springs, and the application of a new springs monitoring risk assessment and adaptive management plan (encompassing measures to limit disturbance from surface activities). These are both considered critical for guarding against ongoing and future threats to these remarkable springs, and to build on the success of a recent bore capping and infrastructure upgrade program.

Finally, in Chapter 14, Andrew Shugg analyzes the success of different management strategies (including various remediation measures) implemented over a period of decades and designed to address recurring bacterial contamination of cold-water mineral springs in central Victoria, Australia. Through careful analysis of a 30 year time series of bacterial monitoring data and the hydrogeological regime, the dynamics of the contamination problem and limits to successful remediation are identified.

Together, the contributions in each section give readers an up-to-date picture of the threats faced by springs across diverse settings. Relevant information is provided on methods for better understanding the nature and extent of such threats. Furthermore, these chapters describe management tools and approaches that are being used and/or developed in response to the strong desire of many people worldwide to safeguard springs and their immense value into the future.

REFERENCES

Ah Chee, D. (2002). Kwatye: Indigenous peoples connection with kwatye (water) in the Great Artesian Basin. *Environment SA, 9*, 20–25.

Brake, L., Harris, C., Jensen, A., Keppel, M., Lewis, M., & Lewis, S. (2019). *Great Artesian Basin Springs, a plan for the future: Evidence-based methodologies for managing risks to spring values*. Report prepared for Australian Government Department of Agriculture.

Cantonati, M., Fensham, R. J., Stevens, L. E., Gerecke, R., Glazier, D. S., Goldscheider, N., et al. (2020). Urgent plea for global protection of springs. *Conservation Biology*. https://doi.org/10.1111/cobi.13576

Currell, M. J., Werner, A. D., McGrath, C., Webb, J. A., & Berkman, M. (2017). Problems with the application of hydrogeological science to regulation of Australian mining projects: Carmichael mine and Doongmabulla Springs. *Journal of Hydrology, 548*, 674–682.

Cuthbert, M. O., Gleeson, T., Reynolds, S. C., Bennett, M. R., Newton, A. C., McCormack, C. J., et al. (2017). Modeling the role of groundwater hydro-refugia in East African hominin evolution and dispersal. *Nature Communications, 8*, 15696.

Fahey, P. S., Fensham, R. J., Laffineur, B., & Cook, L. (2019). *Chloris circumfontinalis* (Poaceae): A recently discovered species from the saline scalds surrounding artesian springs in north-eastern Australia. *Australian Systematic Botany, 32*, 228–242.

Fensham R. J., Ponder, W. F., & Fairfax, R. J. (2010). *Recovery plan for the community of native species dependent on natural discharge of groundwater from the Great Artesian Basin*. Report to Department of the Environment, Water, Heritage and the Arts, Canberra.

Fensham, R. J., Silcock, J. L., Powell, O., & Haberhehl, M. A. (2016). In search of lost springs: A protocol for locating active and inactive springs. *Groundwater, 54*, 374–383.

Graham, J. P., & Polizzotto, M. L. (2013). Pit latrines and their impact on groundwater quality: A systematic review. *Environmental Health Perspectives, 121*, 521–530.

Hughes, P. J., & Lampert, R. J. (1985). *The assessment of Aboriginal archaeological significance of mound springs in South Australia*. Kinhill Stearns report for SA Department of Planning and Environment, Adelaide.

Katz, B. G. (2020). Nitrate contamination in springs. In *Nitrogen overload: Environmental degradation, ramifications and economic costs*. American Geophysical Union Geophysical Monograph 250. Hoboken, NJ.: John Wiley & Sons, Inc.

Knight, R. L. (2015). *Silenced springs: From tragedy to hope*. High Springs, FL: FSI Press.

Kresic, N., & Stevanovic, Z. (2010). *Groundwater hydrology of springs: Engineering, theory, management, and sustainability*. Oxford, United Kingdom: Butterworth Heinemann.

Llewellyn, G. T., Dorman, F., Westland, J. L., Yoxtheimer, D., Grieve, P., Sowers, T., et al. (2015). Evaluating a groundwater supply contamination incident attributed to Marcellus Shale gas development. *Proceedings of the National Academy of Sciences USA, 112*, 6325–6330.

Martin, R., & Dowling, K. (2013). Trace metal contamination of mineral spring water in an historical mining area of regional Victoria, Australia. *Journal of Asian Earth Sciences 77*, 262–267.

Moggridge, B. J. (2020). Aboriginal people and groundwater. *Proceedings of the Royal Society of Queensland, 126*, 11–27.

Murphy, N. P., Adams, N., & Austin, A. D. (2009). Independent colonization and extensive cryptic speciation of freshwater amphipods in the isolated groundwater springs of Australia's Great Artesian Basin. *Molecular Ecology, 18*, 109–122.

Palmer, L. (2015). *Water politics and spiritual ecology*. Routledge.

Ponder, W.F. (2002). Desert springs of the Australian Great Artesian Basin. *Proceedings of the Conference on Spring-fed Wetlands*.

Powell, O., & Fensham, R. (2016). The history and fate of the Nubian Sandstone aquifer springs in the oasis depressions of the Western Desert, Egypt. *Hydrogeology Journal*, *24*, 395–406.

Powell, O., Silcock, J., & Fensham, R. (2015). Oases to oblivion: The rapid demise of springs in the southeastern Great Artesian Basin, Australia. *Groundwater*, *53*, 171–178.

Rossini, R. A., Arthington, A. H., Jackson, S. E., Tomlinson, M., Walton, C. S., & Flook, S. C. (2020). Springs of the Great Artesian Basin: Knowledge gaps and future directions for research, management and conservation. *Proceedings of the Royal Society of Queensland*, *126*, 305–321.

Rossini, R. A., Fensham, R. J., Stewart-Koster, B., Gotch, T., & Kennard, M. J. (2018). Biogeographical patterns of endemic diversity and its conservation in Australia's artesian desert springs. *Diversity and Distributions*, *24*, 1199–1216.

Wangan and Jagalingou Family Council (2015). *Submission to the United Nations Special Rapporteur of the rights of Indigenous peoples*. October 2, 2015.

Zhu, H., Xing, L., Meng, Q., Xing, X., Peng, Y., Li, C., et al. (2020). Water recharge of Jinan Karst springs, Shandong, China. *Water*, *12*, 694.

Part I
Threats to Springs and Their Values

Part I
Threats to Springs and Spring Values

2

Assessing Pollution and Depletion of Large Artesian Springs in Florida's Rapidly Developing Water-Rich Landscape

Robert L. Knight and Angeline Meeks

ABSTRACT

The area of north and central Florida has one of the largest concentrations of large artesian springs in the world. These springs and the rivers they create, including their productive and highly adapted biota, are dependent upon flows of groundwater from the Floridan aquifer, fed by rainfall recharge over a land area of about 100,000 square miles. The Floridan aquifer is also the source of water used for drinking, irrigation, and commercial/industrial industries by more than 14 million people. Human uses of the region's groundwater are in direct conflict with spring flows resulting in long-term flow reductions for the majority of Florida's springs. In addition to this water quantity issue, human activities, including the use of nitrogen fertilizers and disposal of human and animal wastewaters, have polluted a large portion of the Floridan aquifer with nitrogen, implicated in springs eutrophication evidenced by loss of native vegetation and replacement by filamentous algae. While these impairments have been recognized for more than 20 years and are the subject of state efforts to protect spring flows and water quality, conditions are worsening in most springs. The Blue Water Audit is a GIS-based tool that quantifies the sources of water quantity and quality stresses on a parcel-by-parcel basis to facilitate the prioritization of restoration actions. Web-based dissemination of this "aquifer footprint" information is helping to highlight these issues for the public and its elected representatives.

2.1. THE ENVIRONMENTAL STATUS OF FLORIDA'S ARTESIAN SPRINGS

At 100,000 square miles, the Floridan aquifer system (FAS) is one of the largest freshwater supplies on the planet Earth, encompassing all of Florida, much of Georgia's coastal plain area, and additional areas of South Carolina, Alabama, and Mississippi. Trillions of gallons of freshwater occupy the porous limestone that compose the landward portion of the Florida Platform, an accumulation of marine sediments deposited over the past 50 million years (Bellino et al., 2018). With rising and falling sea levels, some of the limestone has been hollowed out by dissolution due to slightly acidic rainfall.

Florida Springs Institute, High Springs, Florida, USA

The combined void spaces of these eroded cavities as well as porosity of the limestone itself provide storage volume for groundwater. Coastal extensions of the platform are filled with seawater as are the deeper portions of the aquifer under much of the state (Williams & Kuniansky, 2016). But high average annual rainfall of about 50 in. each year continuously replenishes the upper volume of the aquifer and naturally overflows in the form of artesian springs across much of the aquifer's area.

2.1.1. Florida Springs Regional Occurrence and Magnitude

The majority of Florida's artesian springs are located north of I-4, from Tampa on the southwest to Orlando on the east (Fig. 2.1). The areas of Florida that provide freshwater recharge to the FAS, known as Florida's

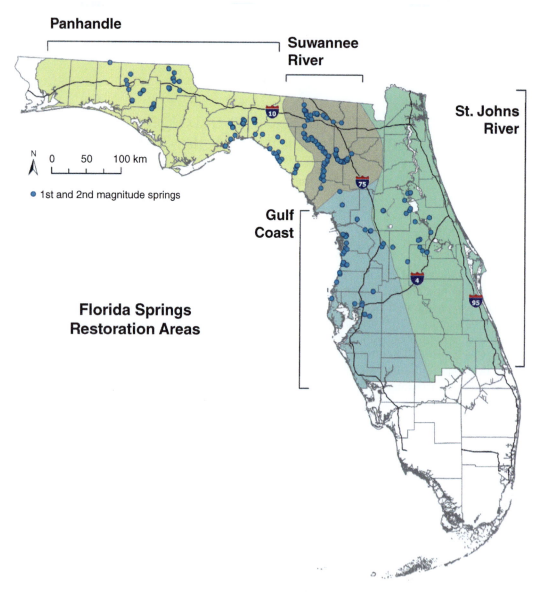

Figure 2.1 Florida Springs Region and four principal Florida Springs Restoration Areas delineated. (Adapted from FSI 2018).

"Springs Region," include about 27 million acres. The total official springs count in Florida's Springs Region is currently 1,090, possibly the largest concentration of artesian springs in the world. Dozens of additional artesian springs, also dependent upon the FAS, are located in South Carolina, Georgia, Alabama, and Mississippi (Williams & Kuniansky, 2016).

2.1.2. Summary of Florida Springs Flow and Water Quality Changes

Florida has the reputation as the state with the highest number of first magnitude springs (median annual discharge over 100 cfs) in the United States and world (Rosenau et al., 1977; Scott et al., 2002). While some of Florida's large springs are river resurgences, at least 27 first magnitude artesian springs are found in the state that predominantly discharge clear groundwater (Table 2.1). Table 2.1 includes reported historic flows from these giants as well as more recent average flows over the past 20 years. In addition to the biggest springs, there are an estimated 70 second magnitude (>10 cfs), 190 third magnitude (>1 cfs), and up to 800 fourth magnitude (>0.1 cfs) or smaller springs in Florida (USGS, 1995).

Prior to modern development, springs were the primary discharge points from the FAS (Bush & Johnston, 1988). Currently, in Florida alone, there are an additional

Table 2.1 First magnitude springs and spring groups recorded in Florida

Spring/spring group name	County	Historical average flow (cfs)	Recent average flows (cfs)
Spring Creek	Wakulla	1,610	255
Crystal River	Citrus	916	465
Silver	Marion	820	519
Rainbow	Marion	763	596
Alapaha Rise	Hamilton	608	392
St. Marks	Leon	519	550
Wakulla	Wakulla	375	461
Wacissa	Jefferson	374	343
Ichetucknee	Columbia	358	310
Holton	Hamilton	289	243
Blue	Jackson	190	108
Manatee	Levy	181	140
Kini	Wakulla	176	150
Weeki Wachee	Hernando	176	136
Homosassa	Citrus	174	93
Troy	Lafayette	166	111
River Sink	Wakulla	164	No data
Hornsby	Alachua	163	51
Blue	Volusia	160	143
Gainer	Bay	159	149
Chassahowitzka	Citrus	138	62
Falmouth	Suwannee	125	49
Blue	Madison	123	102
Silver Glen	Marion	112	99
Natural Bridge	Leon	106	73
Fanning	Levy	102	66
Alexander	Lake	100	100

Source: Historic average flows adapted from Bush and Johnston (1988); recent (2000–2020) flows from Harrington et al. (2010), Florida Springs Institute (2018), and USGS (1995).

30,000 groundwater extraction permits authorizing one or more large wells to withdraw more than 100,000 gal per day from the FAS (Knight, 2015). Many more large wells tap the FAS in Georgia as well as an estimated 1 million smaller self-supply wells in Florida and Georgia (Knight, 2015). In combination, these groundwater extractions have reduced average statewide spring flows by about 32% (Knight & Clarke, 2016).

In stark contrast to this rapid rise in human groundwater extractions, long-term average rainfall totals in Florida's Springs Region have been relatively consistent over the past 100 years (Knight & Clarke, 2016). The anticipated effects of rising air temperatures and sea levels associated with ongoing climate changes are an increase in Florida's precipitation over the long term (EPA, 2016). Current rainfall data have not confirmed this expected trend as annual rainfall totals in Florida's Springs Region continue to range from about 34 to 72 in. per year with an average in the low to mid 50s. While annual average spring flows vary in response to these wet and dry rainfall years, changes in groundwater pumping provide the strongest correlation with long-term spring flow declines (Knight & Clarke, 2016).

Based on historic data, background concentrations of nitrate nitrogen in the FAS and springs were formerly at or near the analytical detection limit of 0.05 mg/L (Harrington et al., 2010). Nitrate concentrations in Florida's groundwaters have been increasing throughout the past 100 years to current levels that are considered to be problematic for springs and downstream receiving waters (Brown et al., 2008). Figure 2.2 provides a summary of observed nitrate-nitrogen increases in 57 sentinel springs identified by the Florida Springs Institute (2018). Average groundwater nitrate concentrations throughout the Florida Springs Region are on the order of 1.0 mg/L, about 20 times the historic baseline (Knight, 2015). About 8% of Florida's springs have nitrate concentrations over 2 mg/L with a few large springs over 5 mg/L, and at least one with more than 40 mg/L in an area of high groundwater contamination by a dairy (Knight, 2015; Florida Springs Institute, 2020).

2.1.3. Observed Ecological Impairments in Florida Springs

Florida springs unimpacted by human development have highly efficient and productive assemblages of aquatic plants and animals. Florida's Silver Springs had average gross primary productivity measured as 17.5 g dry weight/m²/d (57,100 lb/acre/year) in the 1950s (Odum, 1957). Respiratory metabolism was also high, resulting in a well-balanced ecosystem with only a moderate amount of net productivity that was exported downstream. This high rate of ecosystem function was supported by a classic trophic pyramid of primary producers (submerged aquatic plants and attached algae), primary consumers (larval insects and herbivorous turtles and fish), and several levels of consumers, including fish, alligators, and water-dependent birds and mammals (Odum, 1957).

Subsequent research at Silver Springs confirmed the ecological stability of this well-developed natural ecosystem in response to declining flows and increasing nitrate pollution, but within 50 years after Odum's seminal study, it was clear that known and unknown stresses were taking a toll on the healthy balance of the aquatic ecosystem (Munch et al., 2006; Brown et al., 2010). Structural ecological changes observed at Silver Springs included a proliferation of benthic and attached filamentous algae, reduced populations of emerging aquatic insects, reduced and altered fish populations and biomass, and reduced aquatic productivity. Similar and new changes are still occurring at Silver Springs as evidenced by continuing research (Hicks & Holland, 2012; Reddy et al., 2017; Florida Springs Institute, 2019). The damming of the downstream Ocklawaha River, which receives critical tributary flows from the Silver River, not

12 THREATS TO SPRINGS IN A CHANGING WORLD

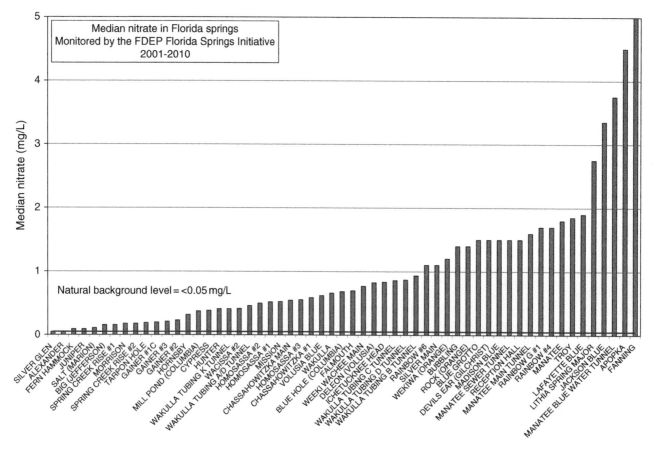

Figure 2.2 Median nitrate-nitrogen concentrations in 57 springs monitored by the Florida Department of Environmental Protection between 2001–2010. Nearly all of Florida's springs have nitrate elevated above natural background and about 80% are above the state numeric nutrient criterion of 0.35 mg/L.

only has resulted in major shifts in fish populations and diversity but also has contributed to explosive expansions of nonnative fish species such as sailfin catfish and blue tilapia (Florida Springs Institute, 2020).

Silver Springs is by no means alone in having documented ecological impairments. More than 20 major spring systems have been determined to have excessive algal proliferation and lost native plant and fish communities (WSI, 2010, 2011; Florida Springs Institute, 2020).

2.2. QUANTIFYING SOURCES OF FLORIDA SPRING/AQUIFER POLLUTION AND DEPLETION

2.2.1. Introduction to the Blue Water Audit

The Florida Springs Institute undertook the Blue Water Audit (BWA) project to quantify groundwater extraction and nitrogen loading throughout the Florida Springs Region (Florida Springs Institute, 2021). Using geographic information system (GIS) technology, a multilayer geospatial analysis provides estimates of the human aquifer footprint for all individual properties greater than five acres throughout the Springs Region.

Data layers include (1) parcels, (2) land use, (3) wastewater systems, (4) atmospheric deposition, (5) aquifer recharge, (6) water use permits, (7) estimated agricultural and landscape irrigation, and (8) springshed boundaries.

Estimates of vertical attenuation rates, fertilizer loading rates, and urban fertilization rates were informed by a detailed review and analysis effort undertaken by the Florida Department of Environmental Protection (FDEP) called the Nitrogen Source Inventory Loading Tool (NSILT) developed by Katz and Eller (2016). While NSILT has been useful for assessing overall loads of nitrogen to the aquifer, it does not provide parcel level detail or estimates of groundwater extractions. The principal advantage of the BWA is that this analysis allows ranking of properties in terms of their estimated individual aquifer footprints, thus facilitating detailed prioritization for implementing groundwater and spring restoration projects.

2.2.2. Nitrogen Estimation

Categories for nitrogen load estimation include atmospheric deposition, crop fertilizers, noncrop fertilizers, wastewater, and livestock waste. Statewide parcel data from the Florida Department of Revenue were used to create the framework for the analysis. FDEP's Statewide Land Cover and Land Use layer and the Florida Department of Agricultural and Consumer Services (FDACS) Florida Statewide Agricultural Irrigation Demand (FSAID) Agricultural Lands Geodatabase were used to represent land use in the nitrogen load estimation process. Septic and sewer data from the Florida Department of Health (FDOH) were used to estimate human wastewater inputs. Data from the National Atmospheric Deposition Program (NADP) were used to estimate atmospheric deposition.

Attenuation, where some nitrogen is removed by subsurface biochemical processes, was estimated using NSILT's attenuation factors. The final step was to adjust the estimates based on recharge level. Detailed recharge maps from individual water management districts (WMDs) were used, as well as a general USGS recharge map, for areas where a more detailed recharge map was not available.

2.2.3. Groundwater Withdrawal Estimation

The components of water use estimation include reported estimates of water use for irrigated agricultural land from the FDACS FSAID Irrigated Lands Geodatabase (ILG), reported water use records from the WMDs, CUP data from WMDs, and general water use estimates from the USGS 2010 and 2015 Florida water use reports (Marella, 2014, 2019). Actual groundwater use data by parcel were available for only a portion of this analysis. Data gaps were estimated based on the FDAC's parcel-specific irrigated crop reports and the detailed water-use estimates provided by Richard Marella.

2.2.4. Validation

The BWA methodology was validated by comparing estimated groundwater extractions and springshed nitrogen loading to actual springs data for flows and nitrogen discharge rates. These validations allowed adjustments to project assumptions for fertilizer loading and attenuation and for regional versus local groundwater pumping impacts on individual spring flows. Detailed discharge and water quality data for eight large springsheds, feeding groundwater and nitrate nitrogen to eight first-magnitude springs or spring groups, were collected and analyzed. Overall agreement between actual nitrogen mass loads from these springs and the BWA water quantity and quality footprint analysis was within 14%.

2.2.5. Floridan Aquifer System Footprints

The BWA analysis is based on spatial data from 2010–2015 and results include detailed nitrogen loading and groundwater withdrawal estimates for each spatial analysis unit, parcels over 5 acres in size as well as generalized land use for urban areas. Additionally, each unit was assigned an overall FAS Nitrogen Footprint. This water quality footprint summarizes the unit's estimated impact on the FAS's water quality. Each unit was also assigned an overall FAS Groundwater Withdrawal Footprint. This Water Quantity Footprint summarizes the unit's estimated impact on the FAS's water quantity. BWA maps were prepared to illustrate the results for the four Florida Spring Restoration Areas.

2.2.6. Nitrogen Loading in the Florida Springs Region

The BWA analysis estimated that roughly 48.2 million lb of nitrogen reach the FAS each year in the Florida Springs Region. With an area of roughly 27 million acres, an average of 1.77 lb of nitrogen per acre is entering the aquifer in the Florida Springs Region each year. Figure 2.3 provides a map of the nitrogen footprint results with the darker colors indicating higher impact on the aquifer. FAS nitrogen loads are considered minimal impact as long as they result in groundwater nitrate concentrations less than the Florida numeric criterion of 0.35 mg/L. Moderate impact loads are expected to result in FAS nitrate concentrations between 0.35 and 2 mg/L. High impact loads are defined as raising groundwater nitrate concentrations between about 2 and 10 mg/L. Severe impacts are defined as nitrogen loading rates to the FAS likely to exceed the drinking water nitrate standard of 10 mg/L.

Of the five nitrogen loading categories, crop fertilizers made up the largest percentage of the total estimated load at 43% or an estimated 20.9 million lb of nitrogen to the aquifer each year in the Florida Springs Region. The remaining percentages are shown in Figure 2.4, where human wastewater is broken into two separate categories, septic and sewer. Table 2.2 provides a breakdown of these loads by springs restoration area to the land surface, attenuated, and reaching the FAS.

A key finding of the BWA is that the bulk of the nitrogen pollution load to the aquifer is coming from a relatively small number of large parcels. For example, parcels with high impact and severe impact Nitrogen Footprints (>3.5 lb N/acre/year to the aquifer) account for only 16% of the parcels analyzed, but they contribute an estimated 72% of the total nitrogen reaching the FAS, or roughly 35 million lb of nitrogen per year.

2.2.7. Groundwater Consumption in the Florida Springs Region

The BWA analysis estimated that an average 2.51 billion gal of groundwater are withdrawn from the FAS each day. Over the 27 million acre Springs Region, the average groundwater withdrawal is about 34,000 gal per acre per

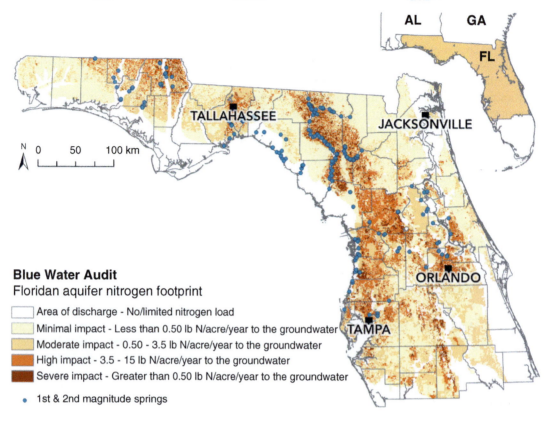

Figure 2.3 BWA nitrogen footprint results for the Florida Springs Region (analysis period 2012–2017).

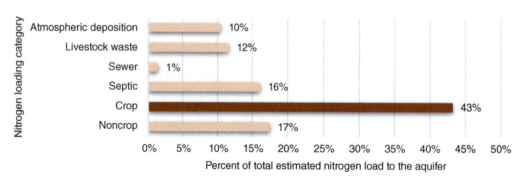

Figure 2.4 BWA estimated nitrogen load in the Florida Springs Region by source category reaching the FAS (analysis period 2012–2017).

Table 2.2 BWA estimated nitrogen loads to the land surface and reaching the FAS within each of the four Springs Restoration Areas (analysis period 2012–2017)

Florida Springs Restoration Areas	Florida Springs Region Area (acres)	Estimated nitrogen load to land surface (lb-N/year)	Estimated nitrogen load to FAS (lb-N/year)
Panhandle	8,098,050	155,497,515	8,813,161
Suwannee	3,624,500	116,771,771	11,353,537
Gulf Coast	5,796,170	250,088,393	14,963,610
St. Johns	9,677,670	304,488,905	13,085,962
Total	27,196,390	826,846,584	48,216,270

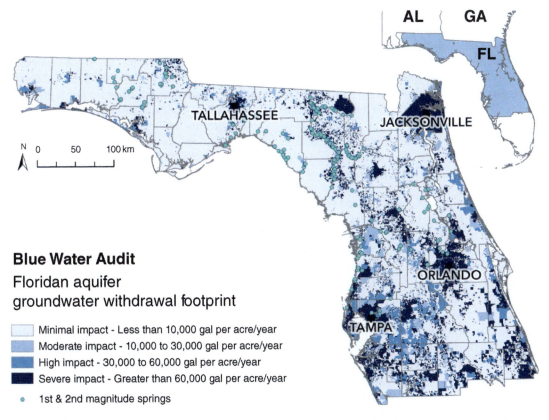

Figure 2.5 BWA estimated FAS groundwater withdrawal footprint in the Florida Springs Region for the period 2012–2017.

Table 2.3 BWA estimated groundwater extraction from the FAS within each of the four Springs Restoration Areas during the analysis period 2012–2017

Florida Springs Restoration Areas	Florida Springs Region Area (acres)	Estimated groundwater withdrawal (MGD)
Panhandle	8,098,050	207
Suwannee	3,624,500	307
Gulf Coast	5,796,170	815
St. Johns	9,677,670	1,183
Total	27,196,390	2,512

year. According to the FAS water balance estimated by Knight and Clarke (2016), this pumping rate has resulted in an average spring flow reduction of 32%. The minimal impact footprint of less than 10,000 gal per acre/year is estimated to reduce spring flows by less than 10%. The severe impact category of 60,000 gal per acre/year, if experienced throughout the Springs Region, might result in a 60% or greater decline in average total spring flows.

Figure 2.5 provides a map of the groundwater withdrawal footprint results. The St. Johns Springs Restoration Area has the highest overall estimated groundwater withdrawal from the aquifer of the four Florida Springs Restoration Areas, with an estimated 1.18 billion gal of groundwater withdrawn from the aquifer each day. Estimated BWA total groundwater withdrawals for each of the four springs restoration areas are shown in Table 2.3.

2.3. UTILIZING WATER AND NUTRIENT MASS BALANCES TO DIRECT SPRINGS PROTECTION AND RECOVERY

Both water and nitrogen are conserved as they are cycled through the environment. Focused monitoring of the components of the water and nitrogen mass balances is necessary to understand and manage these critical resources. Currently, the Florida Springs Region has nearly 100 active flow monitoring sites with many of those on springs and spring runs and spring-fed rivers (see USGS, *Current Water Data for Florida* at https://waterdata.usgs.gov/fl/nwis/rt). Combined with the Florida Automated Weather Network (FAWN) rainfall stations, pan evaporation data, estimated recharge rates, and reported and estimated groundwater withdrawals, existing data are sufficient to construct regional water and nitrogen balances for the FAS and springs. Since groundwater quantity and recharge are limited by rainfall and existing hydrogeology, and because both the human and natural environments utilize the same groundwater for their respective economies, empirical water mass balances offer the most accurate and straightforward method for effectively managing finite water volumes. Similarly, water quality data are routinely collected and reported for many of Florida's springs and spring-fed rivers. Combined with available hydrologic data, nitrogen mass balances offer a practical approach to assessing sources of pollution and achieving regulatory goals.

2.3.1. Simplified Groundwater Mass Balance

Many of Florida's springs have science-based minimum flows and levels (MFL) based on the avoidance of significant harm to their natural ecology. In total, these MFLs allow an average flow reduction between 3% and 15% (WSI, 2021). Based on the documented ecological and aesthetic effects of reduced springs flows, the Florida Springs Institute (2018) has recommended that individual springs MFLs should allow no more than a 5% reduction in average flows. Combined with a regional water mass balance for Florida's Springs Region, it is possible to calculate the allowable groundwater extraction for human uses that will provide on average 95% of historic flows in the springs and rivers. With an overall Florida estimated predevelopment average groundwater inflow of about 11 BGD (Bush & Johnston, 1988), 5% equates to about 550 MGD of net consumption from the FAS available for human uses. Considering that there are often returns of used water to the aquifer, an annual average maximum of 735 MGD based on an assumed 25% return provides a reasonable cap on total average groundwater consumption from the Florida Springs Region.

Based on predevelopment estimates by Bush and Johnston (1988), the total average recharge to the entire FAS was about 14 BGD, with about 2.5 BGD entering Florida as groundwater recharged in Georgia. Current maximum annual pumping from the entire FAS is as high as 4 BGD, including more than 1.5 BGD in Georgia (Williams & Kuniansky, 2016). Without a similar cap on Georgia's groundwater pumping, Florida will need to further lower the estimated 735 MGD cap recommended above to provide restoration of 95% of historic flows in Florida's springs.

2.3.2. Simplified Nitrogen Mass Balance

Based on a review of historic data prior to the utilization of synthetic fertilizers, the natural background nitrate nitrogen concentration in the FAS was about 0.05 mg/L (Harrington et al., 2010). The flow-weighted nitrate nitrogen concentration in Florida's springs over the past decade is about 1.0 mg/L with individual large springs with average concentrations above 5 mg/L. Florida has adopted a regionwide numeric nutrient criterion for nitrate in springs of 0.35 mg/L. Based on the 1.0 mg/L springs average, the simplest assumption is that a lowering of nitrogen loading to the aquifer by >65% will ultimately achieve the numeric springs criterion. Given the BWA estimated average nitrogen attenuation of 94% (see Table 2.2 totals), and that average human nitrogen loads are about 90% of total loads to the land surface (atmospheric inputs average about 10% of the total to the land surface), the estimated allowable human nitrogen loading to the land surface required to achieve the 0.35 mg/L average groundwater nitrate concentration is about 145,000 tons N/year. This mass balance calculation provides a reasonable cap on overall anthropogenic nitrogen loading from fertilizer and wastewater inputs to the land surface in Florida's Springs Region.

2.4. INFORMING THE PUBLIC OF SPRINGS AND AQUIFER HEALTH STATUS

The principal method for publicly sharing the BWA analysis results has been via an interactive map on an educational website created for the BWA project (www.bluewateraudit.org). Interactive maps are useful for showing the spatial aspect of human impacts on the aquifer and where larger impacts are in relation to springs.

One of several ancillary projects to inform the public of these threats is the preparation of maps illustrating the concentration of nitrates in public water systems (PWS). The FDEP defines a PWS as one that provides water to 25 or more people for at least 60 days each year or serves 15 or more service connections (https://floridadep.gov/water/source-drinking-water). These public water systems may be publicly or privately owned and operated. PWS facilities are required to monitor for contaminants under the Safe Drinking Water Act. FDEP makes the PWS facility monitoring data available through its Drinking Water Database (FDEP, 2021). The database currently includes 16 years of monitoring data (2004–2019) with data on various types of contaminants.

Nitrates are known to contribute to excessive algal growth in Florida springs and can also have a negative effect on human health. Since our spring water and drinking water, both come from the FAS, analyzing the concentration of nitrates in drinking water from PWS facilities across the Florida Springs Region provides an independent view into the nitrate contamination of the aquifer.

The PWS data were organized into the following time frames: 2004–2008, 2009–2013, and 2014–2019. An average of results was calculated for each facility for the three time periods. These averages were joined in geographic information systems (GIS) software to a PWS facility location layer. Once the data were joined with the PWS facility layer, a spatial analysis was performed using an inverse distance weighted interpolation technique, which analyzes point data to create a smooth surface (Fig. 2.6). Note the similarity of the spatially distributed

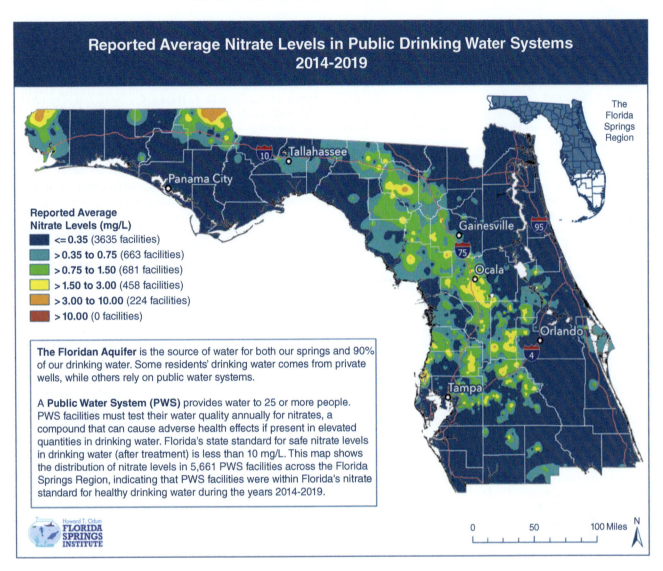

Figure 2.6 Informational summary of the FDEP Public Water System for nitrate-nitrogen contamination of the FAS in the Florida Springs Region. Data *source:* FDEP's Drinking Water Database.

indicators of groundwater nitrate contamination between Figures 2.3 and 2.6 in this chapter.

GIS is often viewed as a scientific tool that is useful for analysis, but spatial data are also key for communicating analysis results. Both static and interactive maps are powerful tools for creating education and outreach materials. The ArcGIS StoryMaps tool was leveraged to create educational content focused on the main components of the BWA analysis, groundwater quality and quantity. StoryMaps allow information to be shared in a narrative form with engaging graphics alongside spatial content. StoryMaps, like the one shown in https://arcg.is/0XPKfz, use maps, engaging photos, and educational content to help describe the complex relationship between the FAS, land use, and water quality.

ACKNOWLEDGMENTS

The authors gratefully acknowledge the support of the staff of the Howard T. Odum Florida Springs Institute. Key analysts who supported this work included Haley Moody and Hillary Skowronski. We also wish to thank Ron Clarke with Wetland Solutions for sharing detailed summaries of North Florida springs data.

REFERENCES

Bellino, J. C., Kuniansky, E. L., O'Reilly, A. M., & Dixon, J. F. (2018). *Hydrogeologic setting, conceptual groundwater flow system, and hydrologic conditions 1995–2010 in Florida and parts of Georgia, Alabama, and South Carolina*. U.S. Geological Survey Scientific Investigations Report 2018-5030.

Brown, M. T., Reiss, K. C., Cohen, M. J., Evans, J. M., Reddy, K. R., Inglett, P. W., et al. (2008). Effects of nutrients on spring organisms and systems. University of Florida Water Institute, Gainesville, Florida.

Bush, P. W., & Johnston, R. H. (1988). *Ground-water hydraulics, regional flow, and ground-water development of the FAS System in Florida and in parts of Georgia, South Carolina, and Alabama*. Professional Paper 1403-C. Reston: United States Geological Survey.

EPA (U.S. Environmental Protection Agency) (2016). *What climate change means for Florida*. EPA 430-F-16-011.

FDEP (2021). *Florida Department of Environmental Protection drinking water database*. https://floridadep.gov/water/source-drinking-water/content/information-drinking-water-data-base

Florida Springs Institute (2018). *Florida Springs conservation plan*. Howard T. Odum Florida Springs Institute. High Spring, Florida.

Florida Springs Institute (2019). *Silver Springs SpringsWatch annual report*. Howard T. Odum Florida Springs Institute, High Springs, Florida.

Florida Springs Institute (2020). *Santa Fe River and springs, environmental analysis*. Phase 3: Final Report Environmental Data. Howard T. Odum Florida Springs Institute. High Springs, Florida.

Florida Springs Institute (2021). Blue Water Project methods summary. Prepared for the Community Foundation of North Central Florida. https://bluewateraudit.org/wp-content/uploads/2021/07/BWA-Methods-Summary_updated_07_27_2021.pdf

Harrington, D., Maddox, G., & Hicks, R. (2010). *Florida Springs Initiative monitoring network report and recognized sources of nitrate*. Florida Department of Environmental Protection. Tallahassee, Florida.

Hicks, R., & Holland, K. (2012). *Nutrient TMDL for Silver Springs, Silver Springs Group, and Upper Silver River (WBIDs 2772A, 2772C, and 2772E)*. Tallahassee: Florida Department of Environmental Protection.

Katz, B. G. & Eller, K. (2016). The Nitrogen Source Inventory and Loading Tool (NSILT) and restoration of water-quality impaired springs. *Florida Scientist*, 79(4), 299–310.

Knight, R. L. (2015). *Silenced springs: Moving from tragedy to hope*. Howard T. Odum Florida Springs Institute. Gainesville, Florida.

Knight, R. L., & Clarke, R. A.. (2016). Florida springs: A water-budget approach to estimating water availability. *Journal of Earth Science and Engineering*, 6, 59–72.

Marella, R. (2014). *Water withdrawals, uses, and trends in Florida, 2010*. Scientific Investigations Report 2014–5088.

Marella, R. (2019). *Water withdrawals, uses, and trends in Florida, 2015*. Scientific Investigations Report 2019–5147.

Munch, D. A., Toth, D. J., Huang, C., Davis, J. B., Fortich, C. M., Osburn, W. L., et al. (2006). *Fifty-year retrospective study of the ecology of Silver Springs, Florida*. Report prepared for the Department of Environmental Protection, Special Publication SJ2007-SP4. Palatka: St. Johns River Water Management District.

Odum, H. T. (1957). Trophic structure and productivity of Silver Springs, Florida. *Ecological Monographs*, 27(1), 55–112.

Reddy, K. R., Dobberfuhl, D., Fitzgerald, C., et al. (2017). *Collaborative research initiative on sustainability and protection of springs (CRISPS) final report (2014–2017)*. Prepared for the St. Johns River Water Management District Springs Protection Initiative. University of Florida Water Institute.

Rosenau, J. C., Faulkner, G. L., Hendry, C. W., Jr., & Hull, R. W. (1977). *Springs of Florida*. Florida Geological Survey Bulletin 31 Revised.

Scott, T., Means, G. H., Means, R. C., & Meegan, R. P. (2002). *First magnitude springs of Florida*. Florida Geological Survey. Open File Report No. 85.

USGS (U.S. Geological Survey) (1995). *Springs of Florida*. U.S. Geological May 1995 Survey Fact Sheet FS-151-95.

Williams, L. J., & Kuniansky, E. L. (2016). *Revised hydrogeologic framework of the FAS system in Florida and parts of Georgia, Alabama, and South Carolina*. U.S. Geological Survey Professional Paper 1807. https://doi.org/10.3133/pp1807

WSI (Wetland Solutions, Inc.) (2010). *An ecosystem-level study of Florida's springs*. Final Report, February 26, 2010. Prepared for the Florida Fish and Wildlife Conservation Commission, Tallahassee, Florida. FWC Project Agreement No. 08010.

WSI (Wetland Solutions, Inc.) (2011). *An ecosystem-level study of Florida's springs, Part II: Gum slough springs ecosystem characterization*. Gainesville, Florida, 2011.

WSI (Wetland Solutions, Inc.) (2021). *Water resource values analysis of outstanding Florida springs and assessment of recreation, aesthetic, and scenic attributes of Florida springs. Task 6: Final report*. Prepared for Suwannee River Water Management District, TWA: 19/20-064.003.

3

Regional Passive Saline Encroachment in Major Springs of the Floridan Aquifer System in Florida (1991–2020)

Rick Copeland, Gary Maddox, and Andy Woeber

ABSTRACT

Due to the awareness of degrading groundwater quality in Florida's freshwater springs and beginning in the early 1990s, the state's water management districts, the Florida Department of Environmental Protection, and the U.S. Geological Survey began efforts to coordinate monitoring of Florida's first- and second-magnitude springs. This study investigates changes in spring discharge and the concentrations of two saline indicators sodium (Na^+) and chloride (Cl^-) from 1991 through 2020 (30 years) in the Floridan aquifer system (FAS). Data were obtained from 32 major springs and three additional discharge gauging stations. Spring discharge was observed to decrease, while concentrations of sodium and chloride increased. As a group, the FAS springs experienced passive saline encroachment. Encroachment occurred not only along Florida's coasts, but also in the interior. Median concentrations of sodium and chloride increased by an estimated range of 7% to 11% per decade. Evidence suggests the major driver is decreasing rainfall and subsequent declines in recharge to the FAS, followed by sea-level rise. The sources of the saline water are from saltwater near Florida's coasts and relict seawater from the deeper portions of the FAS. The observed changes agree with those predicted by the Ghyben-Herzberg principle for coastal, carbonate aquifers.

3.1. INTRODUCTION

Florida has over 1,000 documented springs (Florida Department of Environmental Protection (FDEP), 2016). As Florida's population grows, spring-water quality and quantity changes have been observed (Florida Springs Task Force, 2000). FDEP, Florida's water management districts (WMDs) (Fig. 3.1), and the United States Geological Survey (USGS) standardized efforts to monitor springwater and increased the number of springs being monitored, beginning in the early 1990s. Increasing nutrient concentrations were the initial focus of springwater-quality studies, but later studies expanded the indicator lists to include discharge, along with major ions, including two saline indicators, sodium (Na^+) and chloride (Cl^-). The two indicators are abbreviated Na and Cl respectively. Copeland et al. (2011) discovered that between 1991 and 2003, spring discharge decreased while concentrations of Na and Cl increased in most of the monitored springs A possible driver for the increased saline trend is saltwater encroachment, which has been a documented issue in Florida for decades (Black et al., 1953; Krause & Randolph, 1989; Spechler, 1994; Prinos, 2013: Prinos et al., 2014; and Prinos, 2016). These reasons prompted FDEP leadership to recommend a follow-up study to investigate if the changes have been occurring over a longer period.

The sequel investigation (Copeland & Woeber, 2021) included most of the same springs used in the earlier study but extended the period from 1991 through 2011.

AquiferWatch Inc., Tallahassee, Florida, USA

Threats to Springs in a Changing World: Science and Policies for Protection, Geophysical Monograph 275, First Edition.
Edited by Matthew J. Currell and Brian G. Katz.
© 2023 American Geophysical Union. Published 2023 by John Wiley & Sons, Inc.
DOI:10.1002/9781119818625.ch03

20 THREATS TO SPRINGS IN A CHANGING WORLD

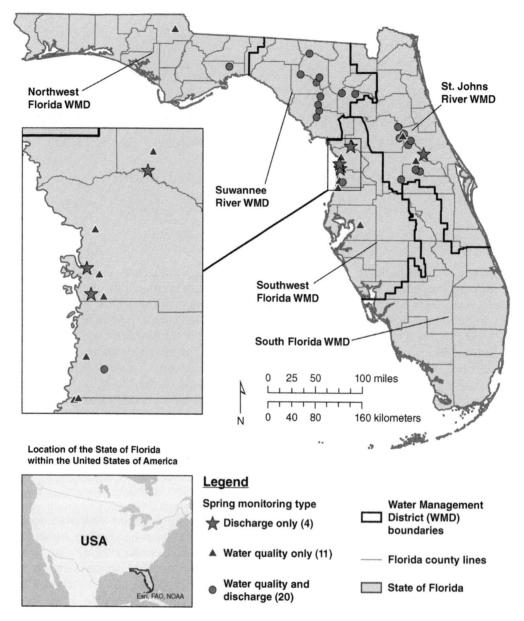

Figure 3.1 Spring monitoring sites used in the study. Florida water management districts and monitoring sites.

During this period, discharge continued to decrease, while concentrations of Na and Cl continued to rise. Copeland and Woeber (2021) postulated the major drivers of the observed changes were: (1) declining rainfall and subsequent declines in recharge, (2) sea-level rise, and (3) groundwater extraction. The major sources of Na and Cl were suspected to be saline groundwater near Florida's coasts and relict seawater from the deeper portions of the Floridan aquifer system (FAS).

Considering the statewide interest in springs and saltwater encroachment, this investigation is the third examination of trends in spring discharge, plus Na and Cl concentrations. The period of study is 1991 through 2020. The springs are primarily in north-central Florida (Fig. 3.1) and occur where the FAS is unconfined or thinly confined (Fig. 3.2). Related to driver (1) above, it should be noted that climatic variability is now recognized as affecting rainfall and river discharge as well as lake chemistry and is due to teleconnections like the Atlantic Multidecadal Oscillation (AMO) and the El Nino southern oscillation (ENSO) (Enfield et al., 2001; Kelly & Gore, 2008; Goly & Teegavarapu, 2014; Canfield et al., 2018). The oscillations affect rainfall, influence spring discharge, and likely influence concentrations of both Na and Cl in springwater.

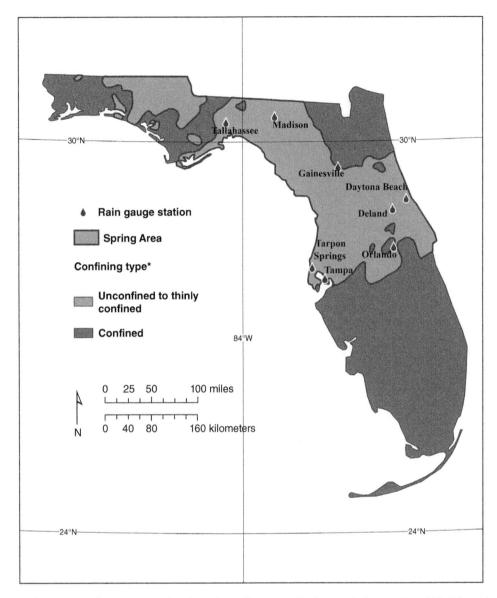

Figure 3.2 Floridan aquifer system in Florida with confinement. Study area is the portion of Florida with unconfined to thinly confined conditions. Solid tear-drop symbols represent rain gauge stations in Spring Area (adapted from Williams & Dixon, 2015).

3.2. FLORIDAN AQUIFER SYSTEM

Florida has three freshwater aquifer systems, from deep to shallow: (1) the Floridan aquifer system (FAS), (2) the intermediate aquifer system and confining unit, and (3) the surficial aquifer system (Southeastern Geological Society, 1986). The largest in terms of areal extent and thickness is the FAS. According to the Southeastern Geological Society (SEGS), it is a thick carbonate sequence, which includes all or parts of Paleocene to early Miocene formations. It can exist under unconfined or confined conditions, depending on the extent of low permeability sediments lying above it. Miller (1986) and Williams and Kuniansky (2016) indicated the FAS is one of the most productive aquifer systems in the world. It underlies all of Florida and portions of South Carolina, Georgia, and Alabama. Klein (1976) mentioned that it can be over 900 m thick in places. Scott (2016), along with Budd and Vacher (2004), mentioned that the FAS is a multiporosity aquifer: fractured and porous where it is confined, and karstic, fractured, and porous where it is unconfined. Scott et al. (2004) indicated that most of Florida's springs are in portions of the state where the FAS is unconfined or thinly confined (Fig. 3.2).

In most places, the FAS can be divided into the Upper Floridan aquifer and the Lower Floridan aquifer, separated by several semiconfining units (Miller, 1986). However, in some places, the two aquifers cannot be differentiated. For this reason, no attempt is made to differentiate the aquifer system, and the undifferentiated term, FAS, is used.

3.3. ENCROACHMENT

Black et al. (1953) mentioned that from the early 1900s through the early 1950s, saltwater had encroached into municipal water supply systems in at least 19 of Florida's coastal counties. Since that report, other authors have reported saline encroachment in Florida. Krause and Randolph (1989) and Spechler (1994) described several possible mechanisms that can drive saltwater encroachment that occurred in northeast Florida. Potential mechanisms included: the landward movement of the freshwater/saltwater interface, the regional upconing of saltwater below pumped wells, and the upward leakage of saltwater from deeper, saline water–bearing zones through confining units. The latter can occur where the units are thin or breached by joints, fractures, collapse features, or other structural anomalies. Movement can also occur because of failed, damaged, or improperly installed well casings, and, as mentioned previously, encroachment can occur by the upward movement of unflushed pockets of relict seawater in the FAS.

In discussions regarding south Florida aquifers, Prinos (2013), Prinos et al. (2014), and Prinos (2016) discussed pathways for saltwater encroachment similar to those mentioned above. However, Prinos (2016) discussed two additional pathways. First, along Florida's coasts, saltwater can flow inland though canals, rivers, boat basins, and coastal marshes, and subsequently leak into the freshwater portions of aquifers. This type of encroachment has been observed in the south Florida aquifers, but can also occur in other areas of Florida where the FAS is under unconfined conditions. Second, Prinos mentioned that encroachment can occur laterally from the coast, moving inland along the base of an aquifer and then upward.

Using conductivity and potentiometric head measurements, Xu et al. (2016) found strong evidence that periodically, under both normal and low-rainfall conditions, seawater has moved inland through cave conduits in the FAS in northwest Florida as much as 11 mi. However, under high-rainfall conditions, when aquifer potentials are high, seawater reverses flow and moves seaward.

As modified from Neuendorf et al. (2005), saltwater encroachment is the displacement of fresh groundwater by the advance of saltwater caused by its greater density. Note, using this definition, there is no requirement distinction that encroachment is due to human activity.

3.4. STUDY AREA, MATERIALS, AND METHODS

The portion of Florida where the FAS is under unconfined to thinly confined conditions (Fig. 3.2) will be referred to as the Spring Area. The springs and discharge sites used in this study are essentially the same ones used in Copeland et al. (2011) and Copeland and Woeber (2021). A total of 32 springs and three stream discharge sites were used (Fig. 3.1). Spring and discharge station names, along with their locational information (WMD, latitude, and longitude) are found in Table 3.1. Of the 35 sites, 31 had sufficient water quality data and 24 had sufficient discharge data, for trend analysis.

Water quality and discharge data were obtained from either the FDEP local WMD or USGS online databases [Suwannee River Water Management District (SRWMD) (2021); St. Johns River Water Management District (SJRWMD) (2021); Southwest Florida Water Management District (SWFWMD) (2021); and the U.S. Geological Survey (2021)]. Data are available by contacting the major author. It should be noted that by Florida Statutes (373.026(2)), state agencies, WMDs, and local agencies are required to cooperate with FDEP in making water quality data available in a central database. Recently, FDEP developed the Water Information Network (WIN) database (Florida Department of Environmental Protection, 2021). At the time of publication, the uploading of groundwater data to WIN is incomplete. However, when complete, efforts to retrieve data in a common format for analyses will be greatly reduced and will increase the efficiency in managing Florida's springs.

All sample collection and all field and laboratory analyses were conducted in accordance with Chapter 62-160, *Florida Administrative Code*. All agencies supplying analytical information did so under FDEP-approved quality assurance project plans. Regarding Na and Cl ions, for the study, laboratory analyses vary between reporting the total and the dissolved species. Total was the most frequently reported and, for this reason, the authors selected the total species to use whenever possible. However, to make the time series as complete as possible, the dissolved species was used whenever the total species was not reported.

Scott et al. (2004) reported on the chemical analysis and discharge of many of Florida's springs. Reiterating comments from an earlier and similar report (Ferguson et al., 1947), Scott et al. indicated the springs in the report represent the major and most important springs in the state. The terms *major* and *most important* have been historically based on discharge. For the springs with available discharge data, the current authors calculated median discharge for each spring and then summed the total discharge of the medians for 92 onshore springs in

Table 3.1 Monitoring sites used in this report

Water management district location and station name	Latitude	Longitude	Water management district location and station namee	Latitude	Longitude
Northwest Florida WMD			St. Johns River WMD (cont'd)		
Jackson Blue[2]	30.7913	−85.1401	Ponce De Leon[1]	29.1343	−81.5294
Wakulla[1]	30.2238	−84.3037	Sanlando[1]	28.6808	−81.3882
Suwannee River WMD			Silver Glen[1]	29.2366	−81.6363
Alapaha Rise[1]	30.4267	−83.0861	Sweetwater[2]	29.2096	−81.6528
Fanning[1]	29.5782	−82.9318	Wekiwa[1]	28.7040	−81.4535
Gilchrist Blue[1]	29.8299	−82.3829	Volusia Blue[3]	29.9387	−81.3319
Hart[1]	29.6660	−82.9482	Southwest Florida WMD		
Hornsby[1]	29.8398	−82.5883	Boat[2]	28.4305	−82.6531
Lafayette Blue[1]	30.1146	−82.2233	BobHill[2]	28.4347	−82.6411
Manatee[1]	29.4804	−82.9736	Buckhorn Main[2]	27.8844	−82.2989
Rock Bluff[1]	29.7889	−82.9149	Catfish[2]	28.8906	−82.5950
Ruth/Little Sulfur[1]	29.9956	−82.9770	Chassahowitzka Main[2]	28.7093	−82.5713
Suwannee Blue[1]	30.0704	−82.9310	Hernando Salt[2]	29.5330	−82.6152
St. Johns River WMD			Hidden River No. 2[2]	28.7691	−82.5835
Alexander[1]	29.0724	−81.5687	Rainbow No. 1[2]	29.1014	−82.4330
12Apopka[1]	28.5593	−81.6745	Weeki Wachee[1]	28.5108	−82.5694
Fern Hammock[1]	29.1745	−82.7013	Chassahowitzka River near Chassahowitzka[3]	28.7150	−82.6064
Marion Salt[1]	29.3411	−81.7257	Homosassa River at Homosassa[3]	28.7850	−82.6181
Palm[2]	28.8437	−81.4501	Rainbow River at Dunnellon[3]	29.0492	−82.4478

Note: The term *spring* is not included in the spring name.
[1] Water quality and discharge (n = 20).
[2] Water quality only (n = 11).
[3] Discharge only (n = 4).

the Scott et al. report [230 m³/s (8,122 ft³/s) (ft³/sec)]. Next, the current authors summed the total median discharge from each of the 24 discharge sites used in this report but restricted it to the 1991– 2020 timeframe, for a calculated total of 89 m³/s (3,124 ft³/s). This represents about 39% of the total discharge; a substantial proportion.

Statewide estimates of groundwater extraction from the FAS are generally reported on 5 or 10 year frequencies by the USGS. Statewide data are available for 1990, 2000, 2010, and 2015, but not 2020. A discussion will be presented later.

Mean annual precipitation data for all of Florida and for selected sites were supplied by the Florida Climate Center (2021). During the study, greater than 99% of all precipitation was rain. For this reason, precipitation will often be referred to as rain or rainfall.

Eight rainfall sites (Fig. 3.2) reported by the climate center are located within the springs area. Figure 3.3 displays annual rainfall and spring discharge for the period (1991–2020). The solid squares represent annual mean precipitation totals for all of Florida (Florida Climate Center, 2021). The solid circles represent annual means of the eight sites in the springs area. Two rainfall Lowess smoothing curves are also in Figure 3.3. The dashed line (upper) curve represents the annual Florida totals, while the dotted line (middle) curve identifies the springs area. They decrease through the first half of the study. Beginning in the late 2000s, and continuing through 2020, both curves increase. The Pearson correlation (r) between the Florida and Spring Area sites is 0.620 (p-value <0.001), and for this reason, only the data from the Spring Area sites are used for the remainder of this report.

Spring discharge annual means either were provided by the monitoring agencies or were calculated by the authors. The annual means are displayed in Figure 3.3 as solid triangles. The Lowess curve (solid line) decreases slightly in the E period and increases during the L period. The correlation (r) between the Spring Area rainfall and spring discharge is 0.309 (p-value = 0.097).

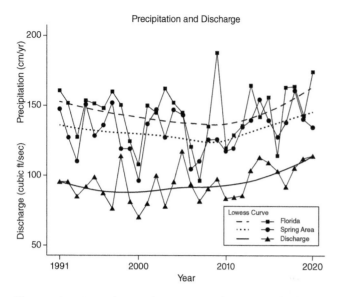

Figure 3.3 Annual precipitation and spring discharge (1991–2020). Solid circles represent annual statewide means (Florida Climate Center, 2021). Solid squares represent annual means of eight rainfall stations and solid triangles represent annual means of 24 discharge sites in Spring Area. Three Lowess curves are (1) dashed line (Florida rainfall), (2) dotted line (Spring Area rainfall), and (3) solid line (spring discharge in Spring Area).

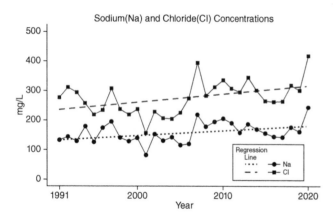

Figure 3.4 Annual sodium and chloride means. Straight lines represent regression lines.

Recall, one of the major drivers of the observed trends for the 1991–2011 time frame is believed to be a decrease in precipitation (and a subsequent decrease in recharge). If so, have increases in rainfall reversed the earlier trends? In part, the current investigation was initiated to address this question. This investigation evaluates trends in precipitation, discharge, and concentrations of both Na and Cl in the Spring Area for the entire period (1991–2020), along with an early (E) period (1991–2011) and a late (L) period (2006–2020). It also revisits the major driving forces of the observed changes. Note, the E period used in this report coincides with the one used by the Copeland and Woeber report (2021). The L period begins in 2006, approximately when rainfall began to increase, and continues through 2020.

For each spring, annual means and medians were calculated for Na and Cl. Figure 3.4 displays the annual means, along with best-fit regression lines. Note, for statistical analyses, medians, rather than means, are used. This topic will be discussed below.

3.5. STATISTICAL METHODS

Most statistical analyses were conducted using the EnvStats, nortest, and rkt packages of the R programming language (R Core Team, 2021). Additional analyses were conducted using the NCSS software (2020).

Data distributions were checked for normality using the Anderson-Darling test (NCSS, 2020). Both rainfall and spring discharge were normally distributed. In addition, data suppliers often provided data for these variables as annual means. For these reasons, the annual means were used for data analysis. Distributions of Na and Cl were strongly and positively skewed (compare their corresponding means to their median values in Table 3.2). Consequently, their annual medians were used for most analytical procedures in this report.

The nonparametric tests, the Mann-Kendall (MK) (Mann, 1945) test and the regional-Kendall (RK) test (Helsel & Frans, 2006), were used for trend analyses. The null hypothesis was of no change in the slope and alpha was preset at 0.10. All tests were two sided. As needed, the Benjamini and Hochberg procedure (1995) was used to test for potential adverse effects of multiple comparisons. None of the tests of tests were adversely affected.

Regarding the RK test, it works best if there is a minimum of 10 years of data available from each site (Helsel & Frans, 2006). For each RK test, this criterion was met. The RK test computes the p-value for the test and a Sen slope (Sen, 1968) for the region.

By using annual means and medians for trend analysis, potential adverse effects of serial autocorrelation (AC), such as seasonality, were reduced. To reduce the effects of spatial AC, a modification of the work completed in the St. Johns River WMD (Fig. 3.1) by Boniol (2002) was used. Boniol determined that spatial AC was sufficiently reduced in groundwater of the FAS in the WMD for Cl at a distance of 15,240 m (50,000 ft). Copeland and Woeber expanded the efforts of Boniol. Using an ArcGIS script tool (Whiteaker, 2015), they generated a coverage of 1,173 equal area hexagons, each with a diameter of 15,240 m, for the entire state. They plotted locations of 56 FAS springs that had been sampled at least once for Cl

Table 3.2 Statistical summaries for rain, discharge, sodium, and chloride

	Units	[1]n	Min	Mean − 1 SD	SD	Mean	Mean + 1 SD	Max
Rain and discharge								
Rain	cm/year	30	96.32	116.48	15.50	131.98	147.48	160.55
Discharge	m^3/s	30	70.16	79.43	15.53	94.96	110.49	146.82
		[2]n	Min	[3]Q1	[3]Q2	Mean	[3]Q3	Max
Na and Cl								
Na	mg/L	815	1.23	4.29	8.51	165.83	125.23	3950
Cl	mg/L	815	3.00	8.00	12.55	280.40	233.75	5960

[1] n = number of annual grand means for the eight rainfall and 24 discharge sites.
[2] n = number of available pairs.
[3] Q1 = 25th percentile, Q2 = 50th percentile = median, Q3 = 75th percentile.

during the period 2005–2011. A spatial join was then performed with the hexagons (polygon layer) containing a unique identifier and the spring locations. The median Cl concentration was determined for each spring. The median Cl values from the 56 springs, including those in this study, were compared to all possible nearest-neighbor springs. To determine nearest neighbors, the point distance tool in ArcMap 10.6 was utilized. This tool was chosen to allow comparisons of feature layers directly. The tool compares the distances between two sets of points. The process involves comparing a point location with all other point locations in the feature layer. This comparison can be performed with either the same layer or different layers, and a search radius can be set to limit processing and search at specific distances for neighboring points. Additional joins are performed to identify the hexagon identifier associated with each point location for further statistical analysis.

At a distance of up to 15,240 m, there were 79 pairs of nearest neighbors. At distances greater than 15,240 m and up to 30,480 m, there were 117 nearest neighbors. For distances greater than 30,480 m up to 45,720 m, there were 1,289 nearest neighbors. Using each set of paired stations, Pearson correlations (Triola & Lossi, 2018) were determined. The correlations for the three distance groupings were 0.273, 0.118, and -.006, respectively. Thus, the effects of spatial AC are reduced considerably at distances greater than 15,240 m. With this in mind, Copeland and Woeber (2021) randomly selected one spring if more than one existed within any single hexagon. As a result, 31 springs were selected for water quality analyses. In addition, 24 discharge stations were located in separate hexagons.

For this investigation, the Sen slope was not used for further statistical analysis unless the RK test inferred the existence of a trend (p-value ≤ 0.10). Nevertheless, inspections of the direction of Sen slopes were used to assist in interpreting causes of observed changes.

Tidal fluctuations can potentially influence trend analyses in springs located near the coast. Annual median or mean values were used for each indicator at each site. Thus, adverse effects of serial correlation are considered minor. In addition, regarding discharge, depending on the site, annual mean data from USGS sites were adjusted for tidal influences.

3.6. RESULTS

Summary statistics for annual means for rain and discharge, and median (Q2) values for Na and Cl are presented in Table 3.2. Note, Na and Cl concentrations vary considerably in the 31 springs, depending on their location relative to the coast. In mg/L, median Na and Cl concentrations range from a minimum 1.23 mg/L and 3.00 mg/L in a spring located in Florida's interior to a maximum of 3,950 mg/L and 5,960 mg/L for the two variables in a spring located near Florida's coast. Based on 815 observations from 31 springs used in this study, in mg/L, the median and mean for Na were 8.51 mg/L and 165.83 mg/L, respectively (Table 3.2). For Cl, the median and mean were 12.55 mg/L and 280.40 mg/L.

Previously, it was mentioned that Pearson correlation between rainfall and discharge was 0.309 (p-value = 0.097). The nonparametric Spearman correlation between Na and Cl was 0.956 (p-value <0.001). The correlations indicate a significant positive correlation between rainfall and discharge and a much stronger positive correlation between Na and Cl. There may be several reasons for the lower rainfall-discharge correlation. First, it was based on a sample size of 30 (annual means), compared with the Na-Cl correlation that was based on the 815 median values (Table 3.2). Second, rainfall sites were not necessarily located close enough to spring discharge sites to have strong correlations. Third, the variances of both rainfall and discharge were slightly greater in the E period, relative to the L period. The coefficient of variation (CV) (standard deviation divided by the mean) was used to make comparisons. For rainfall, the CV in the E period was 0.13 and in the L period was 0.12. For discharge, the CVs were 0.13 and 0.11, respectively.

Figure 3.1 displays the locations of the springs by WMD. With only two springs in the Northwest Florida Water Management District (NWFWMD), they are included with those in the SRWMD. The region is referred to as the NWFWMD and SRWMD region. The remaining regions are the SJRWMD and the SWFWMD. Table 3.3 displays the results of the RK tests for the entire study (1991–2020) for the four variables for the Spring Area and WMD regions. Table 3.4 does the same for both the E and L periods. In both tables, significant p-values are in bold font.

For 1991–2020 for the Spring Area and the WMDs for both rainfall and discharge, there were no statistical trends (Table 3.3) with two exceptions. Discharge decreased in the SJRWMD and rain increased in the SWFWMD. Concentrations of Na and Cl increased significantly in the Spring Area and within each WMD region (p-values <0.001); the most compelling finding of the study.

During the E period (Table 3.4) for the Spring Area and each WMD, there were no trends in rainfall or discharge. Na concentrations increased significantly in the Spring Area, and in each WMD region. Concentrations of Cl did the same, except for the SJRWMD where they did not increase significantly. During the L (Table 3.4) period, rainfall increased in the NWFWMD and SRWMD regions, plus the SJRWMD. Discharge increased in the Spring Area, the NWFWMD and SRWMD regions, plus the SWFWMD. Na concentrations did not change in the Spring Area. They increased in the SJRWMD and the SWFWMD but decreased in the NWFWMD and / SRWMD regions. The decrease in Na in this region plausibly explains why the Spring Area did not experience a significant change. Concentrations of Cl increased in the Spring Area and each WMD.

In the Spring Area, for the study, concentrations of Na and Cl increased by about 0.056 mg/L and 0.135 mg/L per year (Table 3.3). During the E period, Sen slopes for Na and Cl rates increased by 0.086 mg/L and 0.138 mg/L per year, respectively (Table 3.4). During the L period, discharge increased by (0.550 m^3)/(s) per year. Concentrations of Cl increased by 0.135 mg/L per year (Table 3.4).

From 1991 through 2020, the estimated total change in the concentrations of Na and Cl were 1.68 (0.056 x 30) mg/L and 4.05 (0.135 x 30) mg/L (Table 3.3). To estimate the percent rate of change, the total changes for the two variables were compared with the grand median concentrations (8.51 mg/L and 12.55 mg/L, respectively) found in Table 3.2. To one significant figure, the percent annual rates of change for the two indicators were 0.7% and 1.1%.

Table 3.3 Results of regional Kendall tests for springs for the entire study period (1991–2020)

Indicator	Station n[1]	Annual n[2]	Sen slope units	Sen slope	p-value
Rain	8	30	cm/year	–0.049	0.594
Discharge	24	30	(m^3)/(s)/year	–0.045	0.459
Sodium	31	30	(mg/L)/year	0.056	**<0.001**
Chloride	31	30	(mg/L)/year	0.135	**<0.001**
Northwest Florida and Suwannee River Water Management Districts					
Rain	3	30	cm/year	0.119	0.302
Discharge	11	30	(m^3)/(s)/year	0.330	0.281
Sodium	12	30	(mg/L)/year	0.005	**<0.001**
Chloride	12	30	(mg/L)/year	0.135	**<0.001**
St. Johns River Water Management District					
Rain	3	30	cm/year	–0.298	0.377
Discharge	9	30	(m^3)/(s)/year	–0.136	**0.020**
Sodium	10	30	(mg/L)/year	0.111	**<0.001**
Chloride	10	30	(mg/L)/year	0.170	**<0.001**
Southwest Florida Water Management District					
Rain	2	30	cm/year	0.358	**0.014**
Discharge	4	30	(m^3)/(s)/year	0.659	0.224
Sodium	9	30	(mg/L)/year	0.085	**<0.001**
Chloride	9	30	(mg/L)/year	0.248	**<0.001**

[1] Number of stations in region.
[2] Number of years in period.

Table 3.4 Results of regional Kendall tests for the E and L period by region

Period	Station n[1]	Annual n[2]	Sen Slope	p-value	Station n[1]	Annual n[2]	Sen Slope	p-value
		Early				Late		
				Spring Area				
Rain	8	21	−0.237	0.293	8	15	0.270	0.364
Discharge	24	21	−0.670	0.226	24	15	0.550	**0.019**
Na	31	21	0.086	**<0.001**	31	15	0.015	0.244
Cl	31	21	0.138	**<0.001**	31	15	0.135	**<0.001**
		Northwest Florida and Suwannee River Water Management Districts						
Rain	3	21	−0.281	0.129	3	15	1.200	**0.011**
Discharge	11	21	−0.849	0.133	11	15	0.163	**0.007**
Na	12	21	0.083	**<0.001**	12	15	−0.028	**0.024**
Cl	12	21	0.071	**0.002**	12	15	0.075	**0.006**
		St. Johns River Water Management District						
Rain	3	21	−1.104	0.253	3	15	2.460	**0.006**
Discharge	9	21	−0.062	0.497	9	15	0.100	0.131
Na	10	21	0.086	**<0.001**	10	15	0.085	**0.011**
Cl	10	21	0.103	0.164	10	15	0.224	**<0.001**
		Southwest Florida Water Management District						
Rain	2	21	−0.237	0.293	2	15	0.270	0.364
Discharge	4	21	−0.990	0.220	4	15	6.050	**<0.001**
Na	9	21	0.089	**<0.001**	9	15	0.067	**0.006**
Cl	9	21	0.200	**<0.001**	9	15	0.163	**<0.001**

Note: See Table 3.3 for units.
[1] Number of stations in region.
[2] Number of years in period.

For the entire study, the percent changes were 19.7% and 33.0%, respectively.

3.7. DISCUSSION

3.7.1. Conceptual Model

The term *saline* is used to indicate that the source water has greater concentrations of Na and Cl than the receiving groundwater. To assist in understanding the observed changes in spring water, the Ghyben-Herzberg relationship (Freeze & Cherry, 1979) was used. Fetter (2001), and Freeze and Cherry indicated that in the ideal Ghyben-Herzberg relationship, for each meter of drawdown the saltwater/ freshwater interface rises by 40 meters as a sharp line.

Figure 3.5 presents a conceptual model based on the Ghyben-Herzberg relationship. All of Florida's freshwater aquifers and confining units are conceptually lumped together into a freshwater lens. The irregularly shaped lens is generally thickest in the central portion of the state and narrows toward Florida's coastlines. The top part of Figure 3.5 (a) represents the lens during normal times. The bottom part of Figure 3.5 (b) represents long periods of below-normal rainfall. After a lag in rainfall, (aquifer) potentials (Hubbert, 1940), including spring discharge, decline. In addition, the freshwater lens decreases in size (exaggerated in Fig. 3.5). In the FAS, deep groundwater is enriched in carbonate rock-matrix indicators such as calcium (Ca^{+2}), magnesium (Mg^{+2}), potassium (K^{+1}), alkalinity, and sulfate (SO_4^{-2}), along with both Na and Cl (Upchurch et al., 2019; and Sprinkle, 1989). During periods of extended below-normal rainfall, the deep enriched groundwater can migrate horizontally from the edges of the lens and vertically upward from the transition zone at the bottom of the lens.

Krause and Randolph (1989), and Spechler (2001) hypothesized that deep, relict seawater may be a major source for increased saline indicator concentrations in portions of the FAS in northeastern Florida. In an investigation of spring water chemistry in the SRWMD (Fig. 3.1), Moore et al. (2009) observed that upward movement of groundwater from deep within the upper Floridan aquifer of the FAS may, at times, deliver up to 50% of spring discharge. The proportion of deep water is dependent on head gradients within the aquifer. The authors stated that the deep water provides the major source of Na, Cl, potassium, magnesium, and sulfate. Berndt et al. (2005) indicated that spring discharge water can originate from both shallow and deep sources.

28 THREATS TO SPRINGS IN A CHANGING WORLD

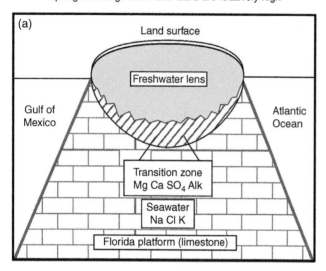

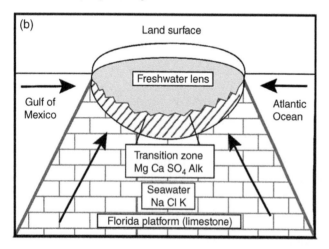

Figure 3.5 Fresh groundwater lens changes over a long dry period: (a) Lens after a long period of average or above average rainfall, (b) lens after a long period of below-average rainfall.

For springs with relatively high TDS concentrations, Berndt et al. speculated that the spring water may have first circulated with deeper groundwater and had a relatively long residence time prior to discharge from springs.

Although not displayed in Figure 3.5, it is implied that if a period of above-normal rainfall prevails, and if recharge exceeds discharge for a long enough time, the lens will increase in size, and concentrations of Na and Cl, along with FAS rock-matrix indicators, will eventually decline. However, as previously stated, some saline water may not be totally flushed (Sprinkle, 1989). Scott et al. (2004) mentioned that during the Pleistocene Epoch, beginning 2.6 million years ago, continental glaciers waxed and waned in the Earth's northern latitudes. In Florida, the potentiometric surfaces and water tables of the aquifers are hypothesized to have dropped with the advancing continental glaciers and then to have risen when the glaciers retreated. The range of sea-level changes may have been up to 140 m (460 ft). Upchurch et al. (2019) indicated that this action would result in saltwater encroachment into the aquifers when sea level rose, and a flushing out of the saltwater when sea level dropped. Although the geological timescale discussed by these authors is considerably different from the decades scale in this study, the processes remain unchanged.

3.7.2. Passive Encroachment

When a well located near the coast in an unconfined aquifer is pumped, the cone of depression around the well can cause upconing of saline groundwater into the well from below in general accordance with the Ghyben-Herzberg principle. Fetter points out that when this type of encroachment occurs, it is an example of active encroachment (Fetter, 2001). Fetter also discussed the term *passive encroachment*. It occurs when some fresh groundwater has been diverted from the aquifer, yet the hydraulic gradient is still sloping toward the saltwater-freshwater boundary. In this situation, the boundary will slowly shift landward until it reaches an equilibrium position based on the new discharge conditions. The mechanisms controlling passive encroachment are the same as active encroachment. However, the rate of encroachment is much slower. Fetter stated that movement is slow. "It may take hundreds of years for the boundary to shift a significant distance" (p. 330). Fetter mentioned that passive encroachment can occur inland as well as in coastal areas. Werner (2017) stated that encroachment can be active, passive, or a combination of the two. Significant increasing trends in the concentrations of both Na and Cl (Tables 3.3 and 3.4) support the hypothesis that passive encroachment is occurring across the Spring Area of Florida.

Recall, during the state's periodic dry periods, when aquifer recharge is reduced and aquifer potentials decline, the freshwater zone shrinks. Younger groundwater is replaced by groundwater with a longer residence time. As a result, groundwater has a greater ionic strength, including increased concentrations of saline indicators such as Na and Cl. This conclusion is supported by Upchurch (1992) and Katz (2004). During times of declining potentials, the likelihood that saline groundwater migrates inland and upward into the freshwater zone is increased.

3.8. POTENTIAL DRIVERS OF THE OBSERVED PASSIVE ENCROACHMENT

Copeland et al. (2011) and Copeland and Woeber (2021) listed several potential drivers of the changes described above. The most significant were (1) below-normal rainfall and subsequent declines in recharge, (2) groundwater extraction, and (3) rising sea level.

3.8.1. Decreasing Rainfall and Consequent Decreases in Recharge

Verdi et al. (2006) mentioned that Florida suffered a severe drought from 1999 to 2002 that affected Florida's water resources. Copeland et al. (2011) indicated the drought was the major driver of change for the period 1991–2003.

During either the E or L periods (Table 3.4), there were no confirmed trends in rainfall and the only significant trend for discharge occurred in the L period. It was upward. Nevertheless, the decreasing Sen slopes of rain and discharge during the E period, and the increasing slopes in the L periods (Fig. 3.3 and Table 3.4) support the concept that rainfall was an important driver of observed changes for the Spring Area.

The decrease in rainfall and followed by an increase are probably related to climatic oscillations (Fig. 3.3). The Atlantic Multidecadal Oscillation (AMO) and the El Niño southern oscillation (ENSO) influence rainfall in Florida (Enfield et al., 2001; Kelly & Gore, 2008; Goly & Teegavarapu, 2014; Canfield et al., 2018). AMO cycles are quasiperiodic, lasting up to 60 and possibly 80 years (Kerr, 2005). The Climate Data Guide (2021) indicated the north Atlantic sea surface temperature (an index for the AMO) increased from the mid-1970s through the late 2000s and has decreased since that time. Of note, the change in the direction of the AMO coincided with the change in rainfall observed in this investigation.

It is important to note the relationships between rainfall and subsequent recharge/discharge in an aquifer system. They can be complex. Theis (1940) and Ponce (2007) stated that under equilibrium conditions in a pristine aquifer, discharge is equal to recharge. To maintain steady-state conditions, Ponce (2007) mentioned that an increase in discharge must be balanced by (1) an increase in recharge to the aquifer from another source (e.g., from an overlying aquifer if the aquifer is confined); (2) a decrease in natural discharge from the aquifer; (3) a loss of storage in the aquifer; or (4) a combination of all three.

Ponce (2007) mentioned that where the aquifer is unconfined, extended periods of below-normal rainfall result in water-table decline. Where it is confined, these conditions can lower the potentiometric surface and groundwater storage. However, below-normal rainfall still can result in additional recharge where the aquifer is thinly confined. The recharge rate will be less than it would be under normal or above-normal rainfall conditions, leading to an overall decrease in storage. In addition, with a sufficient increase in rainfall and recharge for a long enough time, the potentiometric surfaces will eventually increase.

Regarding recharge quantity to the FAS, Bellino et al. (2018) reported the mean annual rate to be 19.0 cm/year. Variations over time were not determined.

3.8.2. Groundwater Extraction

Based on periodic 5 and 10 year summaries by the USGS, groundwater extraction from the FAS in Florida had a net decrease during the study. In units of million m^3/d, in 1990 it was reported as 10.46 m^3/d (Marella, 1992). It rose to 11.72 m^3/d in 2000 (Marella & Berndt, 2005), but then declined to 9.64 m^3/d in 2010 (Marella, 2014), and to 8.85 m^3/d in 2015 (Marella, 2019). Statewide data were not available for 2020. The declines in extraction coincide with efforts by the WMDs to conserve groundwater extraction (U.S. Environmental Protection Agency, 2017a).

Copeland and Woeber (2021) converted the groundwater extraction data from 2010 (Marella, 2014) to a flux in cm/year. They estimated extraction was about 2.4 cm or about 13% of the mean recharge estimate of Bellino et al. (2018). They concluded the effect of extraction was relatively minor, compared with rainfall and subsequent recharge. The net decrease in extraction during this study suggests it was not a major driver of the observed changes in this study.

3.8.3. Sea-Level Rise

In a study of former sea-level rises in Florida, Gully and Florea (2016) indicated that rising sea-levels eventually result in rising aquifer potentials. This can result in a reduction in fresh groundwater in an aquifer (a reduction of the freshwater lens, especially in areas where the FAS is unconfined. As previously noted, as discharge increases, older and more saline groundwater originating from the deeper portions of the aquifer can result in increased concentrations of Na and Cl in discharge water.

Walton (2007) indicated that in Florida, between 1950 and 1999, sea level rose between 8.0 and 23.0 cm. Using linear extraction, an estimate of sea-level increase was between 0.15 and 0.46 cm per year from 1991 through 2011 (E period). The National Oceanic and Atmospheric Administration (NOAA) (2021) stated that between 1993 and 2019 (27 of the 30 years of the current investigation), sea level rose by 8.76 cm. Using the more recent NOAA data, on an annual basis, sea level rose by about 0.32 cm/year. Using this rate for the entire duration of the current study, rising sea levels represent about 2% of the recharge estimates presented earlier (0.32 cm sea-level rise / 19.0 cm recharge). With a limited sea-level rise data set, it is unknown whether the rise in sea level played a significant role in changing the observed indicator concentrations during this investigation. However, Walton predicted sea level in Florida could rise another 25 cm by 2080. By 2100, Wigley and Raper (1992), the U.S. Environmental Protection Agency (2017b), and Lindsey (2022) predict that globally, sea level could rise as much as 1.2 m (4.0 ft), while the NOAA (2021) projects that as an extreme estimate, sea levels could rise as much as 2.5 m. In southeastern Florida, Bloetcher et al. (2011) predicts sea level rise to be between 0.5 and 1.0 m by the end of the 21st century. Unfortunately, as the 21st century proceeds in time, it appears that sea-level rise will play a more important role as a driver of changes in concentrations of Na and Cl.

A question was posed earlier. Although many of the observed changes in this investigation, mostly in the concentrations of Na and Cl, are statistically significant, do the changes have practical significance? The fact that the changes occurred over multiple decades and over an area as large as the spring area (Fig. 3.1) suggests they do.

3.9. UNRESOLVED ISSUES AND NEED FOR ADDITIONAL ENCROACHMENT MONITORING

3.9.1. Unresolved Issues

The changes observed in this report support the conceptual model, but several questions remain unanswered. There was a direct positive correlation between rainfall and spring discharge. During the E period with declining rainfall, spring discharge declined. As predicted by the conceptual model, concentrations of Na and Cl increased. During the L period, when rainfall, discharge, and recharge began to increase, the model predicted a decrease in Na and Cl concentrations would eventually follow. For Na, the slope decreased in the Spring Area and decreased significantly in the NWFWMD and SRWMD region. However, concentrations continued to increase in the SJRWMD and the SWFWMD. Regarding Cl concentrations, they increased in the Springs Area and the three WMD regions. As of 2020, evidence suggests Florida may be experiencing the beginning of the reversal process of encroachment. Unfortunately, this remains uncertain because of continued increases in the Cl concentrations.

Rainfall is the major driver of the observed changes during the time frame of this investigation and may be tied to climatic cycles such as the AMO. Recall, the AMO is a driver of Florida's rainfall and influences surface-water flows (Enfield et al., 2001; Kelly & Gore, 2008; Goly & Teegavarapu, 2014; Canfield et al., 2018). The correlation of rainfall and spring discharge in this study suggests a similar relationship with Florida's spring water.

If the AMO is a major driver of rainfall, then Florida will likely experience increased rainfall for the next several decades. During this period, Floridians will likely be more concerned with surface-water flooding than passive encroachment. Nevertheless, passive encroachment did occur over the course of this study, and as of 2020, encroachment had not abated, at least for Cl. Unfortunately, rainfall will, again, eventually enter a declining stage. When it does, along with the probable increase in the rate of sea-level rise, passive encroachment will likely follow. And again, if the rate of sea-level rise increases, encroachment is likely to be greater than that observed in this study. Floridians would benefit from additional research efforts on the effects that encroachment will have on Florida's groundwater, drinking water, and surface water resources.

3.9.2. Need for Increased Saline Encroachment Monitoring

Passive encroachment observed in this study along with rising sea levels indicate the state needs to continue to monitor spring discharge and saline indicator concentrations. As presented, springs represent good monitoring sites and should be incorporated into saline monitoring efforts whenever possible. It should be noted that the Florida Water Resources Monitoring Council formed a Salinity Network Workgroup in 2011. Key workgroup members include the FDEP, the five WMDs, the USGS, and several counties (Florida Water Resources Monitoring Council, 2019a). One workgroup objective is to improve Florida's ability to monitor for potential saltwater encroachment into major aquifer systems. To this end, the workgroup established a state-wide Coastal Salinity Monitoring Network (Florida Water Resources Monitoring Council, 2019b). It is composed mostly of monitoring wells but does contain a few springs. As sea level continues to rise, it is anticipated that additional springs will be added to the network in the future.

3.10. KEY FINDINGS

At a 90% confidence level, from 1991 through 2020, concentrations of Na and Cl increased in the Florida Spring area. For multiple decades, the region encountered passive saltwater encroachment as defined by Fetter (2001). To the nearest percent, the rates of change for the concentration of Na and Cl were approximately 20% and 33%, respectively, for the duration of the study, or about 7% and 11% per decade.

Evidence suggests the primary driver of the observed changes is below-normal rainfall and a subsequent reduction in recharge to the FAS. Evidence also suggests sea-level rise played a minor role as a driver for changes in Na and Cl concentrations for this investigation. However, several investigators have indicated the rate of sea-level rise is increasing and the rate will continue to increase in the future and therefore become a more important driver of changes in groundwater quality in Florida.

Evidence suggests that an important origin of the saline indicators is from saltwater along Florida's coasts and from saline water located at depth within the FAS. The decrease in spring discharge during the study allowed older and deeper groundwater, located below the freshwater lens and from the coastal regions of Florida, to migrate inward and upward into the springs.

There are several important aspects of this investigation that need emphasis. First, small increases in concentrations of Na and Cl have been observed in major Florida springs for multiple decades. Second, the changes meet the definition of passive saline encroachment. Third, the area of encroachment covers a significant geographical area of the state. And fourth, with increasing rates of sea-level rise predicted in the future, additional monitoring efforts by Florida's water agencies will be needed, including the inclusion of springs.

ACKNOWLEDGMENTS

The authors would like to acknowledge and thank the many individuals who assisted in the production of this chapter. These include Jay Silvanima, Stephanie Sunderman-Barnes, and Chris Sedlacek of the Florida Department of Environmental Protection, and Rick Green and Harley Means of the Florida Geological Survey, all who were involved in editing earlier versions of this work. We would like to thank the many individuals at the U.S. Geological Survey, the Florida Department of Environmental Protection, and Florida's five water management districts who collected data in the field, checked the data for quality assurance, and made available 30 years of data for incorporation into the analysis. Finally, we thank the anonymous reviewers who supplied constructive criticism and suggestions for the chapter.

REFERENCES

Bellino, J. C., Kuniansky, E. L., O'Reilly, A. M., & Dixon, J. F. (2018). *Hydrogeologic setting, conceptual groundwater flow system, and hydrologic conditions 1995–2010 in Florida and parts of Georgia, Alabama, and South Carolina*. U.S. Geological Survey Scientific Investigations Report 2018–5030. https://doi.org/10.3133/sir20185030

Benjamini, Y., & Hochberg, Y. (1995). Controlling the false discovery rate, a practical and powerful approach to multiple testing. *Journal of the Royal Statistical Society, Series B(Methodological)*, 57(1), 289–300. https://doi.org/10.111/j.2517-6161.1995.tb02031x

Berndt, M. P., Katz, B. G., Lindsey, B. D., Ardis, A. F., & Skach, K. A. (2005). Comparison of water chemistry in spring and well samples from selected carbonate aquifers in the United States. In E. Kuniasky (Ed.), *U.S. Geological Scientific Investigations Report 2005-5160* (pp. 74–81).

Black, A. P., Brown, E., & Pearce, J. M. (1953). *Salt water intrusion in Florida, 1953*. Florida State Board of Conservation, Division of Water Survey and Research. Water Survey and Research Paper 9. https://digital.lib.usr.edu/?s62.11

Bloetscher, F., Heimlich, B.N., & Romah, T. (2011). *Counteracting the effects of sea level rise in southeast Florida*. Journal of Environmental Science and Engineering, 5, 1507–1525.

Boniol, D., (2002). *Evaluation of upper Floridan aquifer water quality to design a monitoring network in the St. Johns River Water Management District*. St. Johns River Water Management District Technical Publication SJ2002-1.

Budd, D. A., & Vacher, H. L. (2004). Matrix permeability of the confined Floridan aquifer, Florida, USA. *Hydrogeology Journal* 12(5), 531–549. https://doi.10.1007/s10040-004-0341-5

Canfield, D. E., Hoyer, M. V., Bachmann, R. W., Bingham, S. D., & Ruiz-Bernard, I. (2018). Water quality changes in an outstanding Florida Water: Influence of stochastic events and climate variability. *Lake and Reservoir Management, 32*, 297–313.

Climate Data Guide (2021). *Atlantic multi-decadal oscillation (AMO)*. https://climatedataguide.ucar.edu/climate-data/atlantic-multi-decadal-oscillation-amo

Copeland, R. E., & Woeber N. A. (2021). Changes in groundwater levels, spring discharge, and concentrations of saline and rock-matrix indicators of the Floridan aquifer system, Florida (1991–2011), Tallahassee, FL. *Florida Geological Survey Bulletin, 70*, 275.

Copeland, R. E., Doran, N. A., White, A. J., & Upchurch, S. B. (2011). Regional and statewide trends in Florida's spring and well groundwater quality (1991–2003). Tallahassee Florida. *Florida Geological Survey Bulletin, 69*. http://publicfiles.dep.state.fl.us/FGS/FGS_Publications/B/B69_2009.pdf

Enfield, D. B., Mestan-Nunes, A. M, & Trimble, P. J. (2001). The Atlantic multidecadal oscillation and its relation to rainfall and forever flows in the continental U.S. *Geophysical Research Letters, 28*, 207–208.

Ferguson, G. E., Lingham, C. W., Love, S. K., & Vernon, R. O. (1947). Springs of Florida. Tallahassee, Florida. *Florida Geological Survey Bulletin, 31*. https://ufdc.ufl.edu/UF00094071/00001/207j

Fetter, C. W. (2001). *Applied hydrogeology*. Upper Saddle River, NJ: Prentice-Hall.

Florida Climate Center (2021). *Products & service, data*. https://climatecenter.fsu.edu/products-services/data#Long-Term%20Precipitation%20Data

Florida Department of Environmental Protection (2016). *Geospatial open data, Florida Springs.* https://geodata.dep.state.fl.us/datasets/florida-springs-2016?geometry=-99.274%2C24.985%2C-68.600%2C31.746

Florida Department of Environmental Protection (2021). *WIN/STORET, Watershed Services Program.* https://floridadep.gov/dear/watershed-services-program/content/winstoret

Florida Springs Task Force (2000). *Florida springs: Strategies for protection and restoration.* Tallahassee, Florida Department of Environmental Protection. https://floridadep.gov/sites/default/files/SpringsTaskForceReport_0.pdf

Florida Water Resources Monitoring Council (2019a). Home page. https://floridadep.gov/dear/watershed-monitoring-section/content/fwrmc

Florida Water Resources Monitoring Council (2019b). *Coastal Salinity Monitoring Network.* Salinity Network Workgroup. https://floridadep.gov/dear/watershed-monitoring-section/content/salinity-network

Freeze, R. A., & Cherry, J. A. (1979). *Groundwater.* Englewood Cliffs., NJ: Prentice-Hall. https://www.un-igrac.org/sites/default/files/resources/files/Groundwater%20book%20-%20English.pdf

Goly, A., & Teegavarapu, R. S. V. (2014). Individual and coupled influences of AMO and ENSO on regional precipitation characteristics and extremes. *Water Resource Research, 50,* 4686–4709.

Gully, J. D., & Florea, L. J. (2016). Caves as paleo-water table indicators in the unconfined Upper Floridan aquifer. *Florida Scientist, 79,* 239–256. https://www.jstor.org/stable/44113188?seq=1

Helsel, D. R., & Frans, L. M. (2006). Regional-Kendall test for trend. *Environmental Science and Technology, 40*(13), 4067–407. https://doi.org/10.1021/es051650b

Hubbert, M. K. (1940). The theory of groundwater motion. *Journal of Geology, 8,* 785–944. https://www.journals.uchicago.edu/doi/abs/10.1086/624930

Katz, B. G. (2004). Sources of nitrate contamination and age of water in karstic springs of Florida. *Environmental Geology, 46,* 689–706. https://pubs.er.usgs.gov/publication/70026496. https://doi.org/10.1007/s00254-004-1061-9

Kelly, M. H., & Gore, J. A. (2008). Florida river flow patterns and the Atlantic mutidecadal oscillation. *River Research and Applications, 24*(5), 598–616. https://doi.org/10.1002/rra.1139

Kerr, R. A. (2005). Atlantic climate pacemaker for millennia past, decades hence? *Science, 5,* (309), 43–44.

Klein, H. (1976). *Depth to base of potable water in the Floridan aquifer (revised).* Tallahassee, Florida. Florida Geological Survey Map Series 42. https://ufdc.ufl.edu/UF00099424/00001

Krause, R. E., & Randolph, R. B. (1989). *Hydrology of the Floridan aquifer system in southeast Georgia and adjacent parts of Florida and South Carolina.* U.S. Geological Survey Professional Paper 1403-D. https://pubs.er.usgs.gov/publication/pp1403D. https://doi.org/0.3133/pp1403D

Lindsey, R. (2022). *Climate change: Atmospheric Carbon Dioxide, Climate change: global sea level.* Climate.gov, science and information for a climate smart nation. https://www.climate.gov/news-features/understanding-climate/climate-change-atmospheric-carbon-dioxide

Mann, H. B. (1945). Nonparametric tests against trends. *Econometrica, 13*(3), 245–259. https://doi.org/10.2307/1907187

Marella, R. (1992). *Water withdrawals in Florida during 1990, with trends from 1950 to 1990.* U.S. Geological Survey, Open File Report 92–80. https://pubs.usgs.gov/wri/1992/4140/report.pdf

Marella, R. (2014). *Water withdrawals, use, and trends in Florida, 2010.* U.S. Geological Survey Scientific Investigations Report 22014-5088. https://pubs.er.usgs.gov/publication/sir20145088

Marella, R. (2019). *Water withdrawals, uses, and trends in Florida, 2015.* U.S. Geological Survey Scientific Investigations Report 2019-5147. https://pubs.er.usgs.gov/publication/sir20195147. https://doi.org/10.3133/sir20195147

Marella, R. L., & Berndt, M. P. (2005). *Water withdrawals and trends from the Floridan aquifer system in the southeastern United States, 1950–2000.* U.S. Geological Survey Circular 1278. https://pubs.usgs.gov/circ/2005/1278

Miller, J. A. (1986). *Hydrogeologic framework of the Floridan aquifer system in Florida and parts of Georgia, Alabama, and South Carolina.* U.S. Geological Survey Professional Paper 1403-B. https://pubs.er.usgs.gov/publication/pp1403B

Moore, P. J., Martin, J. B., & Screaton, E. (2009). Geochemical and statistical evidence of recharge, mixing, and controls on spring discharge in an eogenetic karst aquifer. *Journal of Hydrology, 376* (3–4), 443–455. http://citeseerx.ist.psu.edu/viewdoc/download?doi=10.1.1.469.5635&rep=rep1&type=pdf

National Oceanic and Atmospheric Administration (2021). *Ocean service.* https://www.climate.gov/news-features/understanding-climate/climate-change-global-sea-level

NCSS (2020). *Statistical data analysis and graphics. Kaysville, Utah.* https://www.ncss.com/software/ncss/

Neuendorf, K. K. E., Mehl, J. P., & Jackson, J. A. (eds.) (2005). *Glossary of geology.* Alexandria, Virginia: American Geological Institute.

Ponce, V. M, (2007). *Sustainable yield of groundwater.* http://ponce.sdsu.edu/groundwater_sustainable_yield.html

Prinos, S. T. (2013). *Saltwater intrusion in the surficial aquifer system of the Big Cypress Basin, southwest Florida, and a proposed plan for improved salinity monitoring.* U.S. Geological Survey Open-File Report 2013-1088. https://pubs.er.usgs.gov/publication/ofr20131088

Prinos, S. T. (2016). Saltwater intrusion monitoring in Florida. In R. E. Copeland (Ed.), *Florida Scientist, Special Issue, Status of Florida's groundwater resources, 79*(4), 269–278.

Prinos, S. T., Wacker, M. A., Cunningham, K. J., & Fitterman, D. V. (2014). *Origins and delineation of saltwater intrusion in the Biscayne Aquifer and changes in the distribution of saltwater in Miami Dade County, Florida.* U.S. Geological Survey Scientific Investigations Report 2014-5025. http://dx.doi.org/10.3133/sir20145025

R Core Team (2020). *R: A language and environment for statistical computing.* R Foundation for Statistical Computing Report 3-900051-07-0, Vienna, Austria. http://www.R-project.org/

Scott, T. M. (2016). Lithostratigraphy and hydrostratigraphy of Florida. *Florida Scientist, 79*(4), 198–2007.

Scott, T. M., Means, G. H., Meegan, R. P., Means, R. C., Upchurch, S. B., Copeland, R. E., et al. (2004). *Springs of Florida: Florida Geological Survey Bulletin, 66,* 377.

https://ufdcimages.uflib.ufl.edu/UF/00/00/51/66/00001/OFR95revWashingtonCo092209.pdf

Sen, P. K. (1968). Estimates of the regression coefficient based on Kendall's tau. *Journal of the American Statistical Association*, 63(324), 1379–1389. https://doi.org/10.1080/01621459.1968.10480934

Southeastern Geological Society (1986). Hydrogeological units of Florida. *Florida Geological Survey Special Publication, 28*. https://ufdc.ufl.edu/UF00000138/00001

Southwest Florida Water Management District. (2021). *Environmental data portal, external portal, water data viewer (external)*. https://edp.swfwmd.state.fl.us/applications/login.html?publicuser=Guest#waterdata-external.

Spechler, R. M. (1994). *The relation between structure and saltwater intrusion in the Floridan aquifer system, northeastern Florida*. U.S. Geological Survey Water-Resources Investigation Report 01-411.

Spechler, R. M. (2001), The relation between structure and saltwater intrusion in the Floridan aquifer system, northeastern Florida. In E. L. Kuniansky (Ed.), *U.S. Geological Survey karst interest group proceedings: U.S. Geological Survey Water-Resources Investigations* (pp. 25–29). Report 01-4011.

Sprinkle, C. L. (1989). *Geochemistry of the Floridan aquifer system in Florida and in parts of Georgia, South Carolina, and Alabama: Regional aquifer-system analysis: Floridan aquifer system*. U.S. Geological Survey Professional Paper 1403-I. http://citeseerx.ist.psu.edu/viewdoc/download?doi=10.1.1.569.1089&rep=rep1&type=pdf

St. Johns River Water Management District (2021). *Database environmental data retrieval tool*. http://webapub.sjrwmd.com/agws10/edqt/

Suwannee River Water Management District. (2021). *Water data portal*. http://www.mysuwanneeriver.org/portal/waterquality.htm

Theis, C. V. (1940). *The source of water derived from wells essential factors controlling the response of an aquifer to development*. Reprint of American Society of Civil Engineers. U.S. Geological Survey Ground Water Hydraulics Notes No. 34. https://water.usgs.gov/ogw/pubs/Theis-1940.pdf

Triola, M. F., & Lossi, L. (2018). *Elementary statistics*, 13th ed. New York: Pearson.

Upchurch, S. B. (1992). Quality of water in Florida's aquifer systems. In G. L. Maddox et al. (Eds.). *Florida's groundwater quality monitoring program: Background hydrogeochemistry*. Tallahassee, Florida. Florida. Geological Survey Special Publication, 12–63. http://aquaticcommons.org/1307/1/Geochemistry.pdf

Upchurch, S. B., Scott, T. M., Alfieri, M. C., Fratesi, B., & Diobecki, T. L. (2019). *The karst systems of Florida: Understanding karst in a geologically young terrain*. Cham. Switzerland: Springer.

U.S. Environmental Protection Agency (2017a). *Saving water in Florida, factsheet*. https://www.epa.gov/sites/default/files/2017-02/documents/ws-ourwater-florida-state-fact-sheet.pdf

U.S. Environmental Protection Agency (2017b). *Climate change science: Future of sea-level change*. https://archive.epa.gov/epa/climate-change-science/future-climate-change.html#Sea%20level

U.S. Geological Survey (2021). *Streamflow conditions, Florida streamflow table*. https://waterdata.usgs.gov/fl/nwis/current/?type=flow&group_key=basin_cd

Verdi, R. J., Tomlinson, S. A., & Marella, R. L. (2006). *The drought of 1998–2002: Impacts on Florida's hydrology and landscape*. U.S. Geological Survey Circular 1295. https://pubs.usgs.gov/circ/2006/1295/pdf/circ1295.pdf

Walton, T. L. (2007). Projected sea level rise in Florida. *Ocean Engineering, 34*, 1832–1840. https://floridaclimateinstitute.org/images/reports/FSUProjectedSeaLevel.pdf

Werner, A. S. (2017). *On the classification of active and passive seawater intrusion*. American Geophysical Union, Fall meeting, 2017, abstract H11p-01. https://ui.adsabs.harvard.edu/abs/2017AGUFM.H11P.01W/abstract

Whiteaker, T. (2015). Hexagon Theisen tessellation tool. Center for Research in Water Resources. University of Texas at Austin, Austin, Texas. http://tools.crwr.utexas.edu/Hexagon/hexagon/html

Wigley, T. M. L., & Raper, S. C. D. (1992). Implications for climate and sea level of revised IPCC emissions scenarios. *Nature, 357* (6376), 293–300. https://sedac.ciesin.columbia.edu/mva/WR1992/WR1992.html

Williams, L. J., & Dixon, J. F. (2015). Digital surfaces and thicknesses of sselected hydrogeologic units of the Floridan aquifer system in Florida and parts of Georgia, Alabama, and South Carolina. U. S. Geological Survey Data Series, 926.

Williams, L. J., & Kuniansky, E. L. (2016). *Revised hydrogeologic framework of the Floridan aquifer system in Florida and parts of Georgia, Alabama, and South Carolina*. U.S. Geological Survey Professional Paper 1807. https://doi.org/10.3133/pp1807

Xu, Z., Bassett, S., Hu, B., & Dyer, S. (2016). Long distance seawater intrusion through a karst conduit network in the Woodville Karst Plain, Florida. *Scientific Report, 6*, 10. https://www.nature.com/articles/srep32235#citeas. https://doi.org/10.1038/srep32235

4

Karst Spring Processes and Storage Implications in High Elevation, Semiarid Southwestern United States

Keegan M. Donovan[1], Abraham E. Springer[1], Benjamin W. Tobin[2], and Roderic A. Parnell[1]

ABSTRACT

Karst springs and aquifers are significant resources globally yet continue to be poorly understood because of their heterogeneity in porosity and response to climate variability. In semiarid, mountainous regions where total precipitation and groundwater recharge rates will likely decline due to climate change, improved understanding of karst groundwater recharge processes is imperative to plan for future hydrologic responses. The Colorado Plateau is a high-elevation, heavily dissected region of over 2,000 m thickness of sedimentary rock units containing multiple layered karst aquifers. This is the first study to use hydrograph analyses on springs in the uppermost regional Coconino aquifer (C aquifer) of the southern Colorado Plateau in Arizona to detail karst aquifer response to recharge. Coupled hydrograph and stable isotope ($\delta^{18}O$ and δ^2H) analyses document seasonal recharge as well as groundwater mixing and storage processes in the C aquifer. A critical relationship between seasonal snowpack timing and duration of ephemeral spring discharge indicates seasonal buffering and attenuation to drought. Event-scale hydrograph analyses show rapid hydrologic responses to precipitation in two ephemeral karst spring systems and quick drainage without extended seasonal snowmelt contribution. As climate change intensifies throughout the southwestern United States, and other drought-ridden areas of the world, aquifer and spring ecosystem conditions will significantly worsen without mitigation measures. The recharge and groundwater flow processes demonstrated in this study of a complex karst system support informed water resource decision making on the southern Colorado Plateau and other climate-sensitive regions around the world.

4.1. INTRODUCTION

Karst springs and aquifers are significant water resources worldwide and are critical to study to sustainably manage and protect them. Past estimations have claimed that 20%–25% of the world's population depends on karst groundwater in some capacity (Ford & Williams, 2007), and more recent studies have estimated approximately 9.2% of the global population are direct consumers of karst water (Stevanović, 2019). Because of their vulnerable nature, even small shifts in climate and changes in human interaction can lead to serious impacts on the quality and quantity of karst waters (Brinkmann & Parise, 2012).

Karst mountain aquifer systems have widely been neglected because of their inaccessibility and seemingly negligible contribution to water supply. However, a growing body of literature highlights their importance on recharge, seasonal groundwater storage (Bonacci, 2014; Malard et al., 2016; Meeks & Hunkeler, 2015; Shamsi et al., 2018), and underestimation in their river contribution especially during dry seasons (Tobin & Schwartz, 2020). While alpine systems have been shown to be much more noteworthy

[1]*School of Earth and Sustainability, Northern Arizona University, Flagstaff, Arizona, USA*
[2]*Kentucky Geological Survey, University of Kentucky, Lexington, Kentucky, USA*

Threats to Springs in a Changing World: Science and Policies for Protection, Geophysical Monograph 275, First Edition.
Edited by Matthew J. Currell and Brian G. Katz.
© 2023 American Geophysical Union. Published 2023 by John Wiley & Sons, Inc.
DOI:10.1002/9781119818625.ch04

than previous assessments, they are also thought to have higher vulnerability to climate change (Godsey et al., 2014) especially in arid to semiarid regions (He et al., 2019).

Climate models in arid regions commonly predict hotter and drier conditions with longer droughts (Ahmadalipour et al., 2017; Chiew et al., 2011) and a transition from snowpack-driven to rain-dominated recharge (Godsey et al., 2014; Taylor et al., 2013). Researchers expect that resulting declines in aquifer recharge will occur across the southwestern United States and other arid to semiarid regions of the world (Meixner et al., 2016). These effects present an urgency to characterize stressed karst aquifer systems, especially with intense aridification predicted to continue without foreseeable recovery (Hereford & Amoroso, 2020; Tillman et al., 2020). According to Goldscheider et al. (2020), 40.8% of all karst occurs in mountain areas, and 34.2% occurs in arid climates. That is, alpine karst is the most widespread in terms of topography and arid karst is the most prevalent in terms of climate. By detailing present karst groundwater processes in common topographic and climate regimes, extensive management approaches will help to preserve the declining groundwater reserves as climatic shifts directly influence groundwater storage characteristics.

Changes in climate regimes will also affect the dependent ecosystems that springs support. Springs host an immense network of biodiversity and karst springs are hotspots (Goldscheider, 2019). In the Colorado River Basin, springs sustain a disproportionally high percentage of the regional flora (Stevens et al., 2020) including endangered taxa. In the Grand Canyon ecoregion alone, almost 1,000 species populate less than one square kilometer of total springs habitat (Sinclair, 2018). Spring-dependent ecosystems are becoming increasingly reliant on these oases as aridification in the southwestern United States intensifies, warranting concurrent consideration with human water needs (Cartwright et al., 2020; Cantonati et al., 2020). The interconnectedness between climate, springs, and ecosystem health has generated attention in scientific literature but often lacks specific groundwater processes that govern spring discharge. The purpose of this research was to compare the recharge, storage, and drainage processes of high-elevation karst and nonkarst groundwater systems in a semiarid region. Three karst springs (Clover Springs, Hoxworth Springs, and Robber's Roost Springs) and one siliciclastic, nonkarst spring (North Canyon Springs) on the Colorado Plateau guided our scope and research questions. We hypothesized that spring drainage properties vary by hydrostratigraphic unit and associated karstification, with the three karst springs dominated by conduit-flow regimes and the siliciclastic spring controlled by diffuse drainage of the porous media. We predicted that localized recharge, distinct seasonal recharge patterns, and minimal groundwater mixing occur in both karst and siliciclastic springs. The recharge patterns and all associated processes were dependent on the complex distribution of snow and rain in this high-elevation, semiarid region.

4.2. STUDY AREA

The Colorado Plateau extends approximately 210,000 km² across New Mexico, Colorado, Utah, and Arizona. The plateau rises to over 3,700 m in elevation (average of 1,500 m), which causes spatial variability in precipitation (median 300 mm/year) (Hereford et al., 2002). Long-term, century-scale, temporal variability has shown multidecadal periods of drought and wet episodes, while El Niño and La Niña effects control shorter-term variability of approximately 4 to 7 years (Hereford et al., 2002). Fronts originating in the Pacific Ocean (winter precipitation) and tropical rains developing in the Gulf of Mexico and Gulf of California (summer monsoons) govern seasonal precipitation distinctions on the plateau (Hereford et al., 2002).

This study encompasses springs of the Kaibab Plateau and the Mogollon Rim in Arizona on the southern portion of the Colorado Plateau (Fig. 4.1). Both regions experience characteristic bimodal seasonal precipitation, with minimal precipitation between seasons, making studies with hydrograph recession techniques viable. The study areas exhibit classic sinkhole morphology of the epikarst that are predominant recharge locations for the C aquifer (Jones et al., 2017; Huntoon, 1970).

4.2.1. Kaibab Plateau

The Kaibab Plateau is the uplifted region to the north of Grand Canyon National Park, bound to the east by the East Kaibab Monocline and to the west by Crazy Jug Monocline and several faulted segments of the monocline (Fig. 4.1) (Wood et al., 2020; Huntoon, 1970). The plateau reaches a maximum elevation of approximately 2,800 m and receives an average of 610 mm of precipitation per year (PRISM, 2021). The C aquifer on the Kaibab Plateau serves as an important water source for remote cabins, livestock water, firefighting, and backcountry travel (Wood, 2019).

4.2.2. Mogollon Rim

The Mogollon Rim is the rugged erosional escarpment trending northwest-southeast in central Arizona and marking the southern extent of the Colorado Plateau (U.S. Department of Agriculture, 2020). The maximum elevation reaches approximately 2,440 m and shows a vertical relief of 610 m at some cliff faces. With an average of 762 mm of precipitation per year, the Mogollon Rim

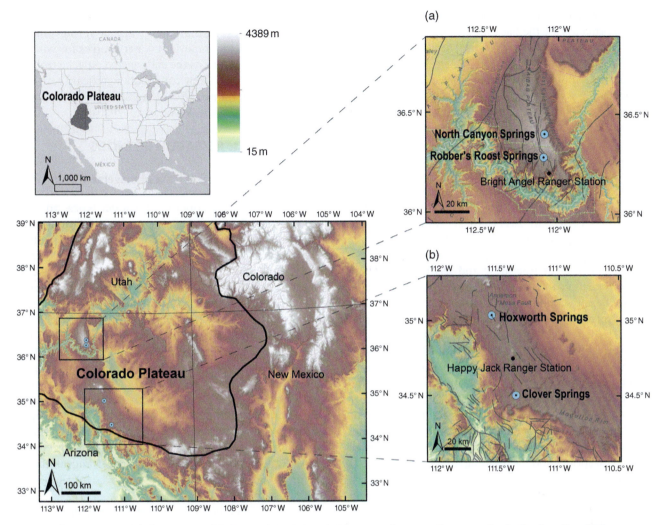

Figure 4.1 Digital elevation model (DEM) of the Colorado Plateau and surrounding areas. Dashed callout lines highlight each study region: (a) Kaibab Plateau springs (North Canyon Springs and Robber's Roost) and the Bright Angel Ranger Station; (b) Mogollon Rim springs (Hoxworth Springs and Clover Springs) and the Happy Jack Ranger Station. Note the mapped faults of both study regions that influence the landscape and spring occurrences.

region is the wettest in Arizona (PRISM, 2021). Deep erosional downcutting of Paleozoic stratigraphy exposes extensive stratigraphic sections, but lacks the soft, soluble gypsiferous Toroweap Formation that is prominent on the Kaibab Plateau (Blakey, 1990; McKee, 1938; Huntoon, 1970). This absence has likely affected the morphology of the karst terrain and subsequent groundwater flow, though the region exhibits substantial karst and epikarst development. Approximately 16 km southeast of Flagstaff, Arizona, the Anderson Mesa-Lake Mary fault system is a major control of surface and groundwater occurrences (Fig. 4.1) (Miller et al., 2007) providing preferential infiltration and groundwater flow paths. Hoxworth Springs lies in the down-dropped block bounded by two normal faults along with over 10 other documented springs in the graben.

4.2.3. Hydrogeology

In central and northern Arizona, the regional aquifer exists as a multilayered karst-siliciclastic system that provides residents with fresh drinking water, aids in sustaining base flow to perennial river systems (Swanson et al., 2020), and supports local tourism, such as over 6 million Grand Canyon visitors each year (Jones et al., 2017). The C aquifer is the uppermost unconfined, perched system in the regional multilayered aquifer and generally consists of three main water-bearing geologic formations within the study area: the Kaibab Formation, the Toroweap Formation (absent in the Mogollon Rim region), and the Coconino Sandstone. These hydrostratigraphic units are primarily recharged through sinkholes and are also infiltrated via fractures, breccia pipe features, and diffuse infiltration

(Jones et al., 2017; Huntoon, 2000). The Hermit Shale and Supai Formation are the siliciclastic confining units in both regions that slow downward, vertical groundwater flow to the deeper Redwall-Muav aquifer (R aquifer) unless it has faulting or fracturing, or it contains vertical conduits.

Because it sources the largest springs in the region (Huntoon, 2000), past hydrogeologic research has focused on the R aquifer (Bills et al., 2007; Crossey et al., 2006; Hill & Polyak, 2010; Huntoon, 1970, 2000; Jones et al., 2017; Schindel, 2015). Still, hydrogeologists have conducted several field studies and modeling efforts on the C aquifer; U.S. Geological Survey reports (Bills et al., 2007; Cooley et al., 1969; Hart et al., 2002; Leake et al., 2005; Pool et al., 2011) have shown highly variable C aquifer properties attributed to porosity variability stemming from lithologic variation (Wood et al., 2020) as well as degree of fracturing and secondary mineralization (Hart et al., 2002). While past C aquifer studies were generally conducted in the context of well yields, karst springs are similarly most productive with increasing effective porosity and can also be subject to the greatest discharge variability (Moniewski, 2015).

Many of the C aquifer field surveys utilized pumping tests to determine aquifer properties in response to induced stress. Spring monitoring offers a nonintrusive, inexpensive way to monitor related hydraulic characteristics and often is the only way to get data in wilderness settings (such as North Canyon Springs). This study analyzed climatic stresses on springs and the resulting drainage properties of the aquifer. While the studied springs are not individually high-volume springs, there are hundreds of analogous springs in both study regions (Springs Online, 2021) that collectively contribute to critical water sources and support ecosystem health (Stevens et al., 2020; Sinclair, 2018).

Clover Springs contributes flow to the Verde River, one of Arizona's last perennial river systems (Mueller et al., 2013). Anderson et al. (2003) and Griffiths et al. (2008) studied the wet meadow geomorphologies at Clover and Hoxworth springs, and this study provides long-term data to assess the success of restoration actions.

North Canyon Springs and Robber's Roost Springs are among the highest discharging C aquifer springs on the Kaibab Plateau, are important water sources for wildlife and fish habitats (such as the endangered Apache trout in North Canyon), and likely contribute significant recharge to the R aquifer through sinking streams and vertical groundwater migration (Wood et al., 2020).

4.3. RESEARCH METHODS

4.3.1. Field Methods and Data Management

We compiled and organized climate data from a combination of manual collection, precipitation models, and existing published data. Field sampling of precipitation isotope data took place at three stations on the Kaibab Plateau following original methods outlined by Wood (2019) and updated by Donovan (2021). Precipitation collectors used mineral oil to eliminate, or substantially minimize, evaporation from bulk snowmelt and monsoon rainwater and represent true meteoric isotopic seasonal signatures. Mogollon Rim precipitation data relied upon published results by Beisner et al. (2016) and Blasch and Bryson (2007). We compiled modeled daily precipitation totals from PRISM (2021) using respective coordinates and elevations at each spring (Table 4.1). Snow depth data and air temperature came from the National Centers for Environmental Information (NOAA, 2021); Bright Angel Ranger Station and Happy Jack Ranger Station provided data for the Kaibab Plateau and Mogollon Rim analyses, respectively.

The study includes continuous monitoring of four springs on the southern Colorado Plateau (Table 4.1). We instrumented each spring with vented InSitu™ pressure transducers to generate continuous records of water pressure at each site. Factory calibration to the 5 Psig range and 3.5 m depth took place for all transducers prior to installation. Each monitoring station sits at or near the spring source to eliminate interference from runoff and only reflect changes representative of springflow, except

Table 4.1 List of springs, hydrostratigraphic unit, spring classification, and monitoring information

Site	Length of record	Spring type	Unit	In situ instrument	Geographic coordinates (WGS84)
Clover Springs	17 June 2010 to 4 September 2020	Hillslope	Kaibab Formation	Level TROLL 500 and Aqua TROLL 600	34.50592, −111.36270 Elevation: 2,093 m
Hoxworth Springs	1 January 2018 to 1 January 2021	Rheocrene	Kaibab Formation	Level TROLL 500	35.04033, −111.47497 Elevation: 2,154 m
Robber's Roost Spring	13 November 2018 to 11 March 2020	Hillslope	Toroweap Formation	Level TROLL 500 and Aqua TROLL 600	36.28051, −112.08891 Elevation: 2,528 m
North Canyon Springs	14 November 2019 to 1 November 2020	Hillslope	Coconino Sandstone	Level TROLL 500	36.39754, −112.08464 Elevation: 2,543 m

Source: Spring classification adapted from Springer and Stevens (2009).
Note: Clover Springs and Robber's Roost Springs have been significantly altered anthropogenically.

for Hoxworth Springs, which has several discharge locations within a stream channel.

We conducted monthly to quarterly site visits to take stable isotope water samples, collect discrete discharge measurements, and download transducer data. We uploaded the transducer data to existing databases and converted water pressure to discharge using modified rating curve methods guided by Sauer (2002). Clover Springs data are published in CUAHSI Hydroshare (Donovan & Springer, 2021) while unpublished data exists on NAU servers. A complete outline of sampling methods, rating curve techniques, base-flow separation analysis, and laboratory procedures is in Donovan (2021).

4.3.2. Hydrograph Analysis

Analysis of the hydrograph recession period provides critical information on drainage properties after peak flows (Doctor et al., 2005), and recession analyses in this study are a proxy for spring behavior during drought conditions. Published research highlights various methods for characterizing complex network geometries (Kovács & Perrochet, 2007), fine-tuning details of hydrograph components (Kovács et al., 2004), and this study provides recession coefficient ranges and associated microregime classification based on statistically significant changes in recession slopes from ANOVA (analysis of variance). Comparisons of the microregimes with previously published recession coefficients, α, in karst terrain (Jones et al., 2017; Amit et al., 2002; Tobin & Schwartz 2016; Bonacci, 1993) confirmed reasonable ranges for this classification scheme.

To characterize the microregimes during drainage, we selected three hydrograph recession periods for Clover Springs and one recession for Hoxworth, Robber's Roost, and North Canyon springs to analyze behavior during monsoon or postmonsoon events. Because Maillet's equation assumes rainfall-dominated precipitation, and because of complex snowmelt variability, we did not include recession analyses for winter recharge.

We selected base-flow recession events based on ideal conditions of long dry periods preceding input from precipitation (Kresic & Stevanović, 2010) and performed analysis using an algebraically manipulated version of Maillet's equation (1905):

$$\alpha = \frac{\log\left(\frac{Q_n}{Q_n+1}\right)}{0.4343(t_{n+1}-t_n)} \quad (4.1)$$

where α is the recession coefficient describing the microregime (in days^{-1}), Q is the discharge (in liters per second, L/s), t is the time (in days), and 0.4343 is a constant relating discharge and time.

Hydrographs used in conjunction with precipitation, snow depth records, and daily air temperature enabled assessment of spring response timing from recharge events. Again, we focused on summer monsoon response times determined by the elapsed time (days) between peak precipitation and peak discharge.

Though analyses from singular recharge events were the focus for flow regime characterization, several other important aquifer indicators illustrate seasonal and annual characteristics. Springflow variability, annual peak discharge timing, total annual discharge, and mean daily discharge are particularly important in regions that have anecdotal evidence of aquifer and spring ecosystem health decline. Extended records of quantitative, evidence-based results provided important general guiding factors to develop comprehensive action.

4.3.3. Regression Modeling

We built two regression models to quantify the role of snowmelt in sustaining seasonal discharge across five winter seasons at Clover Springs. The regression models predict approximate seasonal timing of spring discharge using two metrics: seasonal snowpack duration and peak annual snowpack depth, which is a proxy for total seasonal snow. To normalize the two time series, we defined the duration for all data sets as days elapsed since the start of the respective year. Both models assume that rain contribution is insignificant during snowmelt recharge, and the catchment area does not significantly change across seasons.

4.3.4. Stable Isotope Analysis

Stable isotope ratios, $\delta^{18}O$ and $\delta^{2}H$ (referred to hereafter as delta values), act conservatively in low-temperature groundwater systems and are irreplaceable in groundwater studies (Lee & Krothe, 2001; Kresic, 2013). They are particularly useful for this study region because distinct seasonal delta value signatures reflect the bimodality of precipitation.

We categorized delta values for each spring by the season of collection (Summer monsoon season = June–September; Fall postmonsoon season = October–November; Winter snow season = December–March) and plotted them by respective spring and season. This organization allowed for analysis of recharge seasonality as well as determination of the seasonal mixing and storage within watershed.

4.4. RESULTS

4.4.1. Base-Flow Recession Analysis

Clover and Hoxworth springs showed distinguishable microregimes, indicated by breakthroughs in discharge rates during recession periods, while North Canyon Springs and Robber's Roost Springs displayed only single flow regimes. Though authors have varying recession coefficient

40 THREATS TO SPRINGS IN A CHANGING WORLD

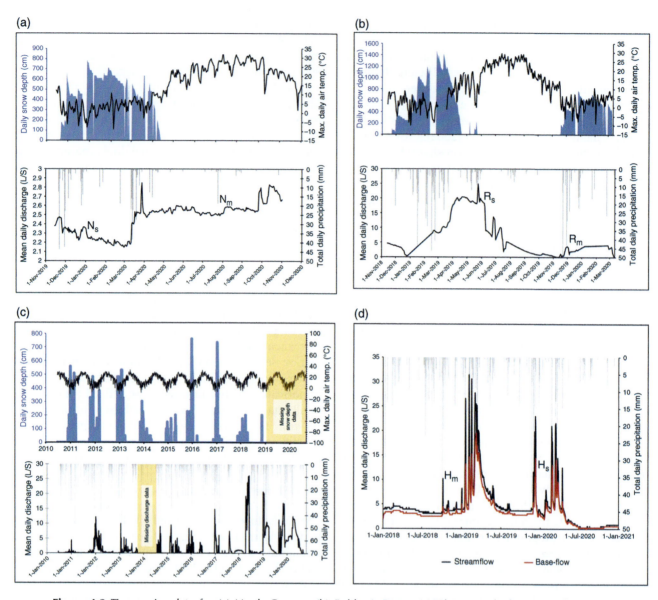

Figure 4.2 Time series data for (a) North Canyon, (b) Robber's Roost, (c) Clover, and (d) Hoxworth springs including hydrographs, hyetographs, snowpack, and daily air temperature. Hoxworth Springs does not include snowpack data because of local insufficient data. Note that Hoxworth Springs discharges in a stream channel (Rheocrene) and necessitated base-flow separation (Donovan, 2021).

ranges for accompanying microregimes, this study provides the following classification: Fast flow (α_1) ranged from 8.5×10^{-2} to 1.1×10^1; intermediate flow (α_2) ranged from 5.0×10^{-2} to 6.0×10^{-2}; and slow flow (α_3) ranged from 1.0×10^{-3} to 10.0×10^{-3}. ANOVA results indicated no statistically significant differences between α_1 values across springs ($F_{1,10} = 16.3$, $p < 0.05$). Though limited in sample size to perform ANOVA, α_3 consistently showed values an order of magnitude lower than α_2 and were used to differentiate between intermediate- and slow-flow regimes.

For Hoxworth, Robber's Roost, and North Canyon springs, the recession identifiers indicate the onset of the recharge and recession period (including winter snowmelt for reference) (Fig. 4.2, Table 4.2). For example, in Figure 4.2, N_m denotes monsoon (and postmonsoon) recharge at North Canyon Springs and N_s signifies snowmelt recharge. A graphic representation of monsoon recession at Clover Springs exemplifies the recession analyses performed for every spring (Fig. 4.3).

4.4.2. Hydrograph Results

Discharge variability ranged from steady at North Canyon Springs to ephemeral at Clover and Robber's Roost springs, highlighting the seasonal buffering and

Table 4.2 Monsoon recession events for each spring

Spring and recession ID	Number of recessions	Recession dates	α_1 (days^{-1})	α_2 (days^{-1})	α_3 (days^{-1})	$1/\alpha_1$ (days)	$1/\alpha_2$ (days)	$1/\alpha_3$ (days)
Clover Springs, C_m	3	7 August 2017 to 16 September 2017	1.1×10^{-1}	5.7×10^{-2}	–	8.7	17.6	–
Hoxworth Springs, H_m	1	7 October 2018 to 14 October 2018	8.9×10^{-2}	5.4×10^{-2}	9.5×10^{-3}	11.2	16.1	105.3
Robber's Roost Springs, R_m	1	30 November 2019 to 11 December 2019	–	5.1×10^{-2}	–	–	19.7	–
North Canyon Springs, N_m	1	9 August 2020 to 17 August 2020	–	–	1.7×10^{-3}	–	–	594.8

Note: Recession coefficients (α) represent the slope of discharge (in logarithmic form) versus time. The inverse ($1/\alpha$) is the total drainage time of the respective microregime uninfluenced by subsequent recharge.

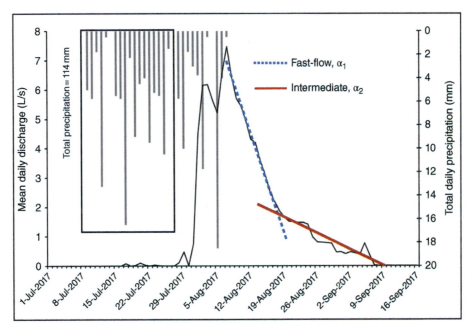

Figure 4.3 Event-scale representation of typical Clover Springs monsoon response and subsequent recession during monsoon recharge. Note the total precipitation preceding a spring response outlined by the solid black box.

drainage mechanisms (Table 4.3). Clover Springs is dry for nearly half of the year.

4.4.3. Regression Modeling

Regression modeling of the snowpack and seasonal springflow emphasized the importance of snowmelt in sustaining discharge at Clover Springs and served as a proxy for ephemeral springs within the study region. The regression model explains the influences that snow has on the ephemeral nature of springs, a critical determination in springs that lack matrix-block drainage or other slow-release mechanisms (Fig. 4.4). The duration of seasonal snowpack and the length of seasonal springflow at Clover Springs showed a significant relationship ($R^2 = 0.8445$, $F_{1,3} = 16.3$, $p < 0.05$) and no statistically significant relationship exists between peak snowpack depth and seasonal springflow ($R^2 = 0.2159$, $F_{1,3} = 0.823$, $p > 0.05$).

4.4.4. Response Timing

Clover and Robber's Roost springs appeared to exhibit flashy (1–2 day) responses from monsoon precipitation when the springs were already flowing (Fig. 4.2b,c). When the springs were dry, however, extended responses occurred, likely varying from antecedent moisture conditions (Schwartz et al., 2013) and evapotranspiration effects. Delayed activation at Clover Springs during the 2017 monsoon season lagged nearly one month with over 100 mm of cumulative precipitation until the spring

Table 4.3 General statistical summary, ephemerality, and discharge variability of Clover, Hoxworth, Robber's Roost, and North Canyon springs

Spring	Mean daily discharge (L/s)	Month of peak discharge	Total days dry per year	Discharge variability	Min daily discharge (L/s)	Max daily discharge (L/s)
Clover Springs	21	January (mode)	129 (average)	Ephemeral	0	26.1
Hoxworth Springs	3.5	March (mode)	Perennial	551%,	0.05	19.5
Robber's Roost Springs	5.2	May	70	Ephemeral	0	24.9
North Canyon Springs	2.5	March	Perennial	28%,	2.1	2.8

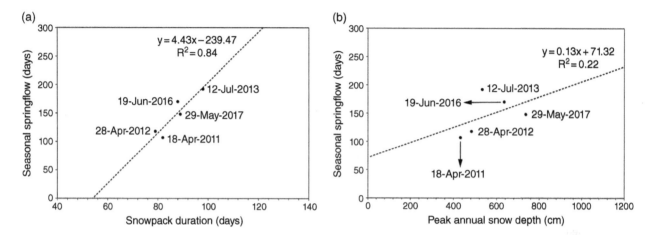

Figure 4.4 Linear regression models at Clover Springs of (a) relationship between seasonal duration of snowpack versus seasonal duration of springflow and (b) peak snowpack (as a proxy for total snow) versus seasonal duration of springflow. Units for both variables are in days elapsed from the start of the year (1 January). Annotations represent the date of seasonal drying of the spring.

responded. It's important to note that there is a complex set of feedback loops that govern spring responses, and further work needs to be conducted to determine the variables driving response timing at these springs.

Hoxworth Springs also showed lagged responses during monsoon season. Cumulative summer monsoon precipitation in 2018 totaled 420 mm before the spring responded (Fig. 4.2d). A postmonsoon 45 mm rain event, H_m, triggered a response within 4 days, and increased discharge by over 2 L/s within a single day.

North Canyon Springs does not appear to respond to singular precipitation events and only responded seasonally to meteoric input. While a 2 week series of March 2020 rain-on-snow events did trigger a response in approximately 15 days, this was a seasonal response; over the course of 2 weeks, the snow depth melted completely (from over 500 cm) and sustained discharge through monsoon season with minimal monsoon recharge (Fig. 4.2a).

4.4.5. Stable Isotope Analysis

The $\delta^{18}O$ and δ^2H clustering shown in Figure 4.5 illustrates distinctions in recharge zones between study areas. Clover Springs and Hoxworth Springs likely experience recharge from similar elevations on the Mogollon Rim based on their isotopic similarities. Likewise, North Canyon Springs and Robber's Roost Springs appear to receive recharge from similar elevations on the Kaibab Plateau.

Because winter snow and summer rain have distinct isotopic compositions, they can be used as natural environmental tracers to assess the effect of seasonality on spring discharge. The seasonal separation of delta values shown in Figure 4.6 highlights generalized groundwater mixing and storage based on time lags between the season of collection and their seasonal signatures. It's important to note a potential limitation for isotope analyses: several snowmelt and sublimation fractionation processes influence isotope composition in addition to seasonal sources of moisture and elevational fractionation (Sharp, 2017). However, the coupled hydrograph analysis results strengthen these interpretations.

4.5. DISCUSSION

Regional seasonal estimates specifically predict decreased groundwater infiltration from April to June through the end of century (Tillman et al., 2020). These

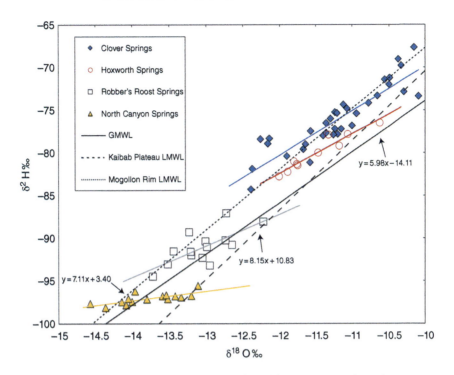

Figure 4.5 Stable isotope delta values for Clover, Hoxworth, Robber's Roost, and North Canyon springs during the study period and associated best-fit lines for spring samples. Also included is the Global Meteoric Water Line (adapted from Craig, 1961) and Local Meteoric Water Lines for the Kaibab Plateau (adapted from Wood et al., 2020) and Mogollon Rim region (adapted from Beisner et al., 2016).

months are critical recharge periods at ephemeral springs on the southern Colorado Plateau and pose a serious threat to sustained seasonal springflow and resulting ecosystem health decline. Warming temperatures, increased evaporation, and declining total groundwater recharge suggest that spring discharge thresholds will increase and consequently decrease the duration and magnitude of spring discharge.

Hydrograph recession analyses, regression modeling, and isotopic analyses built upon these predictions to create an understanding of the complex relationship between climatic input and spring discharge in the study regions. Several generalized explanations may serve as a guiding analog for other similar spring systems across the Colorado Plateau and are used to fill data gaps from Meixner et al. (2016).

Snowmelt-driven spring systems on the southern Colorado Plateau will experience altered seasonal responses with continued shifts to rain-dominated weather systems. Because snowmelt is a critical seasonal storage component (especially in ephemeral karst springs that lack a slow-flow buffer), seasonal springflow durations will be shortened. This model largely occurs because, contrary to delayed and diffuse-flow in siliciclastic spring systems, recharge timing is commonly flashy in conduit-driven karst networks of the southern Colorado Plateau.

These results are important considerations for assessing the cause and duration of water availability from epikarst springflow. The karst systems (Clover, Hoxworth, and Robber's Roost springs) all showed rapid responses to recharge events once springflow initiation occurred. However, the occurrence and magnitude of all responses appeared to depend on antecedent conditions (Schwartz et al., 2013), storm intensity (Liu et al., 2007), and subsequent infiltration. Several prolonged monsoon events did not initiate springflow at all, which is likely due to increased evaporation and soil water recharge, a function of the semiarid climate (Taylor & Greene, 2008). Though this research did not directly evaluate soil-water interactions, past studies have highlighted the influence of both soil (Tooth & Fairchild, 2003; Hu et al., 2015) and vegetation zones on karst aquifers and spring discharge (Schwartz et al., 2013; Tobin et al., 2021).

Countless other research publications and reviews have assessed the role of epikarst and unsaturated zone hydrologic processes on karst aquifer dynamics (Pronk et al., 2009; Bakalowicz, 2004; Polk et al., 2013; Barbel-Périneau et al., 2019; Arbel et al., 2010; Zhang et al., 2013; Williams, 2008; Aquilina et al., 2006; Trček, 2006). Though these efforts generally focus the research questions to address specific phenomena, collectively they highlight the importance of the epikarst and its variety

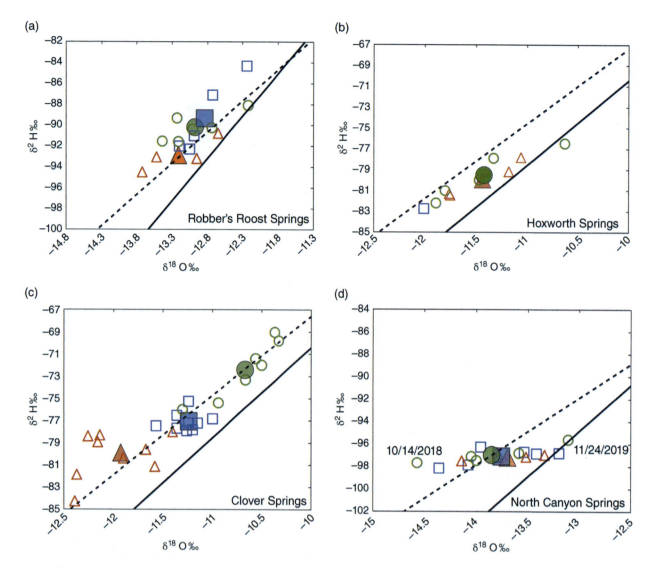

Figure 4.6 Stable isotope ratios for (a) Robber's Roost Springs, (b) Hoxworth Springs, (c) Clover Springs, and (d) North Canyon Springs classified by the season of collection. The symbol scheme is unfilled triangles = winter samples, unfilled squares = summer samples, and unfilled circles = fall samples. The filled symbols represent the mean values of the corresponding season. Also included are the Global Meteoric Water Line (adapted from Craig, 1961) and Local Meteoric Water Lines for the Kaibab Plateau (a and b) (adapted from Wood et al., 2020) and Mogollon Rim region (c and d) (adapted from Beisner et al., 2016).

of functions; the epikarst can rapidly infiltrate water, recharge water to deeper aquifers, store groundwater in voids, or discharge water through springs. The variability of the epikarst results from differences in permeability, porosity, storage capacity, and preceding hydrological conditions. A recent study by Tobin et al. (2021) employed a holistic approach to evaluate the multitude of variables that govern epikarst processes. The conceptual model suggests that fast- and slow-flow components (often conduit-flow and matrix-flow, respectively) can either offer storage or promote flow, which vary based on prior conditions. Along with these storage versus flow dynamics, recharge processes are governed by a set of feedback systems that vary across year, season, and even single events.

4.5.1. Interpretation of Drainage Properties

Each hydrograph record demonstrated that springs have unique drainage characteristics largely controlled by the physical mechanisms and effective porosity of the local aquifer system. Like other shallow karst conceptual models (Gremaud et al., 2009; Goldscheider, 2005), stratification and structural controls were important in

Figure 4.7 Annotated photograph of epikarst outcrop near Clover Springs showing soil-filled conduits (encircled) as well as notable fractures (vertical lines) and bedding planes (horizontal lines).

governing flow patterns. Clover Springs demonstrated both fast-flow and intermediate-flow microregimes but lacked a slow-flow component. Based on interpretations from outcrops, the active mechanisms translate to groundwater transport through conduits (fast flow) and fracture-bedding planes (intermediate flow) (Fig. 4.7). Robber's Roost Springs similarly depended on mixed conduit- and fracture-flow while lacking a third, slow-flow mechanism of drainage. North Canyon Springs and Hoxworth Springs were the only systems sustained by slow flow through matrix-block drainage. Hoxworth Springs also experienced activation of its conduit-fracture networks during monsoon recharge, while North Canyon Springs lacked any sign of fracture or conduit contribution.

While outcrop interpretations and similar published papers support these translations, the microregime distinctions are not absolute. For example, during fast flow (interpreted as drainage through conduits), there is potentially simultaneous contribution from fractures and matrix drainage as well. Further hydrograph component separation is attainable through high-resolution storm sampling and geochemical analyses as performed by Lee and Krothe (2001). However, because of temporally limited geochemical data sets, we were not able to build mixing models for discharge separation on an event basis.

4.5.2. Seasonal Water Storage

The overlap of spring delta values with local meteoric water signatures suggests the springs are all localized systems dependent on precipitation that falls in the local catchment and lacks contribution from regional groundwater flow paths. Springer et al. (2016) and Solder et al. (2020) have supporting explanations that deep aquifer recharge results from mixed regional and local flow systems and indicate an influential modern groundwater component that likely originates in plateau regions. Results from this research highlight the contribution of local flow regimes in high-elevation plateau settings.

Common in arid to semiarid regions, isotopic trends signify an evaporative influence in all four springs (Fig. 4.5) (Sharp, 2017). North Canyon Springs has the gentlest slope, indicating the strongest evaporative trend, and may suggest sublimation of a persistent seasonal snowpack. The North Canyon Springs catchment holds snow into May (Fig. 4.2), and personal observations indicate that in previous years the snowpack remained through June.

Hoxworth Springs delta values all plot below the Local Meteoric Water Line (LMWL), indicating a strong evaporative signal due to its sphere of discharge in a rheocrene environment. The most depleted delta value embodies a summer monsoon sample but is likely a result of evaporation rather than a sign of seasonality. Because of uncertainties in evaporative conditions during the time of sampling in the rheocrene environment, it is difficult to draw conclusions on processes with the given isotope record.

Hydrograph analyses support these isotope interpretations and show that there is a relationship between mixing, storage, evaporation, and drainage times. Clover Springs responds quickly to meteoric input and shows quick epikarst drainage following peak discharge. However, unless the system overcomes complex threshold processes, the spring either does not respond or lags significantly in response. Distinct seasonal isotope signatures suggest that groundwater is neither stored nor mixed between seasons (Fig. 4.5) but is rather flushed through conduits and fractures when the spring responds. Robber's Roost appears to experience some mixing within fractures from overlapping seasonal isotope signatures (Fig. 4.5) and may store water within solutional cavities. North Canyon Springs demonstrates the greatest ability to store and mix water based on overlapping seasonal signatures and homogenization of delta values across seasons during slow, diffuse drainage. In addition, the end members of fall delta values, suggest seasonal long-flow paths, which further substantiates these interpretations.

4.5.3. Snowpack

Seasonal analysis of winter snowmelt highlighted the seasonality of ephemeral springflow and showed that the longevity of seasonal discharge depended on the timing of snowmelt. In the ephemeral karst systems, the drainage period after meteoric input was less than a month through conduits and fractures, while the perennial systems sustained springflow through matrix-block drainage for up to nearly 20 months. That is, without additional recharge events during these drainage periods, springs will dry, as seen at Clover Springs and Robber's Roost Springs seasonally. Snowmelt serves as the slow-flow microregime in these conduit- and fracture-dominated systems, extending seasonal springflow duration and sustaining ecosystem health, and thus is a critical storage and recharge component of its own. Models in Chen et al. (2018) corroborate these interpretations by highlighting similar short-term climate sensitivities in karst aquifers through shifts in snow storage and aquifer recharge.

The release of stored water in the snowpack is important to sustain springflow before the ephemeral systems seasonally dry (Fig. 4.4a). The relationship is statistically significant and is an informative metric for predicting spring discharge timing. The slope of the relationship specifies that for every additional day that snowpack persists, Clover Springs flows for an additional 4.43 days. Because this model inherently includes important water budget components (sublimation and evapotranspiration), they are not explicitly included. For example, increased sublimation fundamentally decreases the snowpack duration in each season, and the sloped relationship is likely a function of this phenomenon. When sublimation and evapotranspiration are not actively depleting the snowpack, snowmelt slowly infiltrates into the spring system, rather than being transferred to the atmosphere.

In contrast, there is no statistically significant relationship between peak snow depth and duration of seasonal springflow (Fig, 4.4b). That is, total snowfall alone is not a quality metric for estimating the duration of ephemeral karst springflow on the southern Colorado Plateau. Sublimation and evapotranspiration have variable influence across seasons and should be explicitly included in models when not intrinsically accounted for.

Hydrograph analyses and the spring isotope record corroborate these observations found in karst systems. Increasing daily air temperatures and subsequent declining snowpack depth resulted in discharge response during every snowmelt season at each spring where data existed (Fig. 4.2). Depleted winter isotope signatures appeared in the karst spring records until the final leg of the snowmelt recession, which validates the notion that snowmelt recessions are a result of same-season snowmelt recharge and justifies the basis for developing regression models. While the sample size of these models is relatively small (n = 5), the results are promising and worth collecting subsequent years of data to verify the model. Replicated studies for other karst systems may further substantiate the results.

4.6. SUMMARY

The coupled analyses conducted in this research provide important clarification on the recharge timing, drainage properties, and storage implications of a complex karst spring network on the southern Colorado Plateau. Though limited to four significant springs in the vast C aquifer, these springs are vital measures of aquifer status in data-poor regions. Moreover, these springs are individually critical for water supply, and they support wetland meadow environments and sustain diverse ecological habitats in a high elevation semiarid region (Sinclair, 2018; Stevens et al., 2020). Although various aquifer tests could further quantify aquifer properties, these spring analyses use nonintrusive, cost-effective approaches for characterizations that are arguably more representative of natural stresses. These low-cost methods can translate to other climate-sensitive regions to guide upland watershed policy and management; water management agencies could similarly implement regular water collection programs and continuous discharge monitoring as indicators for spring and aquifer responses to drought.

In addition, results of this study inform and update groundwater models and data gaps (Leake et al., 2005; Pool et al., 2011; Meixner et al., 2016) on the Colorado Plateau with higher resolution and farther spatial extents previously limited by lack of geophysical or well data.

ACKNOWLEDGMENTS

Funding for this research came from the National Park Service (Task Agreement #P20AC00684 under Master Cooperative Agreement P17AC00624) and U.S. Forest Service (Agreement #19-CS-11030700-010). Thank you to Hannah Chambless, Max Evans, and Zev Axler who all contributed significant time and positive energy to field monitoring trips, and to Ronni Chavez for compiling preanalysis data sets at Hoxworth Springs. Also, thank you to Alex Wood and Riley Swanson for establishing the Kaibab Plateau spring network and for their shared knowledge of spring monitoring.

REFERENCES

Ahmadalipour, A., Moradkhani, H., & Svoboda, M. (2017). Centennial drought outlook over the CONUS using NASA-NEX downscaled climate ensemble. *International Journal of Climatology*, 37(5), 2477–2491. https://doi.org/10.1002/joc.4859

Amit, H., Lyakhovsky, V., Katz, A., Starinsky, A., & Burg, A. (2002). Interpretation of Spring Recession Curves. *Ground Water*, 40(5), 543–551. https://doi.org/10.1111/j.1745-6584.2002.tb02539.x

Anderson, D. E., Springer, A. E., Kennedy, J., & DeWald, L. (2003). Verde River Headwaters Restoration Demonstration Project. U.S. Forest Service.

Aquilina, L., Ladouche, B., & Dörfliger, N. (2006). Water storage and transfer in the epikarst of karstic systems during high flow periods. *Journal of Hydrology*, 327(3–4), 472–485. https://doi.org/10.1016/j.jhydrol.2005.11.054

Arbel, Y., Greenbaum, N., Lange, J., & Inbar, M. (2010). Infiltration processes and flow rates in developed karst vadose zone using tracers in cave drips. *Earth Surface Processes and Landforms*, 35(14), 1682–1693. https://doi.org/10.1002/esp.2010

Bakalowicz, M. (2004). The epikarst, the skin of karst. *Karst Waters Institute Special Publication*, 9, 16–20.

Barbel-Périneau, A., Barbiero, L., Danquigny, C., Emblanch, C., Mazzilli, N., Babic, M., et al. (2019). Karst flow processes explored through analysis of long-term unsaturated-zone discharge hydrochemistry: A 10-year study in Rustrel, France. *Hydrogeology Journal*, 27(5), 1711–1723. https://doi.org/10.1007/s10040-019-01965-6

Beisner, K. R., Paretti, N. V., & Tucci, R. S. (2016). *Analysis of stable isotope ratios (δ18O and δ2H) in precipitation of the Verde River watershed, Arizona 2003 through 2014*. Open-File Report. https://doi.org/10.3133/ofr20161053

Bills, D. J., Flynn, M. E., & Monroe, S. A. (2007). *Hydrogeology of the Coconino Plateau and adjacent areas, Coconino and Yavapai Counties, Arizona*. U.S. Geological Survey Scientific Investigations Report 2005–5222, 1.1, 101. https://dx.doi.org/10.3133/sir20055222v1.1

Blakey, R. (1990). Stratigraphy and geologic history of Pennsylvanian and Permian rocks, Mogollon Rim region, central Arizona and vicinity. *Geological Society of America Bulletin*, 102(9), 1189–1217. https://doi.org/10.1130/0016-7606

Blasch, K. W., & Bryson, J. R. (2007). Distinguishing sources of ground water recharge by using d2H and d18O. *Ground Water*, 45(3), 294–308. https://doi.org/10.1111/j.1745-6584.2006.00289.x

Bonacci, O. (1993). Karst springs hydrographs as indicators of karst aquifers. *Hydrological Sciences Journal*, 38(1), 51–62. https://doi.org/10.1080/02626669309492639

Bonacci, O. (2014). Karst hydrogeology/hydrology of dinaric chain and isles. *Environmental Earth Sciences*, 74(1), 37–55. https://doi.org/10.1007/s12665-014-3677-8

Brinkmann, R., & Parise, M. (2012). Karst environments: Problems, management, human impacts, and sustainability: An introduction to the special issue. *Journal of Cave and Karst Studies*, 74(2), 135–136. https://doi.org/10.4311/2011jcks0253

Cantonati, M., Fensham, R. J., Stevens, L. E., Gerecke, R., Glazier, D. S., Goldscheider, N., et al. (2020). Urgent plea for global protection of springs. *Conservation Biology*, 35(1), 378–382. https://doi.org/10.1111/cobi.13576

Cartwright, J. M., Dwire, K. A., Freed, Z., Hammer, S. J., McLaughlin, B., Misztal, L. W., et al. (2020). Oases of the future? Springs as potential hydrologic refugia in drying climates. *Frontiers in Ecology and the Environment*, 18(5), 245–253. https://doi.org/10.1002/fee.2191

Chen, Z., Hartmann, A., Wagener, T., & Goldscheider, N. (2018). Dynamics of water fluxes and storages in an alpine karst catchment under current and potential future climate conditions. *Hydrology and Earth System Sciences*, 22(7), 3807–3823. https://doi.org/10.5194/hess-22-3807-2018

Chiew, F. H. S., Young, W. J., Cai, W., & Teng, J. (2011). Current drought and future hydroclimate projections in southeast Australia and implications for water resources management. *Stochastic Environmental Research and Risk Assessment*, 25(4), 601–612. https://doi.org/10.1007/s00477-010-0424-x

Cooley, M. E., Harshbarger J. W., Akers, J. P., & Hardt, W. F., (1969). *Regional hydrogeology of the Navajo and Hopi Indian reservations, Arizona New Mexico, and Utah, with a section on vegetation*. Geological Survey Professional Paper 521-A. U.S. Government Printing Office. http://dx.doi.org/10.3133/pp521A

Craig, H. (1961). Isotopic variations in meteoric waters. *Science*, 133(3465), 1702–1703. https://doi.org/10.1126/science.133.3465.1702

Crossey, L. J., Fischer, T. P., Patchett, P. J., Karlstrom, K. E., Hilton, D. R., Newell, D. L., et al. (2006). Dissected hydrologic system at the Grand Canyon: Interaction between deeply derived fluids and plateau aquifer waters in modern springs and travertine. *Geology*, 34(1), 25. https://doi.org/10.1130/g22057.1

Doctor, D. H., Alexander, E. C., & Kuniansky, E. L. (2005). Interpretation of water chemistry and stable isotope data from a karst aquifer according to flow regimes identified through hydrograph recession analysis. *U.S. Geological Survey*, 82-92. https://www.sciencebase.gov/catalog/item/51715dc5e4b0c560b7054fd5

Donovan, K. (2021). *Karst Spring processes and storage implications in high elevation, semiarid southwestern United States*. Master's thesis, Northern Arizona University, ProQuest Dissertations Publishing.

Donovan, K., & Springer, A. E. (2021). Clover Springs discharge, HydroShare. https://doi.org/10.4211/hs.054f6eb3459d43a890ce1e9a6db65f60

Ford, D., & Williams, P. D. (2007). *Karst hydrogeology and geomorphology (Rev. ed.)*. Hoboken, NJ: Wiley. https://doi.org/10.1002/9781118684986

Godsey, S. E., Kirchner, J. W., & Tague, C. L. (2014). Effects of changes in winter snowpacks on summer low flows: Case studies in the Sierra Nevada, California, USA. *Hydrological Processes*, 28(19), 5048–5064. https://doi.org/10.1002/hyp.9943

Goldscheider, N. (2005). Fold structure and underground drainage pattern in the alpine karst system Hochifen-Gottesacker. *Eclogae Geologicae Helvetiae*, 98(1), 1–17. https://doi.org/10.1007/s00015-005-1143-z

Goldscheider, N. (2019). A holistic approach to groundwater protection and ecosystem services in karst terrains. *Carbonates and Evaporites*, 34(4), 1241–1249. https://doi.org/10.1007/s13146-019-00492-5

Goldscheider, N., Chen, Z., Auler, A. S., Bakalowicz, M., Broda, S., Drew, D., et al. (2020). Global distribution of carbonate rocks and karst water resources. *Hydrogeology Journal*, 28(5), 1661–1677. https://doi.org/10.1007/s10040-020-02139-5

Gremaud, V., Goldscheider, N., Savoy, L., Favre, G., & Masson, H. (2009). Geological structure, recharge processes and underground drainage of a glacierised karst aquifer system, Tsanfleuron-Sanetsch, Swiss Alps. *Hydrogeology Journal*, 17(8), 1833–1848. https://doi.org/10.1007/s10040-009-0485-4

Griffiths, R. E., Anderson, D. E., & Springer, A. E. (2008). The morphology and hydrology of small spring-dominated channels. *Geomorphology*, 102(3–4), 511–521. https://doi.org/10.1016/j.geomorph.2008.05.038

Hart, R. J., Ward, J. J., Bills, D. J., & Flynn, M. E. (2002). *Generalized hydrogeology and ground-water budget for the C Aquifer, Little Colorado River Basin and parts of the Verde and Salt River Basins, Arizona and New Mexico*. Scientific Investigations Report 2005-4026. https://doi.org/10.3133/wri024026

He, X., Wu, J., & Guo, W. (2019). Karst Spring protection for the sustainable and healthy living: The examples of Niangziguan Spring and Shuishentang Spring in Shanxi, China. *Exposure and Health*, 11(2), 153–165. https://doi.org/10.1007/s12403-018-00295-4

Hereford, R., & Amoroso, L. (2020). Historical and prehistorical water levels of Mormon Lake, Arizona, as a measure of climate change on the southwest Colorado Plateau, USA. *Quaternary Research*, 1–20. https://doi.org/10.1017/qua.2020.92

Hereford, R., Webb, R. H., & Graham, S. (2002). Precipitation history of the Colorado Plateau region, 1900–2000. https://pubs.usgs.gov/fs/2002/fs119-02/

Hill, C. A., & Polyak, V. J. (2010). Karst hydrology of Grand Canyon, Arizona, USA. *Journal of Hydrology*, 390(3–4), 169–181. https://doi.org/10.1016/j.jhydrol.2010.06.040

Huntoon, P. W. (1970). The hydro-mechanics of the ground water system in the southern portion of the Kaibab Plateau, Arizona. University of Arizona. https://repository.arizona.edu/handle/10150/190965?show=full

Huntoon, P. W. (2000). Variability of karstic permeability between unconfined and confined aquifers, Grand Canyon region, Arizona. *Environmental and Engineering Geoscience*, 6(2), 155–170. https://doi.org/10.2113/gseegeosci.6.2.155

Hu, K., Chen, H., Nie, Y., & Wang, K. (2015). Seasonal recharge and mean residence times of soil and epikarst water in a small karst catchment of southwest China. *Scientific Reports*, 5(1). https://doi.org/10.1038/srep10215

Jones, C. J. R., Springer, A. E., Tobin, B. W., Zappitello, S. J., & Jones, N. A. (2017). Characterization and hydraulic behaviour of the complex karst of the Kaibab Plateau and Grand Canyon National Park, USA. *Geological Society, London, Special Publications*, 466(1), 237–260. https://doi.org/10.1144/sp466.5

Kovács, A., & Perrochet, P. (2007). A quantitative approach to spring hydrograph decomposition. *Journal of Hydrology*, 352(1–2), 16–29. https://doi.org/10.1016/j.jhydrol.2007.12.009

Kovács, A., Perrochet, P., Király, L., & Jeannin, P. Y. (2004). A quantitative method for the characterisation of karst aquifers based on spring hydrograph analysis. *Journal of Hydrology*, 303(1–4), 152–164. https://doi.org/10.1016/j.jhydrol.2004.08.023

Kresic, N. (2013). *Water in karst: Management, vulnerability, and restoration*. New York: McGraw-Hill Education.

Kresic, N., & Stevanović, Z. (2009). Groundwater hydrology of springs: Engineering, theory, management and sustainability. Oxford, U.K.: Butterworth-Heinemann.

Leake, S. A., Hoffmann, J. P., & Dickinson, J. E., (2005). *Numerical ground-water change model of the C aquifer and effects of ground-water withdrawals on stream depletion in selected reaches of Clear Creek, Chevelon Creek, and the Little Colorado River, Northeastern Arizona*. Scientific Investigations Report 2005-5277. http://dx.doi.org/10.3133/sir20055277

Lee, E. S., & Krothe, N. C. (2001). A four-component mixing model for water in a karst terrain in south-central Indiana, USA. Using solute concentration and stable isotopes as tracers. *Chemical Geology*, 179(1–4), 129–143. https://doi.org/10.1016/s0009-2541(01)00319-9

Liu, Z., Li, Q., Sun, H., & Wang, J. (2007). Seasonal, diurnal and storm-scale hydrochemical variations of typical epikarst springs in subtropical karst areas of SW China: Soil CO_2 and dilution effects. *Journal of Hydrology*, 337(1–2), 207–223. https://doi.org/10.1016/j.jhydrol.2007.01.034

Maillet, E. T. (1905). Essai d'hydraulique souterraine et flu-viale.

Malard, A., Sinreich, M., & Jeannin, P.-Y. (2016). A novel approach for estimating karst groundwater recharge in mountainous regions and its application in Switzerland. *Hydrological Processes*, 30(13), 2153–2166. https://doi.org/10.1002/hyp.10765

Mckee, E. D. (1938). Original structures in Colorado River flood deposits of Grand Canyon. *SEPM Journal of Sedimentary Research*, 8. https://doi.org/10.1306/d4269003-2b26-11d7-8648000102c1865d

Meeks, J., & Hunkeler, D. (2015). Snowmelt infiltration and storage within a karstic environment, Vers Chez le Brandt, Switzerland. *Journal of Hydrology*, *529*, 11–21. https://doi.org/10.1016/j.jhydrol.2015.06.040

Meixner, T., Manning, A. H., Stonestrom, D. A., Allen, D. M., Ajami, H., Blasch, K. W., et al. (2016). Implications of projected climate change for groundwater recharge in the western United States. *Journal of Hydrology*, *534*, 124–138. https://doi.org/10.1016/j.jhydrol.2015.12.027

Miller, C., Brewer, D. G., & Covington, W. W. (2007). *Forest structure and surface runoff in the Upper Lake Mary watershed, Arizona*. https://repository.arizona.edu/handle/10150/296679

Moniewski, P. (2015). Seasonal variability of discharge from selected springs in central Europe. *Episodes*, *38*(3), 189–196. https://doi.org/10.18814/epiiugs/2015/v38i3/005

Mueller, J. M., Swaffar, W., Nielsen, E. A., Springer, A. E., & Lopez, S. M. (2013). Estimating the value of watershed services following forest restoration. *Water Resources Research*, *49*(4), 1773–1781. https://doi.org/10.1002/wrcr.20163

NOAA (2021). *National Centers for Environmental Information (data set). Digitally formatted environmental data archived at NCEI*. https://www.ncei.noaa.gov/

Polk, J., Vanderhoff, S., Groves, C. G., Miller, B., Bolster, C., Filippi, M., et al. (2013). Complex epikarst hydrogeology and contaminant transport in a south-central Kentucky karst landscape. *Proceedings of the 16th International Congress of Speleology, Brno, Czech Republic. 21–28 July 2013*, 110–115. http://www.speleogenesis.info/directory/karstbase/pdf/seka_pdf13460.pdf

Pool, D. R., Blasch, K. W., Callegary, J. B., Leake, S. A., & Graser, L. F. (2011), *Regional groundwater-flow model of the Redwall-Muav, Coconino, and alluvial basin aquifer systems of northern and central Arizona*: U.S. Geological Survey Scientific Investigations Report 2010-5180. https://pubs.usgs.gov/sir/2010/5180/

PRISM Climate Group, Oregon State University. (2021). *Northwest Alliance for Computational Science & Engineering data set*. https://prism.oregonstate.edu/

Pronk, M., Goldscheider, N., Zopfi, J., & Zwahlen, F. (2009). Percolation and particle transport in the unsaturated zone of a karst aquifer. *Ground Water*, *47*(3), 361–369. https://doi.org/10.1111/j.1745-6584.2008.00509.x

Sauer, V. B. (2002). *Standards for the analysis and processing of surface-water data and information using electronic methods*. Water-Resources Investigations Report 2001-4044. https://pubs.er.usgs.gov/publication/wri20014044

Shamsi, A., Karami, G. H., Hunkeler, D., & Taheri, A. (2018). Isotopic and hydrogeochemical evaluation of springs discharging from high-elevation karst aquifers in Lar National Park, northern Iran. *Hydrogeology Journal*, *27*(2), 655–667. https://doi.org/10.1007/s10040-018-1873-4

Sharp, Z. (2017). *Principles of stable isotope geochemistry (2nd ed.)*. Hoboken, NJ: Prentice Hall. https://doi.org/10.25844/h9q1-0p82

Schindel, G. M. (2015). *Determining groundwater residence times of the Kaibab Plateau, R-aquifer using temperature, Grand Canyon National Park, Arizona*. ProQuest Dissertations Publishing. https://www.proquest.com/docview/1707872245/B0B5A792BB624C0FPQ/1?accountid=12706

Schwartz, B. F., Schwinning, S., Gerrard, B., Kukowski, K. R., Stinson, C. L., & Dammeyer, H. C. (2013). Using hydrogeochemical and ecohydrologic responses to understand epikarst process in semiarid systems, Edwards plateau, Texas, USA. *Acta Carsologica*, *42*(2–3). https://doi.org/10.3986/ac.v42i2-3.670

Sinclair, D. A. (2018). *Springs geomorphology influences on physical and vegetation ecosystem characteristics, Grand Canyon Ecoregion, USA*. Master's thesis, Northern Arizona University.

Solder, J. E., Beisner, K. R., Anderson, J., & Bills, D. J. (2020). Rethinking groundwater flow on the south rim of the Grand Canyon, USA: Characterizing recharge sources and flow paths with environmental tracers. *Hydrogeology Journal*, *28*(5), 1593–1613. https://doi.org/10.1007/s10040-020-02193-z

Springer, A. E., & Stevens, L. E. (2009). Spheres of discharge of springs. *Hydrogeology Journal*, *17*(1), 83–93. https://doi.org/10.1007/s10040-008-0341-y

Springer, A. E., Boldt, E. M., & Junghans, K. M. (2016). Local vs. regional groundwater flow delineation from stable isotopes at western North America springs. *Groundwater*, *55*(1), 100–109. https://doi.org/10.1111/gwat.12442

Springs Online (2021). *The springs and springs-dependent species database (data set)*. http://springsdata.org/

Stevanović, Z. (2019). Karst waters in potable water supply: A global scale overview. *Environmental Earth Sciences*, *78*(23). https://doi.org/10.1007/s12665-019-8670-9

Stevens, L. E., Jenness, J., & Ledbetter, J. D. (2020). Springs and springs-dependent taxa of the Colorado River basin, southwestern North America: Geography, ecology and human impacts. *Water*, *12*(5), 1501. https://doi.org/10.3390/w12051501

Swanson, R. K., Springer, A. E., Kreamer, D. K., Tobin, B. W., & Perry, D. M. (2020). Quantifying the base flow of the Colorado River: Its importance in sustaining perennial flow in northern Arizona and southern Utah (USA). *Hydrogeology Journal*. https://doi.org/10.1007/s10040-020-02260-5

Taylor, J. C., & Greene, E. A. (2008). Hydrogeologic characterization and methods used in the investigation of karst hydrology. In *Field techniques for estimating water fluxes between surface water and ground water* (pp. 71–114). CreateSpace Independent Publishing Platform. http://citeseerx.ist.psu.edu/viewdoc/download?doi=10.1.1.400.2187&rep=rep1&type=pdf

Taylor, R. G., Scanlon, B., Döll, P., Rodell, M., van Beek, R., Wada, Y., et al. (2013). Ground water and climate change. *Nature Climate Change*, *3*(4), 322–329. https://doi.org/10.1038/nclimate1744

Tillman, F. D., Gangopadhyay, S., & Pruitt, T. (2020). Trends in recent historical and projected climate data for the Colorado River Basin and potential effects on groundwater availability. *Scientific Investigations Report*, 1–24. https://doi.org/10.3133/sir20205107

Tobin, B. W., & Schwartz, B. F. (2016). Using periodic hydrologic and geochemical sampling with limited continuous monitoring to characterize remote karst aquifers in the Kaweah River Basin, California, USA. *Hydrological Processes*, *30*(19), 3361–3372. https://doi.org/10.1002/hyp.10859

Tobin, B. W., & Schwartz, B. F. (2020). Quantifying the role of karstic groundwater in a snowmelt-dominated hydrologic system. *Hydrological Processes*, *34*(16), 3439–3447. https://doi.org/10.1002/hyp.13833

Tobin, B. W., Polk, J. S., Arpin, S. M., Shelley, A., & Taylor, C. (2021). A conceptual model of epikarst processes across sites, seasons, and storm events. *Journal of Hydrology*, *596*, 125692. https://doi.org/10.1016/j.jhydrol.2020.125692

Tooth, A. F., & Fairchild, I. J. (2003). Soil and karst aquifer hydrological controls on the geochemical evolution of speleothem-forming drip waters, Crag Cave, southwest Ireland. *Journal of Hydrology*, *273*(1–4), 51–68. https://doi.org/10.1016/s0022-1694(02)00349-9

Trček, B. (2006). How can the epikarst zone influence the karst aquifer hydraulic behaviour? *Environmental Geology*, *51*(5), 761–765. https://doi.org/10.1007/s00254-006-0387-x

U.S. Department of Agriculture, Forest Service (2020). Mogollon Rim Ranger District. https://www.fs.usda.gov/recarea/coconino/recarea/?recid=54886

Williams, P. (2008). The role of the epikarst in karst and cave hydrogeology: A review. *International Journal of Speleology*, *37*(1), 1–10. https://doi.org/10.5038/1827-806x.37.1.1

Wood, A. J. (2019). *Hydrogeology of the Coconino Aquifer, Kaibab Plateau, Grand Canyon*, Arizona. Master's thesis, Northern Arizona University, ProQuest Dissertations Publishing. https://www.proquest.com/docview/2303837624/fulltextPDF/E58D2A284ED74B78PQ/1?accountid=12706

Wood, A. J., Springer, A. E., & Tobin, B. W. (2020). Geochemical variability in karst-siliciclastic aquifer spring discharge, Kaibab Plateau, Grand Canyon. *Environmental and Engineering Geoscience*, *26*(3), 367–381. https://doi.org/10.2113/eeg-2345

Zhang, Z., Chen, X., Chen, X., & Shi, P. (2013). Quantifying time lag of epikarst-spring hydrograph response to rainfall using correlation and spectral analyses. *Hydrogeology Journal*, *21*(7), 1619–1631. https://doi.org/10.1007/s10040-013-1041-9

5

Nitrogen Contamination and Acidification of Groundwater Due to Excessive Fertilizer Use for Tea Plantations

Hiroyuki Ii

ABSTRACT

Previously Japan used over 1,100 kgN/ha/year on its numerous green tea plantations. The groundwater from catchment areas of some tea plantations is highly acidic (pH 4) and heavily contaminated with NO_3-N (80 mg/l) and other heavy metals, particularly Al (50 mg/l). Although the Japan Ministry of Environment recommended reducing nitrogen fertilizer use, approximately 500 kgN/ha/year is still deemed necessary by growers to ensure better tasting, higher quality tea leaves. As a result, NO_3-N, total Al, and Mn concentrations of lower stream spring waters remains over the established Environmental Standard for drinking water of 10, 0.2, and 0.05 mg/l, respectively. Under the volcanic loam soils of the tea plantations, phosphorus was absorbed by soils resulting in low P concentrations in spring and river water. As a result, even though the NO_3-N concentration was maintained high, phosphorus deficiency prevented eutrophication downstream. In this study, the NO_3-N concentration of river or spring water for tea plantation catchments under alternative scenarios was estimated from the amount of nitrogen fertilizer used and the amount of nitrogen absorbed by plants. The amount of nitrogen fertilizer used, or nitrogen fertilizer used minus plant absorption, was found to vary linearly with the NO_3-N concentrations of river or spring water for the catchment. As such, the NO_3-N concentrations of river or spring water was readily estimated by land use and fertilizer nitrogen and plant nitrogen absorption. A balance of land use for the catchment based on these findings is considered necessary for preventing fertilizer contamination in future agriculture.

5.1. INTRODUCTION

5.1.1. Excess Nitrogen Fertilizer is Used for Quality Enhancement and Production of Tea Leaves

Japanese have traditionally relished green tea, enjoying both its delicate aroma and sweet taste. Green tea, or *ocha* as it is known in Japanese, is an important commodity crop with the price of green tea varying by orders of magnitude as determined by its taste. Green tea's delicate taste is intimately related to the amount of nitrogen fertilizer used to grow it (Morita, 2000; Chen et al., 2015). In particular, the taste of nascent tea sprouts is far better with a higher nitrogen concentration of 3% to 6% in the tea leaves. Both the taste and price of tea are determined by its total nitrogen and free amino acid concentrations. In recent years, total concentrations of nitrogen and free amino acid in tea leaves have increased along with the annual amounts of applied nitrogen fertilizer finally attaining 900 kgN/ha/year (Morita, 2000). To produce superior tasting higher priced tea, Japanese tea growers favored extensive nitrogen fertilizer use. Previously applied amounts of approximately 1,000 kgN/ha/year of nitrogen fertilizer grew to as much as 2,000 kgN/ha/year at some tea plantations (Hirono et al., 2009; Morita, 2000).

Faculty of Systems Engineering, Wakayama University, Wakayama, Japan

Threats to Springs in a Changing World: Science and Policies for Protection, Geophysical Monograph 275, First Edition.
Edited by Matthew J. Currell and Brian G. Katz.
© 2023 American Geophysical Union. Published 2023 by John Wiley & Sons, Inc.
DOI:10.1002/9781119818625.ch05

According to research conducted by the Japan Agricultural Cooperative (Morita, 2000; Hajiboland, 2017), to produce superior tasting, quality *ocha*, tea plants require at least more than 500 kgN/ha/year of nitrogen fertilizer.

In Taiwan, when comparing the quality of green and black teas, significant positive relationships were observed between the nitrogen content of tea flushes and the quality of green tea in all seasons, regions, and nitrogen fertilizer types, whereas a negative correlation was observed between nitrogen content and the quality of black tea (Chen et al., 2015). From this it was thought that extensive nitrogen fertilizer was essential for producing high quality green tea. In Taiwan, the annual amount of nitrogen fertilizer used at tea plantations reached 736 kgN/ha/year (Chen & Lin, 2016). In China, the world's largest tea producing country, nitrogen fertilizer application reached 1,200 kgN/ha/year with an average value of 553 kgN/ha/year (Wang et al., 2020c). If a N supply of up to 600 kgN/ha/year of fertilizer is applied, the total amount of harvested tea produced yields 5,800 to 6,400 kg (Hajiboland, 2017). Tea plants require a high N supply, with the current fertilization rate ranging from 450 to 1,200 kgN/ha/year, exceeding that being used in other artificial ecosystems. Recent research suggests that the excessive application of chemical fertilizer in Chinese tea gardens is a major problem with more than 30% of the tea garden area being oversprayed (Tang et al., 2020; Wang et al., 2020b).

In 2017, Egypt was the heaviest user of nitrogen fertilizer for cropland at 367 kgN/ha/year, the Netherlands was at 243 kgN/ha/year, and 219 kgN/ha/year for China (Roser & Ritchie, 2021). Nitrogen fertilizer for cropland in Japan and the United States was 86 and 73 kgN/ha/year, respectively. Therefore, tea plantations consumed a lot of nitrogen fertilizer relative to other crops. On an international scale, the amount of nitrogen fertilizer used per unit area at Japanese tea plantations is extremely high although recent values have decreased. In some cases, 1,200 kgN/ha/year in China and 2,000 kgN/ha/year in Japan were recorded.

In 2015, Japan had a total planted area of 446,000 ha with tea plantations accounting for 45,900 ha (Ministry of Agriculture, Forestry and Fisheries in Japan (MAFF); https://www.maff.go.jp). Global tea production reached 300 million tons in 2009 with leading producers like China, India, Kenya, Sri Lanka, and Japan producing 136 million, 98 million, 31 million, 29 million, and 9 million tons, respectively (MAFF; https://www.maff.go.jp). In 2013, global tea production exceeded 5 million tons, and its consumption continues to increase (Tang et al., 2020).

Tea production accounts for the largest amount of nitrogen fertilizer use in the world and excessive nitrogen fertilizer used at tea plantations is an ongoing issue of environmental concern. Therefore, the pollution of surface water and groundwater around tea plantations caused by excess nitrogen fertilizer remains a typical example of fertilizer pollution. In Japan, groundwater nitrogen fertilizer pollution was found only at tea plantation areas and downstream areas from spring and house wells (Ii et al., 1997, 1998, 2000; Tanaka et al., 2001). The appropriate use of nitrogen fertilizer at tea plantations is now a serious issue among environmentalists.

5.1.2. Nitrogen Fertilizer Compounds and Their Toxicity

Nitrogen fertilizer has several kinds of chemical forms, which include nitrate (NO_3^-), nitrite (NO_2^-), and ammonia (NH_4^+). Popular among nitrogen fertilizers is ammonium sulfate, $(NH_4)_2SO_4$, which easily undergoes nitrification in soil or on the ground according to the following equation:

$$(NH_4)_2 SO_4 + 4O_2 = 4H^+ + 2NO_3^- + 2H_2O + SO_4^{2-}. \quad (5.1)$$

Ammonium sulfate is widely applied at tea plantations because tea plants are readily able to absorb both ammonia and nitrate. Although nitrite is not a predominant nitrogen fertilizer at tea plantations, small amounts of nitrite are produced from ammonia and nitrate in soil under some conditions. Nitrate and nitrite are toxic substances with nitrite being known to cause methemoglobinemia, or a lack of oxygen in the blood, particularly in infants, with some nitrate changing into nitrite in the body (MAFF; https://www.maff.go.jp).

Ingestion of nitrite may produce a nitroso compound, which is carcinogenic. As a result, nitrite and nitrate are thought to be carcinogenic although there is little actual proof (World Health Organization (WHO;https://www.who.int), Food and Agriculture Organization of the United Nations (FAO; http://www.fao.org/home/en/). Because of this, environmental standards for nitrate and nitrite concentrations in drinking water sources, groundwater, and river water have been established globally. The nitrate and nitrite concentrations set out in the Environmental Standard (WHO; https://www.who.int) are 50 mg/l (nitrate N concentration of 11 mg/l) and 3 mg/l (nitrite N concentration of 0.9 mg/l).

5.1.3. Acidic Groundwater Pollution Caused by Excess Nitrogen Fertilizer Used for Tea Plantation

Soil acidification in the United States arising from over 20 years of nitrogen fertilizer use on plants was documented to have reduced pH values from 5.5 to 4.0 (Schroder et al., 2011). Long-term fertilization brought out a decrease of cation exchange capacity (CEC) for the soil of sweet corn planted in Wisconsin with the application of ammoniacal N fertilizer (Barak et al., 1997). The nitrogen fertilizer, which was regularly used, was $(NH_4)_2SO_4$, which easily changes into H^+, NO_3^-, and SO_4^{2-} as shown in the nitrification equation (5.1). From equation (5.1), soil water around tea plantations becomes strongly acidic. As groundwater soluble concentration depends on temperature, Eh, pH, and soluble element concentrations of the soil, Ca^{2+}, Mg^{2+}, K^+, Al^{3+}, and Mn^{2+} concentrations were influenced, in particular, Al^{3+} and Mn^{2+} concentrations increased with a decrease of pH values, and Ca^{2+}, Mg^{2+}, and K^+ concentrations increased with pH values (Tian & Niu, 2015).

Previous study of tea plantations in Japan by Nakasone and Yamamoto (2004), found that soil water pH values for the tea plantations were below 3.5. In particular, Al^{3+} concentrations for groundwater around the tea plantations reached an abnormal value of 50 mg/l with a decrease in pH values. Fish living in rivers and lakes along the catchment were found to have been killed by the spring water, which had had high Al^{3+} concentrations and low pH values (Nakasone & Yamamoto, 2004). H^+ and Al^{3+} damaged fish gills and fish died as a result because of oxygen deficiency (Haines, 1981; Schofield & Trojnar, 1980). For humans, the allowable Al concentration in drinking water, as designated by the Environmental Standard for Japan and USEPA, are 0.2 mg/l and 0.05 to 0.2 mg/l (MAFF, https://www.maff.go.jp; WHO, https://www.who.int; United States Environmental Protection Agency). The Al concentration specified by the WHO is 0.2 mg/l. Al is a main component of soil and its solubility depends on pH, therefore under acidic conditions, monitoring Al is required for maintaining environmental conditions. Lakes with high transparency and rivers depleted of fish and low concentrations of plankton were found along the catchment (Nakasone & Yamamoto, 2004). Therefore, the acidification caused by excess nitrogen fertilizer use at tea plantations and its accompanied high Al concentration were significant environmental issues. Groundwater acidification and Al pollution caused by excess nitrogen fertilizer was found only at tea plantation areas and downstream areas from spring and house wells (Ii et al., 1997, 1998, 2000; Tanaka et al., 2001).

Tea (*Camellia sinensis* L.) plants have an optimal pH range of 4.5 to 6.0 and prefer ammonium (NH_4^+) over nitrate (NO_3^-). Therefore, strong soil acidification and nitrification are detrimental to their growth (Wang et al., 2020a). As a countermeasure, rapeseed cake organic fertilization may effectively increase pH and better preserve soil C and N pools (Xie et al., 2021). Runoff was also reduced in tea plantations that adopted this technique, which greatly reduced regional soil acidification and water eutrophication trends. The Japanese government has also been advocating conservation measures such as reducing chemical fertilizers and synthetic pesticides. As a result, organic forms of nutrients were believed to be good for the environment and less harmful than chemical fertilizers. However, the incorrect application of nutrients in any form may contribute to nitrogen pollution (Kuroda et al., 2016).

5.1.4. Absorption of Nitrogen Components by Tea Plants (Tea Plantation)

The absorption of nitrogen fertilizer components by tea plants is relevant to groundwater nitrogen pollution. The nitrogen from nitrogen fertilizer for a tea plantation after fertilization migrates into the air (nitrous oxide emission loss, N_2O), tea plants, soil water, groundwater, and river water (Sun et al., 2020). The amount of nitrogen fertilizer absorbed by the tea plantation area can be estimated from the amount of nitrogen fertilizer used, the tea plantation area, and the nitrogen concentration in soil or groundwater and river water. Various data exist about the relation between the nitrogen absorption of tea and nitrogen fertilizer used. The amount of applied nitrogen fertilizer increased from 540 to 1,080 kgN/ha/year with the absorption of nitrogen by tea plants being 210 to 240 kgN/ha/year with the annual difference being a negligible 30 kgN/ha/year (Morita, 2000). Because tea production was uniform under the condition of 300 to 1,200 kgN/ha/year, nitrogen absorption by tea plants did not increase despite changes of nitrogen fertilizer used (Morita, 2000). However, nitrogen absorption by tea plants did increase from 105 to 625 kgN/ha/year when nitrogen fertilizer used ranged from 450 to 1,001 kgN/ha/year (Yamazaki et al., 2011).

Under the amounts of nitrogen fertilizer used from 500 to 1,000 kgN/ha/year, tea plant nitrogen absorption for Yamazaki et al. (2011) showed values varying from 105 to 625 kgN/ha/year, however, the value for Morita (2000) obtained from the Shizuoka tea plantation area was quite uniform, from 210 to 240 kgN/ha/year. Yamazaki et al. (2011) covered all Japanese tea plantation areas and the reasons for variable values were attributed to precipitation and the underestimation of land use contribution from small areas, which are easily overlooked during the calculation of land use area. The main source of nitrogen pollution is infiltrated nitrogen from fertilizer excluding nitrogen absorbed by tea plants from soil. Although absorption by tea plants is variable, the difference between nitrogen fertilizer applied and the absorption by tea has been found to be large with most of the nitrogen fertilizer component directly moving into groundwater (Ii et al., 1997; Nakasone & Yamamoto, 2004).

In Taiwan, 736 kgN/ha/year of nitrogen fertilizer was applied with 18.3% of the nitrogen being absorbed by tea plants and 30% lost due to storm runoff and 52% stored in soil and groundwater (Chen & Lin, 2016). In China, because excessive nitrogen fertilizer was used for vegetable crops (Zhang et al., 1996), high nitrogen concentrations in groundwater exceeding the environmental standard were widely distributed. In Kenya, nitrogen fertilizer application was found to have contributed to an increase in nitrate levels in groundwater although NO_3-N concentrations were 4.9 to 8.2 mg/l and still below the environmental standard (Maghanga et al., 2013). In summary, many countries other than Japan use excessive amounts of applied nitrogen fertilizer on tea plantations and are equally at risk of subsequent environmental damage.

5.1.5. Reducing N_2O Emission Gas from Excess Fertilizer Helps Prevent Global Warming

Excessive nitrogen fertilizer application at tea plantations induces N_2O gas emission from soil. Human activity has increased concentrations of atmospheric nitrous oxide (N_2O) over recent decades. The global warming potential of N_2O is approximately 300 times that of carbon dioxide. Fortunately, new lime nitrogen and dicyandiamide fertilizer can effectively reduce N_2O emission from tea field soils through the nitrification of dicyandiamide (Hirono & Nonaka, 2014; Bhatia et al., 2010). This new nitrogen fertilizer also increases pH values due to its lime component and impedes the acidification of soil and groundwater. The combination of relatively high soil C content, high fertilizer N use, acidic soils, and high rainfall is known to promote N_2O formation in soils via nitrification, denitrification, and chemodenitrification (Wang et al., 2020b). Results from mainly China and Japan suggest that annual N_2O emissions from soils of global tea plantations are on average 17.1 kgN/ha (or 8,008 kg CO_2-eq/ha), being substantially greater than those reported for cereal croplands (662–3,757 kg CO_2-eq/ha) (Wang et al., 2020b). Incidentally, N_2O emission and synthetic nitrogen fertilizer used at tea plantations were effectively reduced by employing biofertilizer made from plant waste including sweet potato starch waste and rice straws containing *Trichoderma viride* (Xu et al., 2017).

5.2. METHODS AND MATERIALS

In Japan, tea trees are lined up in parallel as shown in Figure 5.1. For field collection of water samples from the studied plantations at Shizuoka prefecture, central Japan (Fig. 5.2), pH, ORP values, and temperature were measured in the field using portable sensors. A combination of house well, spring, river, and pond waters were directly collected from sites indicated on Figure 5.2. Porous cups were buried between each tea tree line in the plantations. Soil water was sampled using porous cups. The depth of the porous cups was from 30 to 210 cm and the soil water at each depth was collected. After collection, water samples were preserved in plastic or glass bottles. Dissolved ions in water were measured by ion chromatography after being filtered through a 0.45 μm membrane. Total metal concentrations for water were measured by ICP-AES after acidification.

5.3. SOIL WATER CHEMISTRY AT TEA PLANTATIONS LOCATED ON VOLCANIC LOAM (FIELD RESULT 1)

Tea plantation soil water was sampled at Shimizu district in Shizuoka Prefecture, central Japan which experiences 2400 mm/year of precipitation and a 16°C annual average temperature as shown in Figure 5.2. Shizuoka Prefecture is a typical tea plantation area and one of the largest green tea production regions in Japan. From 0 to 10 cm in depth, the soil layer was brownish black and consisted almost entirely of plant residue (andosol). From 10 to 150 cm in depth, the soil layer mainly consisted of brownish volcanic loam (Ogawa et al., 2005). Figure 5.2 shows the study area and the monthly precipitation. Figure 5.3 shows pH and the concentrations of components (NO_3-N, SO_4^{2-}, and Total P) of soil water, which were affected by fertilizer applications in the tea plantation in Shimizu, Shizuoka, Japan. From 2003 to 2004, monthly precipitation reached 700 mm in August and conditions were very wet except during winter. Very low pH values of 3.2 to 4.0 were recorded each season from surface soils to 210 cm in depth with the difference in pH at each depth being small. Twice a year, 600 kgN/ha was applied, first from January to February and then again from July to September (Ogawa et al., 2005). In January and between July and October, pH values were very low during nitrogen fertilizer application. However, pH values for April and June were high due to the very wet conditions in 2004 after nitrogen fertilizer application (Ogawa et al., 2005). Tea plant roots cannot survive if the pH falls below 5, which indicates that excess fertilizer actually destroyed tea plant roots and compromised tea plant health (Morita, 2000).

$(NH_4)_2SO_4$ is a very common nitrogen fertilizer and maximum SO_4^{2-} concentrations in soil water reached an extremely high value of 1,600 mg/l. SO_4^{2-} concentrations varied with depth from 100 mg/l at 30 cm to a very high 400 to 800 mg/l at 210 cm (Ogawa et al., 2005). These results showed that components of infiltrated nitrogen fertilizer were still high even beyond 2 m depth (Nakasone & Yamamoto, 2004; Hirono & Nonaka, 2014; Ogawa et al., 2005) and this was thought to indicate likely groundwater contamination with fertilizer. From equation (5.1), chemical fertilizer, $(NH_4)_2SO_4$ changes into NO_3^-. As a result, NO_3-N concentrations in surface soil water were highly variable, from 10 to 240 mg/l, and even reached 50 to 130 mg/l at depths of 210 cm where infiltrated NO_3-N concentrations remained high (Fig. 5.3). Furthermore,

Figure 5.1 Tea plantation in Yame and Shimizu districts. Many fans protect for frost. Tea leaves are cut by machine.

NO_3-N concentrations in soil water were high in August and low between October and January following nitrogen fertilizer application, although the NO_3-N concentrations in soil water still exceeded the Japanese environmental standard to a depth of 210 cm.

Next, soil nitrogen concentration was analyzed by Ogawa et al. (2005). Average NH_4-N concentrations in the 0 to 10, 10 to 30, and 30 to 50 cm layers of soil were 280, 19, and 9 µg/g throughout the year. NO_3-N concentrations in the 0 to 10, 10 to 30, and 30 to 50 cm layers of soil were 153, 62, and 59 µg/g and NO_3-N concentration remained high at 40µg/g in the 130 to 150 cm layer. NO_3-N concentration was high during August and February and low during April following nitrogen fertilizer application (Ogawa et al., 2005). Therefore, the same seasonal changes were seen in both NO_3-N concentrations of soil and soil water.

Soil water total P concentrations at 30 cm in depth were extremely high, 10 to 55 mg/l, however, concentrations in soil water below this depth were extremely low. P was

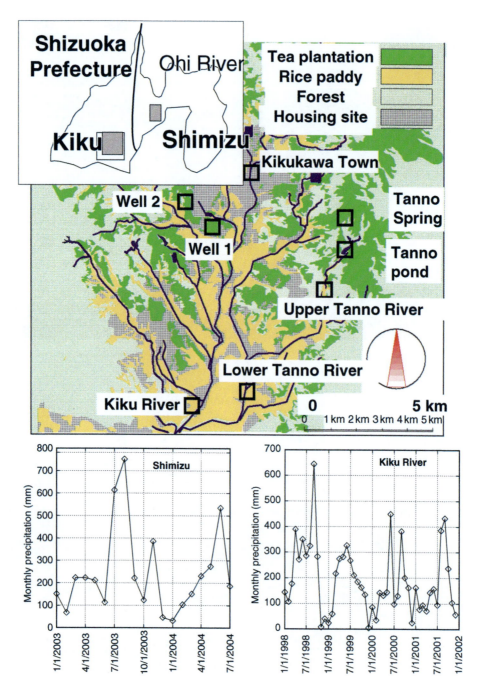

Figure 5.2 Shimizu and Kiku study areas and land use of the Kiku River catchment in 1997 and monthly precipitation for both areas.

thought to have been absorbed by soil due to high CEC (Nakasone & Yamamoto, 2004; Barak et al., 1997) and volcanic loam. Volcanic loam contains active Al and fixes phosphoric acid (Nanzyo et al., 1993). Therefore, P did not move and remained only in the surface of the soil. Other fertilizer components, however, were easily transported farther in depth with infiltrating rain.

Cu and Ni are not usually present in high concentrations in nitrogen fertilizers, but they are naturally present in uncontaminated soils and can be mobilized as pH decreases thus accumulating in groundwater. Figure 5.4 shows the concentrations of trace elements (total Al, Mn, Zn, Ni, and Cu) of soil water for tea plantations in Shimizu City, Shizuoka, Japan. Al is a main soil

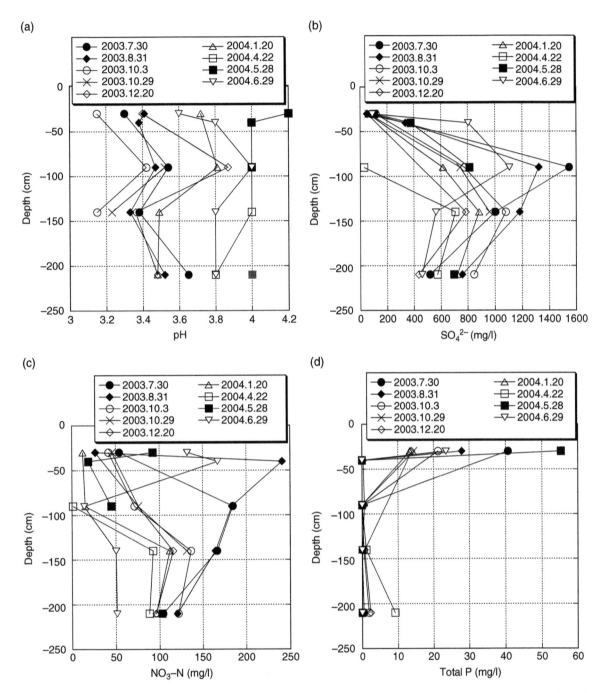

Figure 5.3 (a) pH and concentrations of component, (b) SO_4^{2-}, (c) NO_3-N, and (d) Total P, of soil water affected by the fertilizer application for a tea plantation in Shimizu, Shizuoka, Japan.

component and its concentration depends on pH values. Al concentrations in soil water increased with a decrease in the pH buffering phase (Tian & Niu, 2015). Al concentrations in soil water were also extremely high and the concentration distribution with depth was very similar to SO_4^{2-}. Al concentration was 200 mg/l at 100 to 150 cm in depth. Even at 210 cm in depth, Al concentrations in soil water were 50 to 150 mg/l (Ogawa et al., 2005).

As the allowable Al concentration for drinking water set forth in the Japanese environmental standard is 0.2 mg/l (MAFF; https://www.maff.go.jp), high Al concentration in soil water was a significant concern because this may also lead to high Al concentrations in groundwater and spring water.

Mn concentrations in soil water increased from 1 to 6 mg/l with depth and were from 3 to 6 mg/l at 210 cm in

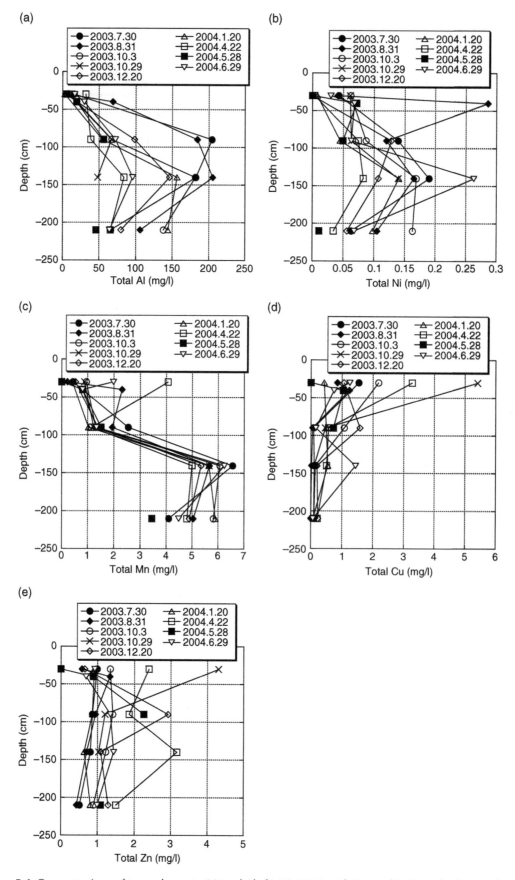

Figure 5.4 Concentrations of trace elements, (a) total Al, (b) Ni, (c) Mn, (d) Cu, and (e) Zn, of soil water for a tea plantation in Shimizu, Shizuoka, Japan.

depth (Ogawa et al., 2005). The permissible Mn concentration for drinking water in Japan is 0.05 mg/l (MAFF; https://www.maff.go.jp). An increase of Mn concentration with depth and high Mn concentrations relative to the Japanese standard were thought to lead to higher levels of Mn in groundwater and spring water. Thus, consistent monitoring of Mn concentration for well and spring water around tea plantations is considered necessary. Zn concentration in soil water varied from 0 to 4 mg/l at the surface and 0.5 to 1.5 mg/l at 210 cm in depth (Ogawa et al., 2005). The Japanese Environmental Standard for drinking water Zn concentration is 1.0 mg/l (MAFF; https://www.maff.go.jp) and the sampled soil water Zn concentration was near the standard. Ni concentrations in soil water were over 0.05 mg/l from 30 to 210 cm in depth. The allowable Ni concentration for drinking water according to the Japanese Environmental Standard is 0.01 mg/l (MAFF; https://www.maff.go.jp), therefore high Ni soil water was thought to lead to Ni contamination in groundwater, well water, or spring water around tea plantations. Cu concentration in soil water decreased from 5 to 0.2 mg/l with depth and was less than 0.2 mg/l at 210 cm in depth. The Japanese environmental standard for drinking water Cu concentration is 1.0 mg/l (MAFF; https://www.maff.go.jp). Al, Mn, Ni, and Zn concentrations in soil water were very high and exceeded the environmental standard under low pH conditions.

5.4. WATER CHEMISTRY FOR THE KIKU TEA PLANTATION, SHIZUOKA, EXHIBITING A LOW PHOSPHORUS CONCENTRATION SPRING (FIELD RESULT 2)

Figure 5.2 shows the land use of the Kiku River catchment and its monthly precipitation on the south slope of the Makinohara Plateau. This catchment in Shizuoka experiences an annual precipitation of 2,000 mm/year with an average temperature of 16°C. Summer is the wet season, with precipitation of approximately 400 mm/month, whereas winter is the dry season. Land use of the Kiku River catchment is composed of tea plantations on the plateau as well as ponds or water reservoirs on the slope and rice fields at the bottom. The Makinohara Plateau is located in the center of a tea plantation area making it part of one of the biggest tea plantation areas in Japan. The top soil of the tea plantation is volcanic loam while the plateau is mainly composed of Pliocene to Pleistocene marine silt to sandstone. The high CEC of the soil is thought to have been caused by volcanic loam and young marine silt (Nakasone & Yamamoto, 2004). On the plateau there are many wells located in a very wide tea plantation area. Down the plateau, spring or river water, derived from groundwater on the plateau, flows down and through the Tanno Pond, the Tanno River, and finally to the Kiku River. Since 1998, total nitrogen fertilizer use has decreased in order to reduce nitrogen contamination (Ii et al., 2000; Yamano et al., 2001; Tanaka et al., 2001, 2002; Nakasone & Yamamoto, 2004; Hirono et al., 2009). Before 1997, about 1,100 kgN/ha/year was applied, but from 1998, it decreased and has been at approximately 550 kgN/ha/year since 2006.

Figure 5.5 shows pH values, NO_3-N and trace element concentrations (total Al, Mn, Zn, and Ni) along the catchment (Ii et al., 2000; Tanaka et al., 2001, 2002). The pH values of wells around the tea plantation on the plateau, and spring water were 4 to 4.5. These values are consistent with soil pH values, ranging from 3.2 to 4.0 (shown in Fig. 5.3). Downstream, pH values increased 5 to 7 for the Tanno Pond and 7 to 8 for rivers. Although nitrogen fertilizer use decreased, pH values did not change after 1998 except for the Tanno Pond. In contrast, NO_3-N and total Al concentrations decreased along with nitrogen fertilizer use. By June 2001, NO_3-N concentrations at the upper stream of the catchment were 20 to 30 mg/l over the Japanese Environmental Standard (MAFF; https://www.maff.go.jp). NO_3-N concentrations at the most downstream point were less than 5 mg/l and under the Japanese environmental standard. Also, in June 2001, total Al concentrations for wells and springs were 1 to 30 mg/l even after reducing the amount of nitrogen fertilizer used. Downstream, total Al concentration decreased with an increase in pH values. Although the Al concentration of the Kiku River at the bottom of the catchment decreased to 0.1 mg/l, which was under the Japanese environmental standard for drinking water, the Al concentration for most of the water along the catchment was over the standard. Fish in the Tanno Pond were reported in the media to have died before nitrogen fertilizer amounts were reduced (Nakasone & Yamamoto, 2004). The death of the fish is thought to have been due to low pH values and/or high Al concentration. An increase in pH from spring water to the Tanno Pond was thought to have contributed to making the pond color change to blue by precipitating minute $Al(OH)_3$ particles. Similarly, the Mn and Zn concentrations in water along the catchment decreased going downstream. After reducing nitrogen fertilizer amounts, Mn and Zn concentrations for wells and spring water decreased. Notably, while all Zn concentrations for the catchment were under the standard, most Mn concentrations were over the standard. Ni concentrations for wells also decreased after nitrogen fertilizer decreased, however Ni concentrations of well water were still over the environmental standard (Ii et al., 2000; Tanaka et al., 2001, 2002).

The rice fields at the bottom of the catchment require considerable amounts of water, therefore both river water

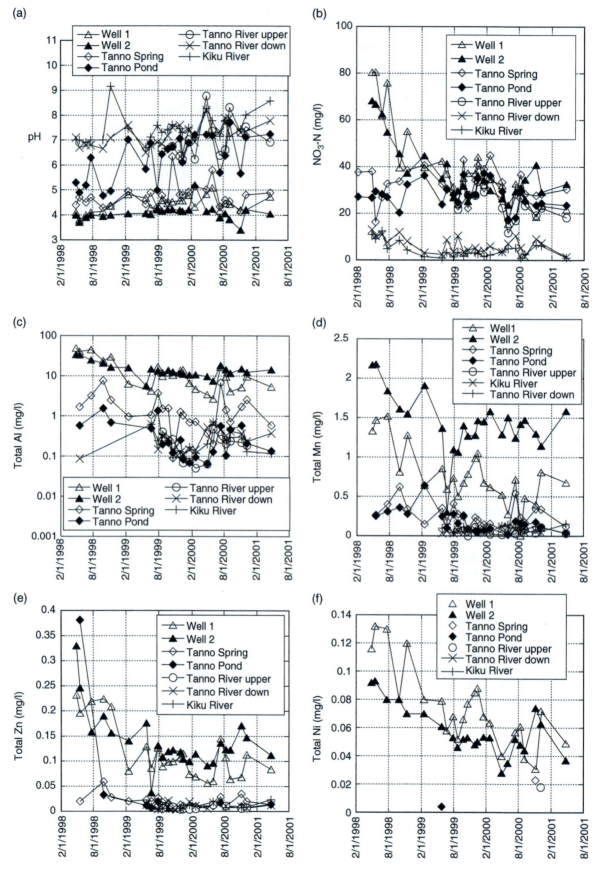

Figure 5.5 (a) pH values and (b) NO$_3$-N and trace elements, (c) total Al, (d) Mn, (e) Zn, and (f) Ni, concentrations along the Kiku River catchment. The logarithmic scale for Al (all other scales are linear).

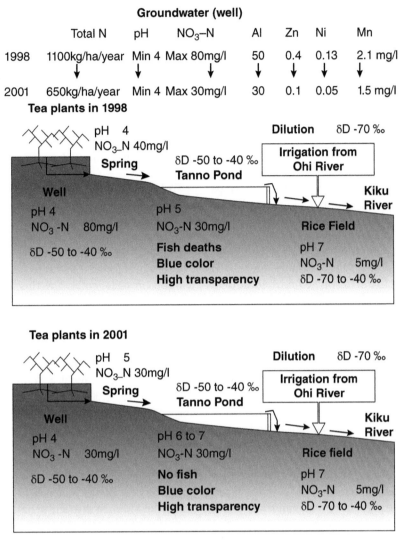

Figure 5.6 Water chemistry change for groundwater and surface water caused by decreased fertilizer use from 1998 to 2001 on tea plantations along the Kiku River catchment.

and the water reservoir on the slope were important water sources. However, because of water shortages, Ohi River irrigation water from another big catchment was utilized for rice fields. As Ohi River water was sourced from a mountain 3,000 m in height and the tea plantation area is near the sea and only 200 m in height, the oxygen and hydrogen stable isotope data for water are quite different. The isotope values for Ohi River (δD -70‰) are lower than those of both well waters and springs (δD -50 to -40‰) on the plateau. Isotope data for the downstream (δD -70 to -40‰) of the catchment were changeable with season; these values were thought to change due to the mixing with Ohi River irrigation water (Yamano et al., 2001). As a result, NO_3-N and total Al concentrations were diluted with Ohi River irrigation water and these values were thought to fall below the Japanese environmental standard as a result of this dilution. Therefore, reducing amounts of nitrogen fertilizer used at tea plantations to 650 kgN/ha/year in 2000 is insufficient for passing the environmental standard of catchment water. Dilution by irrigation or precipitation, however, was effective in reducing nitrogen concentrations.

Figure 5.6 shows the change in water chemistry for groundwater and surface water caused by decreased amounts of fertilizer from 1998 to 2001 on tea plantations along the Kiku River catchment. With decreasing the amounts of total nitrogen fertilizer used for tea plantations from 1,100 kgN/ha/year to 650 kgN/ha/year, NO_3-N and metal concentrations (Al, Zn, Ni, and Mn) of groundwater (house wells) decreased from 80 to 30 mg/l, 50 to 30 mg/l, 0.4 to 0.1 mg/l, 0.13 to 0.05 mg/l, 2.1 to 1.5 mg/l although pH values along the catchment were little changed. However, the NO_3-N concentration of the Tanno Pond remained high at over 30 mg/l.

Dilution by Ohi River irrigation water decreased the NO_3-N concentration of river water at the catchment. A reduction in fertilizer to 650 kgN/ha/year for tea plantations was insufficient to bring the NO_3-N concentration of river water below the environmental standard.

5.5. WATER CHEMISTRY FOR A CATCHMENT CONTAINING A TEA PLANTATION AND OTHER LAND USES (FIELD RESULT 3)

As shown in Figure 5.2, the Shimizu district experiences 2,400 mm/year of precipitation and has an average annual temperature of 16°C. Although dilution by irrigation or precipitation was effective for reducing nitrogen concentrations, and reduction in fertilizer amount helped to reduce nitrogen contamination, tea plants are thought to require at least 500 kgN/ha/year to achieve the desired quality. Therefore, effective dilution requires additional water, which is difficult to supply from other catchments. According to tea plantation soil water chemistry results, the NO_3-N concentration of groundwater and spring water was extremely high, over 30 mg/l, and their pH values were very low (approximately 4). Therefore, direct use of groundwater and spring water derived from the tea plantation was impractical for either irrigation or drinking water.

It is very difficult with only tea plantation land use to come within the limitations set in the environmental standard so the next evaluation was performed using a combination of catchment land utilization. The Shimizu district is an intensive tea plantation area. In 1978, when a tea plantation was started, 50% of the land was forest and orange orchards composed another 30%. Land use for tea plantation increased with time and, in 1997, the tea plantation area reached over 20% of land used while land used for orange orchards decreased to less than 10%. NO_3-N concentrations in one particular river at the catchment increased from 5 mg/l in 1980 to 15 mg/l in 1992. After 1999, it decreased correspondingly with reduced amounts of nitrogen fertilizer. This catchment was divided into 26 smaller catchments. Land use for tea plantation, orange orchards, and forest for the 26 catchments varied from 0% to 40%, 0% to 60%, and 0% to 95%, respectively. River water nitrogen concentration was influenced by the amount of nitrogen fertilizer used for tea and orange orchards. During this time, the amounts of nitrogen fertilizer used were 540 and 240 kgN/ha/year for tea and orange plants respectively (Nishio et al., 2011).

Figure 5.7 shows the relationship between NO_3-N concentrations and the amounts of nitrogen fertilizer used for 26 small catchments between 2008 and 2010 as calculated from land use and fertilizer use. NO_3-N concentrations of river increased with the amount of nitrogen fertilizer used in the catchment. When the amount of nitrogen fertilizer used was less than 250 kgN/ha/year from this relation, the NO_3-N concentration of river water for each catchment was less than 10 mg/l, which met the environmental standard. The amount of nitrogen fertilizer used for the catchment was useful for estimating NO_3-N concentrations in river water and evaluating contamination. The relation between NO_3-N concentration and the amount of nitrogen fertilizer used per catchment area was not linear. The reason for this is that the relation did not evaluate plant absorption, which may change the relation.

To highlight plant absorption, we showed the relation between NO_3-N concentration of river for the catchment and the amount of nitrogen fertilizer used minus the amount of nitrogen absorption by plants. In this catchment, there were mainly three land uses: tea, oranges, and forest. Only nitrogen fertilizer was used for tea and oranges. Following this, we showed the relation between NO_3-N concentration and the amount of nitrogen fertilizer used minus the amount of nitrogen absorption by tea plants and orange trees considering the nitrogen absorption of tea plants and oranges. The actual amount of nitrogen absorption for tea was variable. As was noted in section 5.1.4, under nitrogen fertilizer use from 500 to 1,000 kgN/ha/year, tea plant nitrogen absorption ranged from 105 to 625 (Yamazaki et al., 2011) or from 210 to 240 kgN/ha/year in Japan (Morita, 2000). In Taiwan, 2,700 kg/ha/year of fertilizer was applied with only 18.3% of applied nitrogen (total nitrogen fertilizer was 736 kg/ha/year) and 5.5% of applied phosphate being utilized by tea plants (Chen & Lin, 2016).

For the Shimizu district, the amount of nitrogen fertilizer used for tea plantations was 540 kgN/ha/year and was under the condition of low nitrogen fertilizer for both results. Then, the tea plant nitrogen absorption value was used for 210 kgN/ha/year because the study area of Morita (2000) was also Shizuoka and the absorption rate, 21% (210/1,000 kg/ha/year), agreed with Taiwan values. The amount of nitrogen fertilizer used and plant absorption for oranges was 240 kgN/ha/year and 200 kgN/ha/year (Nishio et al., 2011; MAFF, https://www.maff.go.jp).

Figure 5.7 shows the relationship between NO_3-N concentrations and the amount of nitrogen fertilizer used minus the amount of nitrogen absorption by plants for 26 small catchments between 2008 to 2010. In Figure 5.7, the river nitrate ion concentration is determined by the amount of nitrogen fertilizer used minus the nitrogen absorbed by the tea. The relation considering absorption was linear relative to the results obtained when not considering absorption and is deemed more convenient for estimating river nitrogen concentrations for the catchment estimated from the linear relation (Nishio et al., 2011). When the amount of nitrogen fertilizer used minus the

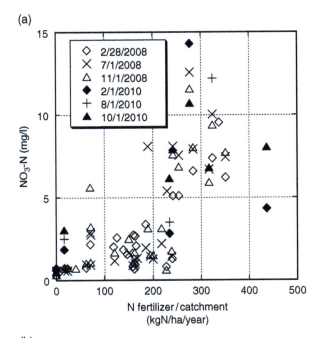

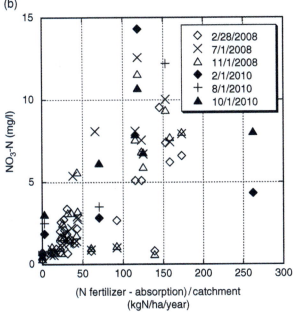

Figure 5.7 The relationship between (a) NO_3-N concentrations and amounts of nitrogen fertilizer use and (b) nitrogen fertilizer minus absorption for small catchments from 2008 to 2010.

amount of nitrogen plant absorption per catchment was less than 100 kgN/ha/year, the NO_3-N concentration of river water for each catchment was less than 10 mg/l and in line with the environmental standard. The difference between nitrogen fertilizer used and nitrogen absorption for the tea plantation was 300 kgN/ha/year (540 -240 kgN/ha/year). When land uses for a tea plantation and forest (where no nitrogen fertilizer is used) are less than one third and over two thirds, respectively, the NO_3-N concentration of river water will be within the environmental standard.

5.6. DISCUSSION

From field results and background information, nitrogen contamination for excess nitrogen fertilizer use at tea plantations was discussed and methods to control nitrogen contamination in the future will be proposed.

5.6.1. Groundwater Chemistry Changes Occurring from the Soil Surface to Spring in Tea Plantation (Comparing Results 1 and 2)

Result 1 showed soil water chemistry under the condition of about 600 kgN/ha/year nitrogen fertilizer used in 2003 and 2004. Result 2 showed water chemistry from house well water (less than 10 m in depth) to spring water along the tea plantation catchment. The amount of nitrogen fertilizer used for result 2 decreased from 1,100 (before 1998) to about 650 (after 2000) kgN/ha/year. Comparing soil water in 2003 and 2004 from approximately 2 m in depth (results 1) with house well water and spring water in tea plantation in 2000 and 2001 (result 2) under the condition of about 600 kgN/ha/year nitrogen fertilizer used, pH values increased from 3.5 to 4.0 of soil water to 4 to 4.5 of well and spring water going downstream. NO_3-N concentration decreased from 50 to 100 mg/l in soil water to 20 to 40 mg/l in well and spring water. Similarly, total Al concentration decreased from 50 to 150 in soil water to about 1 to 10 mg/l in well and spring water. Total Mn concentration decreased from 3-6 in soil water to about 0.1–1.5 mg/l in well and spring water. Total Zn concentration decreased from 0.5–1.5 in soil water to 0.01–0.1 mg/l in well and spring water. Total Ni concentration decreased from 0.01–0.1 in soil water to less than 0.04 mg/l in well and spring water.

Even after a reduction in nitrogen fertilizer use, the NO_3-N, total Al, and Mn concentrations of lower stream spring water were still over the 10, 0.2, and 0.05 mg/l Japanese Environmental Standard for drinking water. Generally, Al concentration is controlled by pH values (Ii et al., 1998). Figure 5.8 shows the relation among pH values and NO_3-N and trace element concentrations of groundwater for a tea plantation. Al concentration clearly increased with a decrease in pH values from soil water, well water, to spring water for tea plantations in the Yame, Shimizu, and Kiku study areas. The Yame area in Kyushu in western Japan is also famous for its tea plantations. Al was controlled by pH values. Similarly, total Mn and Zn concentrations increased with a decrease in pH values, however, the Mn concentration of the Yame district was 10 times higher than that of the Shimizu and Kiku

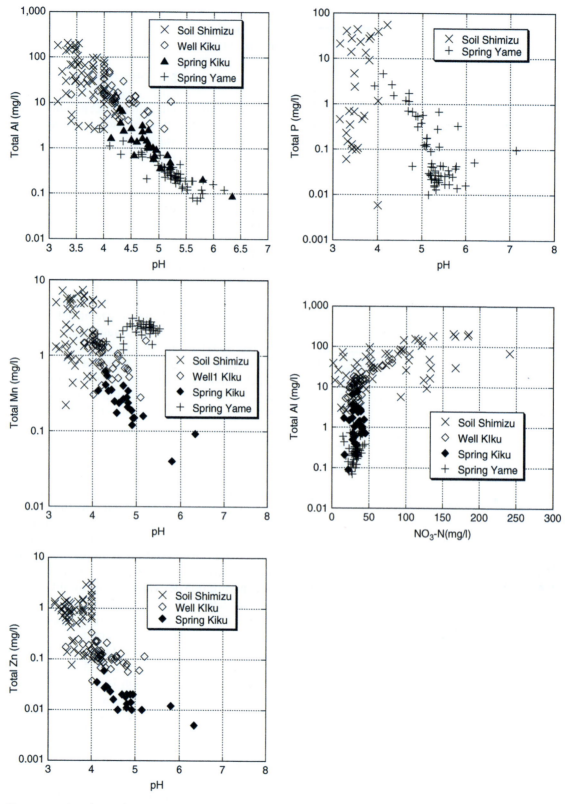

Figure 5.8 The relationship among pH values and NO_3-N and trace element concentrations of groundwater for a tea plantation.

districts. Under the same geological condition (Shimizu and Kiku district), total Mn and Zn concentrations linearly related with pH values. The Yame area is composed of pelitic schist, known as the Sangun Formation, derived from deep ocean floor sediments with no volcanic loam (Geological Survey of Japan, 2021). Basically, Al, Mn, and Zn were controlled by pH values, however, some other factors were also considered.

Total P concentration in soil water at Shimizu had no relation with pH values but total P concentration in the spring at Yame linearly decreased with pH values (Fig. 5.8). From Figure 5.3, P in soil water at Shimizu was concentrated at the soil surface and depended on depth. However, total P concentration in the spring water of Yame depended on pH values. The marked difference for total P with pH was thought to have been caused by soil character as Shimizu soil contains volcanic loam. Volcanic loam absorbed P at the surface and the P concentration of soil water was very low. Total P concentration was related with depth as seen in Figure 5.3. NO_3-N concentration increased with total Al concentrations for soil water. However, Al concentrations of well and spring waters in the Kiku district changed from 0.1 to 100 mg/l even under the same NO_3-N concentration condition (Fig. 5.8). Therefore, total Al concentration only increased with the decrease in pH values.

5.6.2. Effect of Reducing the Amount of Nitrogen Fertilizer Used from Result 2

Amounts of total nitrogen fertilizer decreased from 1,100 kgN/ha/year before 1998 to 650 kgN/ha/year after 2001, resulting in NO_3-N and metal concentrations (Al, Zn, Ni, and Mn) in groundwater and spring water decreasing going downstream. However, the NO_3-N concentration of the Tanno Pond remained high at over 30 mg/l. Ni concentrations of well water were still over the environmental standard and the Al concentration for most of the water along the catchment was also over the standard (Ii et al., 2000; Tanaka et al., 2001, 2002). Dilution by Ohi River irrigation water facilitated a decrease in NO_3-N concentration of river water at the catchment. A reduction to 650 kgN/ha/year for tea plantations was insufficient to bring the NO_3-N concentration of river water within the environmental standard.

In light of this, the Japanese government has been advocating conservation measures such as reducing the use of chemical fertilizers and synthetic pesticides. However, to produce superior tasting, quality Japanese tea, tea plants require more than 500 kgN/ha/year of nitrogen fertilizer according to research conducted by the Japan Agricultural Cooperative (Morita, 2000; Hajiboland, 2017). For this reason, the amount of nitrogen fertilizer used at tea plantations in Japan still remains over 500 kgN/ha/year. At this level of application, the nitrogen concentration of groundwater and spring water at tea plantations exceeds the Japanese Environmental Standard. In summary, when land use is only a tea plantation in the catchment, reducing the fertilizer in the tea plantation alone did not satisfy the environmental standard value, so it is necessary to consider land use other than the tea plantation in the catchment and try to reduce the amount of fertilizer applied to the entire basin. Thus, the next evaluation was performed for land use combination of catchment.

5.6.3. Nitrogen Contamination for Combination Land Use (Result 3)

Nitrogen fertilizer use changed with land use, therefore the total amount of nitrogen fertilizer used for catchment is calculated using the area and amount of nitrogen fertilizer used for each land use. As a result, the NO_3-N concentration of river water for the catchment increased with the amount of nitrogen fertilizer used for catchment. The relation between NO_3-N concentration and the amount of nitrogen fertilizer used per catchment area was not linear. The reason for this being that the relation did not evaluate plant absorption and the relation itself may change with plant absorption.

Next, considering plant absorption, the relation between NO_3-N concentration of river for catchment and the amount of nitrogen fertilizer used minus the amount of nitrogen absorption by plants were clarified. This relation was linear and easily estimated. River NO_3-N concentration was estimated from this relation and when the amount of fertilizer used minus the amount of plant absorption per catchment was less than 100 kgN/ha/year, NO_3-N concentration of river water for catchment was less than 10 mg/l, which is the environmental standard. As a result, NO_3-N concentrations of river for catchments can be managed in order to come within the environmental standard by combining land use for the catchment, even if the NO_3-N concentration of groundwater under the tea plantation always exceeds the environmental standard. Therefore, a combination and balance of land use for the catchment is necessary for protecting fertilizer contamination in future agriculture.

5.7. CONCLUSION

Previously, Japan used the largest amount of nitrogen fertilizer of any country for tea plantations, and excess nitrogen fertilizer utilization (over 1,100 kgN/ha/year) causes high nitrogen and Al concentrations (50 mg/l) in groundwater. Decreasing the nitrogen concentration of groundwater is possible by reducing the amount of applied nitrogen fertilizer, so the Japan Ministry of

Environment subsequently recommended a decrease in the amount of nitrogen fertilizer used. However, in order to produce superior tasting, quality Japanese tea, the amount of nitrogen fertilizer used at tea plantations in Japan continues to remain approximately 600 kgN/ha/year. Under this continuing high nitrogen fertilizer use condition, pH values increased from 3.5–4.0 to 4–4.5. NO_3-N concentration decreased from 50–100 mg/l to 20–40 mg/l in soil water from 200 cm in depth to spring water for the tea planation catchment. Total Al concentration decreased from 50–150 to about 1–10 mg/l. Total Mn concentration decreased from 3–6 to about 0.1–1.5 mg/l. Total Zn concentration decreased from 0.5–1.5 to 0.01–0.1 mg/l. Total Ni concentration decreased from 0.01–0.1 to less than 0.04 mg/l. However, the NO_3-N, total Al, and Mn concentration of lower stream spring water still remained above the Japanese Environmental Standard for drinking water of 10, 0.2, and 0.05 mg/l.

As the soil of Shizuoka is high CEC with volcanic loam and a tertiary marine silt, soil was thought to absorb most of the P derived from fertilizer. Therefore, the P concentration of deep soil water and spring water was thought to decrease to 0.001–0.01 mg/l. Subsequently, photosynthesis did not occur in the downstream lake and river due to lack of P. As a result, pH values of downstream lake and river waters remained low with high transparency while nitrogen and metal concentrations remained high.

For the Kiku River catchment, Ohi River irrigation diluted contamination and thus pH increased. NO_3-N, Al, and Mn concentrations of downstream water decreased to within the Japanese Environmental Standard. Irrigation water is thus considered important in reducing NO_3-N concentration and increasing pH values. Effective dilution requires additional catchment water, however this water is not always easy to supply from other catchments. For the Kiku River catchment, the NO_3-N concentration of groundwater and spring water was extremely high at over 30 mg/l and their pH values were very low at 4. Therefore, direct use of groundwater and spring water derived from the tea plantation was impractical for either irrigation or drinking water. It is very difficult under only tea plantation use for a catchment to come within the environmental standard so the next evaluation was performed for a combination of land use in the catchment.

NO_3-N concentrations of river water for the catchment changed with the amount of nitrogen fertilizer used and plant nitrogen absorption for catchment. The relation between the NO_3-N concentration of river and the amount of nitrogen fertilizer used minus the amount of nitrogen absorption by plants considering land use of catchment was linear. This way, the NO_3-N concentration of catchment river water was easily estimated by land use using nitrogen fertilizer used and plant nitrogen absorption.

From this estimation, the planning of combination land use can be performed, which meets the environmental standard. Therefore, a combination and balance of land use for the catchment is necessary for protecting fertilizer contamination in future agriculture. As the scale of global food production increases, so does the amount of nitrogen fertilizer being applied. This fertilizer is not only being used for tea crops but also for other food plants. Land use for catchments is thus important for controlling nitrogen contamination.

ACKNOWLEDGMENTS

Thanks are due to Professor Emeritus Tatemasa Hirata of Wakayama University and Professor Masataka Nishikawa of Tokyo University of Science for introducing research fields and Toyokazu Tanaka, Yohei Nishio, Yoshie Fukuoka, and Kenichi Yamano for assistance in preparing field data.

REFERENCES

Barak, P., Jobe, B. O., Krueger, A. R., Peterson, L. A., & Laird, D. A. (1997). Effects of long-term soil acidification due to nitrogen fertilizer inputs in Wisconsin. *Plant and Soil, 197,* 61–69. https://www.jstor.org/stable/42948199

Bhatia, A., Sasmal, S., Jain, N., Pathak, H., Kumar, R., & Singh, A. (2010). Mitigating nitrous oxide emission from soil under conventional and no-tillage in wheat using nitrification inhibitors. *Agriculture, Ecosystems and Environment, 136,* 247–253. https://doi.org/10.1016/j.agee.2010.01.004

Chen, C. F., & Lin, J. Y. (2016). Estimating the gross budget of applied nitrogen and phosphorus in tea plantations. *Journal of Sustainable Environment Research, 26,* 124–130. https://doi.org/10.1016/j.serj.2016.04.007

Chen, P. A., Lin, S. Y., Liu, C. F., Su, Y. S., Cheng, H. Y., Shiau, J. H., et al. (2015). Correlation between nitrogen application to tea flushes and quality of green and black teas. *Scientia Horticulturae, 181,* 102–107. https://doi.org/10.1016/j.scienta.2014.10.050

Geological survey of Japan (2021). *Geological map of Japan 1:200,000.* https://www.gsj.jp/Map/JP/geology

Haines, T. A. (1981). Acidic precipitation and its consequences for aquatic ecosystems: A review. *Transactions of the American Fisheries Society, 10,* 669–701. https://doi.org/10.1577/1548-8659(1981)110<669:APAICF>2.0.CO;2

Hajiboland, R. (2017). Environmental and nutritional requirements for tea cultivation. *Folia Horticulturae, 29*(2), 199–220. https://www.sciendo.com/article/10.1515/fhort-2017-0019

Hirono, Y., & Nonaka, K. (2014). Effects of application of lime nitrogen and dicyandiamide on nitrous oxide emission from green tea fields. *Soil Science and Plant Nutrition, 60*, 276–285. https://doi.org/10.1080/00380768.2014.890015

Hirono, Y., Watanabe, I., & Nonaka, K. (2009). Trends in water quality around an intensive tea-growing area in Shizuoka, Japan. *Soil Science and Plant Nutrition, 55*, 783–792. https://doi.org/10.1111/j.1747-0765.2009.00413.x

Ii, H., Hirata, T., Matsuo, H., Nishikawa, M., & Tase, N. (1997). Surface water chemistry, particularly concentrations of NO_3^- and DO and $\delta^{15}N$ values, near a tea plantation in Kyushu, Japan. *Journal of Hydrology, 202*, 341–352. https://doi.org/10.1016/S0022-1694(97)00076-0

Ii, H., Hirata, T., Matsuo, H., Tase, N., & Nishikawa, M. (1998). pH and chemistry of nitrogen, phosphate, sulfur and aluminum in surface water near tea plantation in Kyushu, Japan. *Journal of Japan Society of Civil Engineering, 594*(vii-7), 57–63. In Japanese with English summary.

Ii, H., Hirata, T., Tanaka, T., Nishikawa, M., Nakajima, M., & Umehara, T. (2000). Spring, pond and river water chemistries derived from a tea plantation in central Shizuoka prefecture. *Annual Journal Hydraulic Engineering, 44*, 1155–1160. In Japanese with English summary.

Kuroda, H, Lin X. L., Kitamura, T., Oouchi, T., & Sugaya, K. (2016). Accumulated nitrogen from organic fertilizer affects river nitrogen pollution. *Proceedings of the 2016 International Nitrogen Initiative Conference, "Solutions to improve nitrogen use efficiency for the world,"* 4–8 December 2016, Melbourne, Australia. https://www.ini2016.com/pdf-papers/INI2016_Kuroda_Hisao.pdf

Maghanga, J. K., Kituyi, J. L., Kisinyo, P. O., & Ng'etich, W. K. (2013). Impact of nitrogen fertilizer applications on surface water nitrate levels within a Kenyan tea plantation. *Journal of Chemistry,* Article ID 196516. https://doi.org/10.1155/2013/196516

Morita, A. (2000). Studies on nutritional physiological responses of tea plant and nitrogen dynamics in tea fields under heavy fertilizer application. *Shizuoka Tea Experiment Station, Technical Bulletin, No.1*. 1–81. In Japanese with English summary.

Nakasone, H., & Yamamoto, T. (2004). The impact of the water quality of the inflow water from tea fields on irrigation reservoir ecosystems. *Paddy and Water Environment, 2*, 45–50. https://doi.org/10.1007/s10333-004-0039-2

Nanzyo, M., Dahlgren, R., & Shoji, S. (1993). Chemical characteristics of volcanic ash soils. In S. Shoji, M. Nanzyo, & R. Dahlgren (Eds.), *Volcanic ash soils genesis: Properties and utilization* (pp145–187). Developments in Soil Science, 21. Amsterdam: Elsevier Science Publishers.

Nishio, Y., Ii, H., & Hirata, T. (2011). Relation between amount of nitrogen fertilizer used and water quality in tea plantation, Shimizu district, Shizuoka city. *Annual Journal of Hydraulic Engineering, 55*, 1279–1284. In Japanese with English summary.

Ogawa, Y., Nakasugi, O., Nishikawa, M., Hirata, T. & Ii, H. (2005). Behavior of soil nitrogen and leaching of metal elements from arable land. *International Journal of Environmental and Analytical Chemistry, 85*(3), 209–221. https://doi.org/10.1080/03067310500050767

Roser, M., & Ritchie, H., (2021). Fertilizer. In *Our world in data.* https://ourworldindata.org/fertilizers#nitrogen-fertilizer

Schofield, C., & Trojnar, J. (1980). Aluminum toxicity to fish in acidified waters. In T. M. Miller & P. Morrow (Eds.), *Polluted rain.* New York: Plenum Press.

Schroder, J., Zhang, H., Girma, K., Raun, W., Penn, C., & Payton, M. (2011). Soil acidification from long-term use of nitrogen fertilizer on winter wheat. *Soil Fertility and Plant Nutrition, 75*(3), 957–964. https://doi.org/10.2136/SSSAJ2010.018

Sun, C., Chen, L., Zhai, L., Liu, H., Wang, K., Jiao C., & Shen, Z. (2020). National assessment of nitrogen fertilizers fate and related environmental impacts of multiple pathways in China. *Journal of Cleaner Production, 277*, 123519. https://doi.org/10.1016/j.jclepro.2020.123519

Tanaka, T., Ii, H., Hirata, T., Nishikawa, M., Nkajima, T., Umehara, K., et al. (2001). Groundwater chemistry near a tea plantation in the central of Shizuoka prefecture, Japan. *Annual Journal Hydraulic Engineering, 45*, 355–360. In Japanese with English summary.

Tanaka, T., Ii, H., Hirata, T., Nishikawa, M., Umehara, K., & Ogawa, Y. (2002). Groundwater metal and nitrogen contaminations caused by nitrogen fertilizer in tea plantation catchment, the center of Shizuoka, Japan. *Journal of Hydroscience and Hydraulic Engineering, 20*(1), 37–47. https://www.jsce.or.jp/publication/e/book/book_jhhe.html#201

Tang, S., Liu, Y., Zheng, N., Li, Y., Ma, Q., Xiao, H., et al. (2020). Temporal variation in nutrient requirements of tea (*Camellia sinensis*) in China based on QUEFTS analysis. *Scientific Reports, 10*, 1745. https://doi.org/10.1038/s41598-020-57809-x

Tian D., & Niu, S. (2015). A global analysis of soil acidification caused by nitrogen addition. *Environmental Research Letters, 10*, 024019. http://dx.doi.org/10.1088/1748-9326/10/2/024019

Wang, J., Tu, X., Zhang, H., Cui, J., Ni, K., Chen, J., et al. (2020a). Effects of ammonium-based nitrogen addition on soil nitrification and nitrogen gas emissions depend on fertilizer-induced changes in pH in a tea plantation soil. *Science of the Total Environment, 747*, 141340. https://doi.org/10.1016/j.scitotenv.2020.141340

Wang, Y., Yao, Z., Pan, Z., Wang, R., Yan, G., Liu, C., et al. (2020b). Tea-planted soils as global hotspots for N_2O emissions from croplands. *Environmental Research Letters, 15*(10), 10418. https://iopscience.iop.org/article/10.1088/1748-9326/aba5b2

Wang, Z., Geng, Y., & Liang, T. (2020c). Optimization of reduced chemical fertilizer use in tea gardens based on the assessment of related environmental and economic benefits. *Science of the Total Environment, 713*, 136439. https://doi.org/10.1016/j.scitotenv.2019.136439

Xie, S., Yang. F., Feng, H., Yu, Z., Liu, C., Wei, C., & Liang, T. (2021). Organic fertilizer reduced carbon and nitrogen in runoff and buffered soil acidification in tea plantations: Evidence in nutrient contents and isotope fractionations. *Science of the Total Environment, 762*(25), 143059. https://doi.org/10.1016/j.scitotenv.2020.143059

Xu, S., Zhou, S., Ma, S., Jiang, C., Wu, S., Bai, Z., et al. (2017). Manipulation of nitrogen leaching from tea filed soil using a *Trichoderma viride* biofertilizer. *Environmental Science and Pollution Research International, 24*, 27833–27842. https://link.springer.com/article/10.1007/s11356-017-0355-x

Yamano, K., Ii, H., Hirata, T., Tanaka, T., Nishikawa, M., & Ogawa, Y. (2001). The effect of fertilizing from tea plantation and water conveyance from Ohigawa River upon nitrogen contamination in Kikugawa basin, central Shizuoka prefecture. *Journal of the Japan Society of Civil Engineering, Environmental Engineering Research, 38*, 197–205. In Japanese with English summary.

Yamazaki, S., Nakayama, D., & Matsuyama, H. (2011). Observational study on the concentration/load of nitrate-nitrogen in small rivers derived from fertilizer. *Journal Japan Society Hydrology and Water Resources, 24*(4), 202–215. https://doi.org/10.3178/jjshwr.24.202

Zhang, W. L., Tian, Z. X., Zhang, N., & Li, X. Q. (1996). Nitrate pollution of groundwater in northern China. *Agriculture, Ecosystems and Environment, 59*, 223–231. https://doi.org/10.1016/0167-8809(96)01052-3

6

Springs of the Southwestern Great Artesian Basin, Australia: Balancing Sustainable Use and Cultural and Environmental Values

Gavin M. Mudd and Matthew J. Currell

ABSTRACT

The Great Artesian Basin (GAB) is one of the world's largest groundwater systems and supports a wide variety of springs, associated ecosystems, and cultural values. Historically, groundwater was extracted from the GAB with little regard for sustainable management, with much of the groundwater from artesian bores wasted through evaporation and seepage. By the late twentieth century, agriculture still dominated groundwater extraction; however, water use for mining and petroleum projects began increasing, a trend that continues. In particular, the Olympic Dam mining project in South Australia has been extracting groundwater since 1983 from a wellfield located on the southwestern margins of the GAB, an area containing a vast array of culturally significant, ecologically unique, and sensitive springs, including iconic mound springs. The extraction rate has increased over time, leading to concerns about impacts on the springs and their associated values. There are plans to expand the mine that would potentially increase the extraction rate further. This chapter reviews the hydrogeological setting of the springs of the southwestern GAB and their cultural and environmental values and synthesizes and analyses the available groundwater monitoring data associated with the wellfield. The case highlights the critical importance of detailed spatial and temporal hydrogeological monitoring, including both spring flow rates and groundwater level/pressure data, and the need to link monitoring and management of such sites to key cultural and environmental values.

6.1. INTRODUCTION

There is a common saying among Indigenous communities the world over: "Water is life" (Jewett & Garavan, 2019). (This is based on the authors' experiences visiting, supporting, and working for Indigenous communities across Australia, New Zealand, southern Africa, Fiji, and North America. Marshall (2017) further elaborates on the inseparability of water, life, and land in Australian Aboriginal nations' cultures and spiritual beliefs.) While streams and lakes are obvious sources of freshwater, groundwater contains 100 times more their volume in storage (Gleeson et al., 2016; Oki & Kanae, 2006), sustains and connects streams and lakes, and finds surface expression through springs, wetlands, and waterholes. In many dry landscapes around the world, springs are fundamental to supporting all types of life. In Australia, most of the continent ranges from semi-arid to hyperarid, meaning that springs have been and remain fundamental to the survival of Indigenous and non-Indigenous peoples and ecological communities. The permanence of springs in otherwise dry landscapes and, in some cases, their disappearance (Fensham et al., 2016) is a great reminder of the need to protect and conserve precious groundwater resources.

The Great Artesian Basin (GAB) is recognized as one of the largest groundwater systems in the world and underlies approximately 1.7 million km^2 (or 22%) of Australia (GABCC, 2014; Habermehl, 2020b). The GAB

School of Engineering, Royal Melbourne Institute of Technology, Melbourne, Australia

Threats to Springs in a Changing World: Science and Policies for Protection, Geophysical Monograph 275, First Edition.
Edited by Matthew J. Currell and Brian G. Katz.
© 2023 American Geophysical Union. Published 2023 by John Wiley & Sons, Inc.
DOI:10.1002/9781119818625.ch06

is host to an enormous reservoir of relatively fresh groundwater and a complex and diverse range of springs, including 13 recognized spring supergroups. Among these are the springs of the southwestern margins of the GAB around Kati Thanda (Lake Eyre), which are often affectionately called "oases in the desert" (Harris, 1980–1981). (For many years, the large salt lake of northern South Australia was called Lake Eyre, but was renamed Kati Thanda in respect of Indigenous peoples in December 2012. We adopt Kati Thanda throughout this chapter in place of Lake Eyre.) Discharge from the GAB supports groundwater-dependent ecosystems (including a wide array of critically endangered species and ecological communities), Indigenous and non-Indigenous heritage, as well as pastoral (agricultural), mining, and petroleum activities and small towns and other community water supplies.

Until the latter parts of the twentieth century, most use of GAB groundwater supported pastoral (mostly cattle) stations. Typically, pastoral water bores experienced (and continue to experience) artesian flows, sometimes reaching >10 ML/d, and water from these was allowed to flow freely across the landscape in bore drains, meaning significant losses, inefficient use, and invasion and spread of weeds and feral animals. This included introduction of water into previously dry ecosystems (or water-remote areas), leading to structural impacts on ecological processes and biodiversity (GABCC, 2014). From the 1970s, the development of underlying petroleum reservoirs (especially in the Cooper Basin) and major mining projects from the 1980s onward have also relied on GAB groundwater. In South Australia, the development of a wellfield in the midst of the Kati Thanda Spring supergroup to supply water to the developing Olympic Dam mine, located ~100 km to the south, has been and remains deeply controversial (AAC, 2020; Boyd, 1990; Mudd, 2000; Read, 2003).

In 2001, the Australian government listed most GAB spring ecosystems as endangered under federal environmental law, reinforcing their national conservation significance (Fensham et al., 2010). There have also been significant community and scientific efforts to see the GAB springs recognized internationally as being of World Heritage significance (e.g., in the late 1990s; see Jenkin, 2001), albeit unsuccessfully.

This chapter focuses on the Kati Thanda Springs, providing a detailed examination of monitoring data and impacts of the wellfield developed for the Olympic Dam mining project. Since the establishment of the mine and initial wellfield, there have been proposals (including as recently as 2020) to expand production and increase groundwater extraction further. As such, the assessment herein and other recent proposals to develop a comprehensive adaptive management plan for GAB springs (Brake et al., 2020) are important and topical issues.

6.2. THE GREAT ARTESIAN BASIN SPRINGS

6.2.1. Cultural and Ecological Values

The GAB springs were and remain fundamental to Australia's Indigenous peoples. Since springs were a permanent and reliable source of freshwater, they were integrated into agricultural purposes, ceremony, dreaming stories, and caring for country (e.g., Harris, 1980–1981; Hercus, 1990; Ransley & Smerdon, 2012). Despite European occupation and development of the lands and groundwaters, including springs, from the late 1800s onward in various ways, indigenous cultural connections to the springs remain deep and significant (Habermehl, 2020a; Lewis & Harris, 2020; Moggridge, 2020).

The GAB springs of the Kati Thanda south region are situated on the unceded traditional lands of the Arabana people, whose ancient and continuing connection to the land and waters of the region, including Kati Thanda, is recognized in a formal Native Title determination. In their submission to a recent national government inquiry into the protection of Aboriginal cultural heritage (brought about by the destruction of the Juukan Gorge rock shelter, documented as having at least 32,000 years of continual human occupation and use), the Arabana explained the significance of the springs for their people and culture:

> To many, much of the country would appear to be desert, but to the Arabana it is a place of rich abundance of life. The thing that sustains that life, are what we call the mound springs, which are seepages at the southern edge of the Lake Eyre Basin. In our country there are over 6,000 of these springs and they are of great significance to the Arabana people. These springs were created by our ancestors whose stories we know and respect, on whose water we depend for life, and on which the birds, plants and animals also depend. (AAC, 2020, p. 1)

Further, the Arabana expressed their concern over the disappearance of springs, under the influence of ongoing groundwater extraction for the Olympic Dam mine:

> Unless something is done by the Commonwealth, our springs will disappear and with them will disappear the unique flora and fauna that live in and around these springs. There are many endemic and endangered species of plants, fish and birds that depend on these springs, and they too will disappear. (AAC, 2020, p.2)

Many studies, including Fensham and Fairfax (2003), and numerous papers included in the recent special issues of the *Proceedings of the Royal Society of Queensland* (Arthington et al., 2020) and *Hydrogeology Journal* (Miraldo Ordens et al., 2020) on the GAB springs, further highlight the ecological significance and biodiversity of the springs. As documented by Rossini et al. (2018), species that are endemic to a particular group of springs, or an individual spring pool or wetland, are characteristic throughout the GAB, including the Kati Thanda Springs. These endemic species encompass dozens of taxa of plants, molluscs, fishes, and crustaceans, noting that the diversity of endemic species is almost certainly underestimated in current official listings (Rossini et al., 2018). It is well documented that GAB spring pools (including those of the southwest GAB), the focus of this chapter, represent evolutionary refugia for plants and animals over geologic timescales, remaining the only permanent sources of water in the landscape during periods of major climatic and landscape change (Davis et al., 2013; Keppel et al., 2012; Murphy et al., 2015). This provided isolated habitats for the genetic evolution of the endemic species, a topic still being actively explored (Murphy et al., 2015).

6.2.2. Geology and Hydrogeology of the Great Artesian Basin

The geology and hydrogeology of the GAB have recently been comprehensively described by Smerdon et al. (2012), Keppel et al. (2013), and Habermehl (2020b). Here, we provide only a very brief description and direct readers to these (and related) studies for further detail. The GAB is composed of three major sedimentary basins, the Eromanga, Surat, and Carpentaria basins, with additional geological basins adjacent and/or underlying these (e.g., Cooper, Bowen, Galilee, Pedirka, Simpson, Adavale, and Warburton basins and Warrabin Trough). The geology consists of multiple layers of sandstones, shales, siltstones, and mudstones deposited from the Jurassic to the Cretaceous, with sandstones acting as aquifers for groundwater storage and flow and shale and other low-permeability units acting as aquitards.

Together, the three GAB subbasins act (to a certain degree) as a connected hydrogeologic system. Recharge and elevated groundwater levels that occur along the eastern margins in the Great Dividing Range in Queensland drive groundwater flows mostly to the south, southwest, and west. Some, however, have questioned the level of regional interconnectivity (e.g., Dafne, 2016; Endersbee, 2001; Mazor, 1995). The elevation in the recharge areas is typically around 400 m relative to the Australian Height Datum (AHD). (For AHD, zero (0) is mean sea level.) Much of the GAB occurs below relatively flat plains leading to artesian conditions across most of the GAB. The connections between the primary basins and adjacent or underlying basins remain relatively poorly understood, contentious (especially in the context of ongoing water extraction and mining and oil and gas development), and subject to ongoing studies (e.g., Moya, 2015; Radke & Ransley, 2020). The GAB is therefore a highly complex and heterogeneous system, with multiple aquifers, aquitards, and local/regional scale flow systems with variable degrees of interaquifer and interbasin connectivity at different scales (Radke & Ransley, 2020). The role of faults and structural features such as unconformities in controlling flow patterns, geochemistry, and the occurrence of springs is well documented (Crossey et al., 2013; Flook et al., 2020; Keppel et al., 2020).

Along the southwest Great Artesian Basin, where the study area is located, there is a major sandstone aquifer comprising the combined Late Jurassic Algebuckina sandstone directly overlain by the Early Cretaceous Cadna-Owie Formation. This aquifer unconformably overlies Proterozoic basement and is overlain by an aquitard (Early Cretaceous Bulldog shale). The Mesozoic sediments dip shallowly northward; the sandstones outcrop around the edge of the basin and around bedrock highs close to the edge of the basin. Potentiometric surface contours define a groundwater flow regime from north to south converging around the Kati Thanda Springs (Priestley et al., 2020).

The GAB is estimated to contain about 6.5×10^{13} m^3 (64,900 million ML) of groundwater (GABCC, 2014). Quantification of a water budget for the GAB is, necessarily, approximate given its vast scale and the uncertainties in key parameters and processes (Smerdon et al., 2012). The most recent assessment of recharge and discharge for the principal Cadna-Owie and Hooray aquifer system is shown in Table 6.1.

6.2.3. Springs

There are a variety of hydrogeological processes or mechanisms that form GAB springs, with excellent reviews by Habermehl (1982), K-S (1984), Boyd (1990), Keppel et al. (2011), Brake et al. (2020), and Habermehl (2020a). From these studies, the most important processes are flow through a fault or fracture zone, aquifer outcrop or subcrop, as well as aquitard thinning. Spring location can also be influenced by topographic position, mineral precipitation (e.g., resulting in long-term accumulation of crusts, which influence permeability), vegetation, microbial activity, land use (e.g., trampling by cattle or tourism), and drawdown caused by pastoral groundwater extraction. The Kati Thanda south region is famous for hosting the classic mound spring type, formed by the evaporation of groundwater and precipitation of carbonate minerals (since the groundwater chemistry is bicarbonate-dominant), which over time builds up a mound (Fig. 6.1). The underlying flow mechanisms are varied, combining most (if not all) of those noted above. Many mound springs contain pools at their surface and the discharge forms a wetland down-gradient of the mound, both of which are crucial microhabitats for biodiversity (e.g., endemic fish, macroinvertebrates, vegetation, snails, frogs, and other taxa). Although it is inherently difficult to measure the total

Table 6.1 Estimated water balance of the Cadna-Owie / Hooray aquifer system (ML/d)

	Groundwater recharge				Spring discharge		Diffuse discharge	
	Cl mass balance[a,b]	Numerical modeling[c]	Conceptual model[a]	Bore extraction[c]	Habermehl[d]	5% PP[a]	10% PP[e]	15% PP[e]
Surat	430 to 808	507	649	636	38	126	252	381
Central Eromanga	444 to 723	452	389	425	68	170	340	507
Western Eromanga	19 to NA	110	19	164	66	101	197	299
Carpentaria	NA to NA	277	1,184	175	NA	NA	NA	NA
Total	893 to 1,532	1,345	2,241	1,400	173	397	789	1,186

Note: Cl = chloride, PP = preferential pathways, NA = not available.
[a] Ransley and Smerdon (2012).
[b] Kellett et al. (2003), Habermehl et al. (2009).
[c] Welsh et al. (2012).
[d] Habermehl (1982).
[e] Harrington et al. (2012).

Figure 6.1 The Bubbler mound spring (Coward springs subgroup, location shown in Fig. 6.2) showing key features of the spring pool, drainage channel, and supported wetland (area of vegetation) as well as the older extinct Hamilton Hill mound (top left) (photo courtesy of Isla and Linda Marks, Eric Miller).

flow from a specific spring vent (e.g., due to vegetation interference, subsurface and diffuse losses, different field methods), it has been found that wetland area correlates well to measured flows (Fatchen & Fatchen, 1993; White & Lewis, 2011; Williams & Holmes, 1978). There are (to our knowledge) limited studies investigating the relationship between artesian pressure and monitored spring flow rates, incorporating relevant geological factors such as aquifer thickness, permeability or structural features, and other potential controls on the pressure-flow relationship (e.g., atmospheric pressure, surface modifications).

6.3. OLYMPIC DAM AND ITS GAB WELLFIELDS

The giant Olympic Dam deposit in central South Australia, ~100 km south of the Kati Thanda Springs complex, was discovered in 1975 and contains copper (Cu), uranium (U), gold (Au), silver (Ag), rare earth elements, cobalt, and tellurium. Only Cu-U-Au-Ag have been extracted since production commenced in August 1988. Olympic Dam is governed by the Roxby Downs (Indenture Ratification) Act 1982, with special provision for water resources, indigenous heritage, radiation protection, environmental regulation, project governance, and other matters that often override the normal legislation for that area. (The Olympic Dam mineral deposit was discovered on the Roxby Downs pastoral station; hence the mining town is named Roxby Downs and the project is sometimes referred to as Roxby Downs, as in the Indenture Act.)

In the 1982 environmental impact statement (EIS) prepared for environmental approvals (KSR, 1982), it was proposed to develop two large wellfields around the southwestern margins of the GAB near Kati Thanda South. The first, Wellfield A, was to be constructed in the midst of the Kati Thanda Springs region, with the second, Wellfield B, to be located farther into the GAB to the northeast. Wellfield A was expected to be able to supply 6 ML/d from five production bores for the life of the project, but was not considered scalable to longer term demand of 33 ML/d, hence the need for Wellfield B. Initial extraction began in August 1983 at about 1.3 ML/d with water trucked to Olympic Dam. Wellfield A was formally licensed in May 1986, with construction proceeding on the 108 km pipeline and concurrent with additional investigations, it was shown that higher yields could be obtained from Wellfield A with additional bores. The wellfield was subsequently expanded in stages to match mine construction (1986–1988) and operating demands to the mid-1990s, reaching an average of 15.1 ML/d in 1995/96 (WMCR-ODC, var.). During this time, there were increasing concerns about the impacts of Wellfield A extraction on the springs, especially those of the Hermit Hill Springs complex (Keane, 1997; Mudd, 2000).

Olympic Dam sought approvals to undertake a major expansion through releasing a new EIS in May 1997 (K-S, 1997), although by this stage concerns about impacts on the springs and increasing water demands had already led to the need for Wellfield B, which was approved under the special water resources provisions of the Indenture Act. There were two important parts to this process: (1) purchase of water entitlements from pastoral stations, effectively switching the GAB extraction from pastoral to mining use, and (2) a major reduction in extraction from Wellfield A with the majority now coming from Wellfield B. A map of the two wellfields is shown in Figure 6.2, including locations of the local and basinwide spring supergroups across the Great Artesian Basin. The history of extraction from both wellfields as well as measured or estimated pastoral bore use is shown in Figure 6.3.

Another major mine expansion was proposed in 2009, including conversion from underground to open-cut mining and a sevenfold increase in processing capacity (BHPB, 2009), however this failed to proceed after uranium prices collapsed following the Fukushima nuclear disaster in Japan in March 2011. More recently, the mine expansion was identified by the Commonwealth government as a potential project it would support to promote economic recovery following the COVID-19 pandemic. The assessment process for a new expansion project was restarted in 2019 but again withdrawn due to marginal economics. If there is a future proposed expansion, there would be significant pressure to develop additional water supply capacity (for example, through an additional wellfield), although there is no specific proposal as of mid-October 2021.

As part of the Special Water Licenses for Wellfield A (and later B), Olympic Dam is required to monitor groundwater levels in a range of monitoring bores, flows from production, and pastoral bores as well as flow and ecological monitoring and assessment of the numerous springs. The monitoring results are compiled annually and have recently become publicly available (AGC, var.; BHP, var.; WMCR-ODC, var.). The compliance limits for Wellfield A are stated as (1) drawdown between observation bores GAB8 and HH2 shall not exceed 4 m; and (2) drawdown at Jackboot bore (at the northern most corner of Wellfield A) shall not exceed 5 m (K-S, 1997). Limitations of a purely drawdown-based monitoring and compliance program for the protection of springs have recently been discussed in Currell (2016), drawing on Theis (1940) and Konikow and Leake (2014).

6.4. ASSESSING WELLFIELD A IMPACTS ON SPRINGS AND BORES

6.4.1. Spring Flow and Groundwater Level Trends Through Time

After significant criticism of the 1982 EIS regarding the lack of technical rigor in the assessment of potential impacts of GAB extraction on the springs (Mudd, 2000), additional extensive field studies were undertaken (K-S, 1984). This work led to the theory that the location of Wellfield A was within a localized subbasin, hydraulically separated from the GAB aquifers feeding the springs by major northwest trending faulting, especially the low permeability Norwest Fault Zone (see Fig. 6.3). It was hypothesized that the subbasin sustaining the important Hermit Hill Springs complex would therefore not be impacted, although some springs and pastoral bores would see significant drawdowns and impacts within the Wellfield A subbasin.

A detailed map of monitoring, production, and pastoral bores and spring locations is given in Figure 6.3, including key geological features (basement and aquifer outcrop, fault zones) and drawdown recorded as of mid-2008. The year 2008 is chosen since recovered groundwater levels had effectively reached steady state by this time (see Fig. 6.4). Subsequent annual reports also

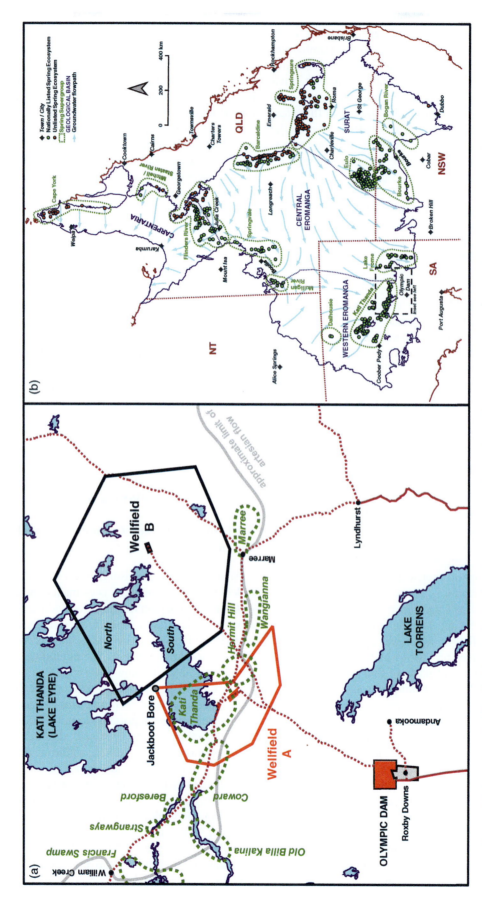

Figure 6.2 Location of (a) Wellfields A and B for the Olympic Dam project and Kati Thanda spring subgroups and location of (b) the Great Artesian Basin, spring supergroups and principal groundwater flow paths (redrawn and synthesized from K-S, 1984; BHPB, 2009; Habermehl, 2020a; Mudd, 2000).

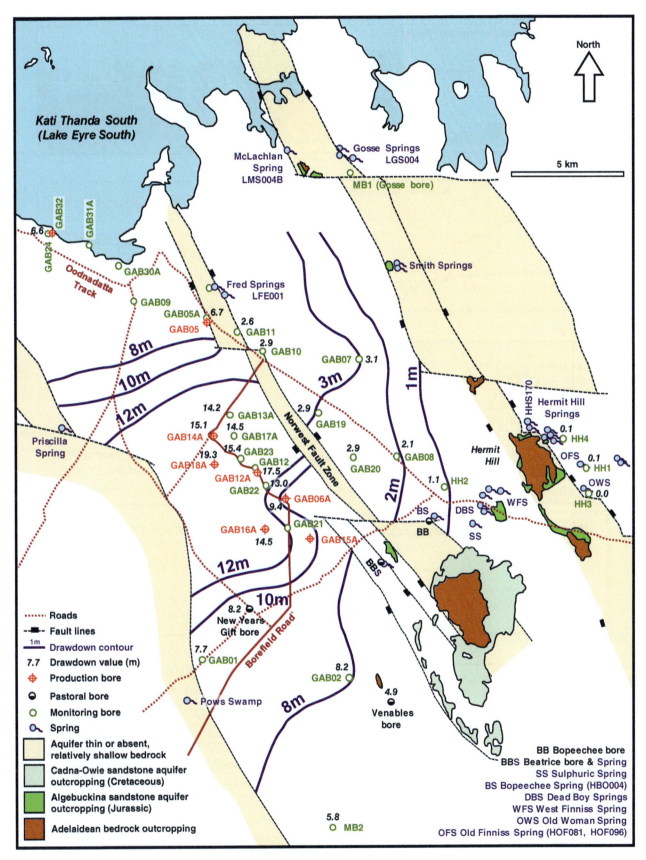

Figure 6.3 Location of production, monitoring and pastoral bores and springs, simplified geology and drawdown at approximately June 2008 (baseline is May 1986, the issue date of the special water license for Wellfield A) (redrawn from BHP (var.), Fig. 5-1, 2007–2008 edition).

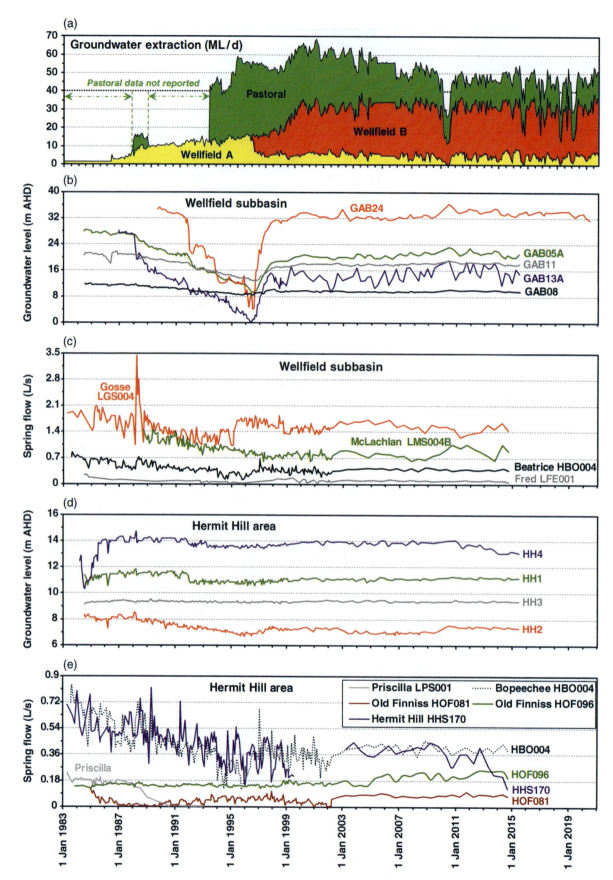

Figure 6.4 (a) Water extraction by Wellfields A and B and pastoral activity, (b, c) groundwater levels and spring flows for the Wellfield A subbasin region, and (d, e) Hermit Hill Springs complex (data synthesized from AGC, var.; BHP, var.; WMCR-ODC, var.).

provide less detail regarding presentation of drawdown surrounding Wellfield A and spring and water level monitoring at individual sites. Generally, drawdown has propagated from the wellfield to the northwest and southeast most markedly, controlled by basement highs aligned in the NW-SE orientation, as predicted. However, some propagation of drawdown to the east across the Norwest Fault Zone (encompassing at least Bopeechee Spring) is also evident. It should be noted that in the northern area of Figure 6.3, drawdown to the west of McLachlan and Gosse springs was between 5 to 15 m by mid-1994 (Fig. 23, AGC, var.), again indicating substantial drawdown propagation across the Norwest Fault Zone in this northern region (see also groundwater levels over time in Fig. 6.4). For the same time (mid-1994), bore GAB07 reported drawdown of close to 5 m, showing drawdown propagation across the Norwest Fault Zone in the center of the region, very close to the critically important Hermit Hills group of springs (including Bopeechee, Dead Boy, Sulphuric, and West Finniss springs, with the Hermit Hill, Old Finniss, and Old Woman springs farther separated by basement outcrop and faulting).

By the time of Wellfield B's development in the mid-1990s, the hydraulically disconnected conceptualization was being questioned, due to perceived and actual impacts on springs outside the Wellfield A subbasin (Mudd, 2000). This was supported by the 1997 EIS acknowledging the inaccuracy of the 1980s impact predictions (Table 6.2) as well as the ongoing monitoring showing drawdown extending across the Norwest Fault Zone. This agrees with the work of Bense et al. (2013), who showed that the hydrogeological behavior of faults is complex and often very difficult to predict from observed physical geology alone.

Since the work of Keane (1997) and Mudd (2000), the full extent of annual wellfield monitoring reports has become publicly available, specifically AGC (var.), WMCR-ODC (var.), and BHP (var.). The data from these reports have been digitized and a selection of the data presented herein. The primary data, such as spreadsheets, are not publicly available. Given that this monitoring is done for statutory compliance purposes, we use the data as presented in these reports; further details on monitoring methods are provided in these reports. All data series are shown as lines, although it must be noted that they are based on point monitoring and do not represent continuous monitoring. We present the spring flow and groundwater level data for the Hermit Hill Springs complex and the Wellfield A subbasin in Figure 6.4, including Wellfields A and B and pastoral water extraction over time. Although not shown here, an overlay of these reproductions on the source graphs clearly captures the key timing and magnitudes of changes in levels and/or flows. From 2016, however, the reporting changed from individual bore and spring data to showing only total flows by spring complex, regional drawdowns, and a few bores, preventing an accurate analysis of spring and bore-level effects and their relationship to extraction and hydrogeological controls.

For the Wellfield A subbasin, groundwater levels declined in unison with the increasing extraction rate from the mid-1980s to the mid-1990s, with a rebound occurring after the reduction in extraction from mid-1996. Although bore GAB24, the northern most bore, has returned to approximately prewellfield levels, bores GAB05A, GAB08, GAB11, and GAB13A still remain significantly lower (i.e., GAB05A ~21 vs. ~28; GAB08 ~9.8 vs. ~11.6; GAB11 ~18.3 vs. ~20.7; GAB13A ~15.8 vs. ~27.6; all m AHD). The pattern for spring flows shows some rebound after the reduction in Wellfield A extraction but still general declines overall (i.e., Gosse ~1.8 vs. ~1.5; McLachlan ~1.3 vs. ~0.92; Beatrice ~0.35 vs. ~0.60; Fred ~0.25 vs. ~0.10; all L/s). For McLachlan, it appears that the extent of flow recovery is minimal, rarely rising above the ~0.7 L/s reached at the peak of Wellfield A extraction (i.e., flows since that time range from 0.7 to 0.8 L/s). These aspects are further discussed below.

For the Hermit Hill region, there was a common pattern of gradual decline in groundwater levels in line with the increasing extraction from Wellfield A, and gradual recovery following the reduction in extraction (i.e., bores HH1 to HH4). Bore HH3 has returned to similar levels prior to Wellfield A extraction, with HH4 showing similar behavior; although in the last few years of available data the level declined substantially (by ~1 m). Both bores HH1 and HH3 show similar behavior to subbasin bores (i.e., a rebound after reduction in wellfield extraction, but still below pre-wellfield levels). Given the declining trend in HH4, it is unfortunate that there appears to be no further level data publicly reported since mid-2015.

Flow trends for the selected spring vents in the Hermit Hill region are variable. For the two Old Finniss Spring vents, HOF081 and HOF096, flows have recovered after the reduction in extraction from Wellfield A with HOF096 showing an increasing trend and greater flows than prior to commencement of Wellfield A extraction, contrasting with HOF081, which remains below prewellfield flows. Venables and Priscilla were rendered extinct, as a result of the wellfield extraction (Table 6.2), ceasing to flow in mid to late 1990, respectively. Bopeechee Spring (HBO004) has shown limited recovery since the reduction in Wellfield A extraction, with flows increasing from the lows of ~0.12 to ~0.35 L/s (1994 to 1996) to ~0.35 to ~0.45 L/s (2002 to 2015), as compared with average flows of ~0.77 L/s in 1983. For the Hermit Hill Spring vent HHS170, flows have mostly followed a similar pattern to Bopeechee; however, since 2013 flows have reduced substantially.

Table 6.2 Comparison of predicted versus actual impacts on key GAB springs by the mid-1990s, assuming a semipermeable or impermeable fault zone

Spring complex	Spring system	Predicted flow reductions (%)		Actual flow reduction (%)
		Impermeable	Semi-permeable	
Hermit Hill	Beatrice	100	100	40
	Bopeechee	<2	20	43
	Hermit Hill	<1	<1	36
	Old Finniss	<2	<2	Marginal increase
	Venables	100	100	100 (extinct May 1990)
Wangianna Lake Eyre	Davenport	<1	<1	Close to 0
	Emerald	3	3	Close to 0
	Fred	6	17	50
	Priscilla	75	60	100 (extinct late 1990)

Source: K-S (1997).

There are two key observations from the analysis of bore and spring flow monitoring data. First is the almost complete lack of reported data since mid-2015, due to the change in reporting structure, unfortunate particularly in light of HHS170's declining trend immediately prior to this time. As such, the ongoing extent and/or cause of this steep decline in flow cannot be resolved. Field observations from local citizens suggest flow remains reduced and wetland area and vegetation are receding. (Observations are from a local community member who has been visiting, measuring flows, and recording observations on the state of the mound springs since 1985; field notes used with permission.) Second is the reduced frequency of monitoring since approximately 2003, which leads to the appearance of flatter hydrograph patterns, which do not capture the significant seasonal and interannual variability previously observable in monitoring data. The reduced monitoring frequency also hampers the ability to assess in greater detail the potential factors that could be driving the measured flow or level variability. Combined, the reduction in the extent of reporting and monitoring frequency reduces confidence in the monitoring program, especially its ability to discern key trends in a timely manner and attribute the major cause(s) of these.

6.4.2. Spring Flows and Groundwater Level Correlation

Surprisingly, there appear to be no studies that have assessed the correlations between Kati Thanda Spring flows and nearby groundwater levels, despite its being established that artesian pressure (or level) is fundamental to driving groundwater flow from a spring (e.g., Currell et al., 2017). Given the availability of ~40 years of (digitized) monitoring data, we explored the relationship between level and spring flow for three sites. We chose the following spring/bore pairs as the closest combinations of available monitoring data (including allowing for faults, which could affect the relationship between level and spring flow):

1. Hermit Hill: key spring vent HHS170 paired with bore HH4, ~2 km apart (Note the earliest monitoring data, March 1983 to July 1985, have been excluded from the regression since these values appear to be significant outliers with no explanation as to why; see later Fig. 6.5.)
2. Bopeechee: key spring vent HBO004 paired with bore HH2, ~1.9 km apart
3. Fred: key spring vent LFE001 paired with bore GAB11, ~2.2 km apart
4. Old Finniss: key spring vents HOF081 and HOF096 each paired separately with bore HH1 or HH3 (~0.8 and ~1.2 km apart, respectively)

The results, Figure 6.5, show a correspondence between bore water level and spring vent flows in some cases, with least-squares regression coefficients of 0.25 (HHS170-HH4), 0.39 (LFE001-GAB11), and 0.52 (HBO004-HH2), respectively. Both pairs of HOF081 and HOF096 show significant scatter of flow rates within a somewhat narrow pressure range (see inset graph). For the Hermit Hill and Bopeechee pairs, the level-flow relationship shows steep gradients (i.e., linear slope of ~0.22), indicating that small reductions in groundwater pressure drive a substantial reduction in flows. For Fred Spring, however, the gradient is much more subdued (~0.017), suggesting a greater pressure reduction would be required to affect spring flows. The least-squares regression gradient (i.e., m from y = mx + c) has the same units as transmissivity (m^2/s), reinforcing the relative difference in permeability or transmissivity between the different cases.

These are perhaps intuitive results, reflecting the basic mechanics of groundwater flow (i.e., Darcy's law); however, there are additional factors to consider. First, Fred Spring and GAB11 are on the eastern side of a fault zone, suggesting that the lower transmissivity of the fault zone may be influencing the flow-level relationship. Second,

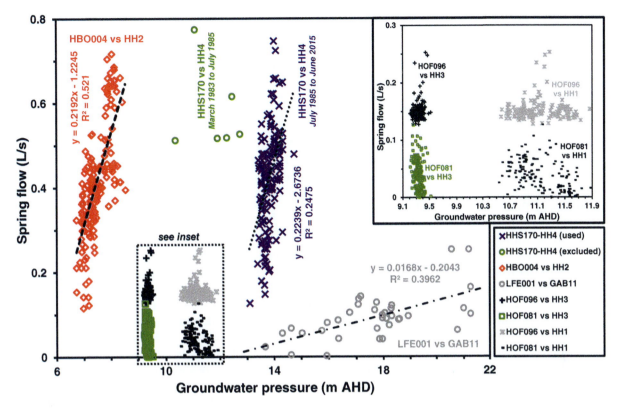

Figure 6.5 Relationship between groundwater levels and spring flows for the Hermit Hill (spring vent HHS170 versus bore HH4), Bopeechee (spring vent HBO004 versus HH2), Fred (spring vent LFE001 versus GAB11), and Old Finniss (spring vents HOF081 and HOF096 versus bores HH1 or HH3) paired sites (data synthesized from AGC, var.; BHP, var.; WMCR-ODC, var.).

based on analysis of the flow-level trend for HHS170 (dotted line, Fig. 6.5), the linear regression suggests that zero flow would occur when the artesian groundwater level falls below ~11.9 m AHD (and ~5.6 and ~12.2 m AHD for Bopeechee and Fred Springs, respectively). Upon visual inspection, it could be argued that, accounting for scatter in the data, this could occur at a higher pressure (i.e., ~13 m). Currell et al. (2017) have previously discussed this simple head/flow relationship for culturally significant springs on the eastern margin of the GAB; however, they noted another critical factor is the geomorphic threshold (i.e., total hydraulic head required in a spring's source aquifer to drive discharge above surface elevation and thereby maintain a spring). There are no sufficiently accurate surface elevation data (within ± 0.1 m) that are publicly available for each spring vent and bore (to our knowledge), meaning it is not possible to determine the extent of artesian pressure at each bore and the elevation of each spring discharge point relative to artesian pressure at monitored bores. An urgent priority for protection of these springs should thus encompass a precise elevation survey at each spring and determination of the geomorphic threshold, below which spring flow will (likely) be unable to be sustained.

Third, the level data for HHS170 and HBO004 occur together within a typically very tight range (i.e., ~13.4–14.3 m AHD for HHS170 and ~6.7–8.4 m AHD for HBO004) but vary more widely for LFE001 (i.e., ~13.7 to 22.7 m AHD). This further reinforces the potential for other factors to be affecting flows at Fred, whereas levels are arguably the primary control for HHS170 and HBO004. Conversely, there appears to be no obvious correlation between flows and levels for either combination of HOF081 and HOF096 with bores HH1 or HH3, suggesting measurement variability and/or minimal drawdown impacts at these locations (their location on the eastern side of Hermit Hill could help explain this behavior). There are a variety of other factors that could possibly affect these relationships, including barometric (atmospheric) pressure differences during flow and level measurements (e.g., Rasmussen & Crawford, 1997), vegetation and algal growth affecting (near-) surface permeability, any change in spring morphology over time (e.g., discharge point being closed due to carbonate precipitation and flow emerging from an alternative or new vent), cattle trampling of springs reducing permeability or causing changes in the point(s) of discharge (e.g., Brake et al., 2020), different measurement methods and personnel over time adopting slightly different measuring points and/or conventions (i.e., human variability in field work), as well as the extent of hydrogeological links between the chosen bore and paired spring vent (i.e., influence of geological structure or other features in mediating the level-flow relationship). This also highlights the need to locate monitoring bores as close as reasonably possible to springs, noting the risks of cultural and environmental damage associated with drilling too close to springs as well as a monitoring regime, which takes account of all of these possible factors to the best extent technically possible. Finally, the monitoring regime must make sure that measurements of flows and levels are taken at very similar times. Overall, these results should be considered exploratory only, with further detailed investigation and analyses performed on the primary monitoring data combined with individual synthesis of geology and hydrogeology for each spring-bore pair. Such detailed assessment would facilitate a more comprehensive understanding of the dynamics of each spring-bore pair.

It should be noted that the requirements of the approved monitoring program have continually changed and evolved over time (see WMCR-ODR, var.; BHP, var.). For example, the frequency of artesian bore monitoring was reduced from monthly to quarterly in February 1999, ostensibly due to the gradual changes in groundwater levels after the rapid recovery due to reduction in extraction rate in Wellfield A in the mid-1990s. Furthermore, spring flow measurements were reduced from monthly to biannual in 2002 in favor of remote sensing to assess wetland area, although these results are rarely included in statutory reporting. From 2015, annual statutory reporting presented results with respect to hydrogeological zones rather than individual bore and spring data. Over the decades, it is clear that the extent of bore and spring data and information being included in annual statutory reporting has been considerably reduced, a critical issue given the need for such data to inform a comprehensive understanding of the hydrogeological, ecological, and cultural health of the springs.

To provide a concise overview of the current extent of impacts due to Wellfield A, the key hydrogeological aspects of levels and flows are synthesized in Table 6.3. We do this by comparing pre-Wellfield A flows or levels, taken as data from 1983 to July 1986 (prior to large scale extraction commencing in August 1986) or the earliest data reported, with flows or levels around 1995 to 1996 (i.e., peak impacts due to extraction by Wellfield A) with an average of levels and flows from 2010 to 2015 (or later if available) as well as the ratio between 2015 to baseline values to assess the extent of recovery after the reduction in Wellfield A extraction rates. Finally, we look at the ratios between baseline values and recent monitoring results as well as the trend of the last 5 years of available monitoring data to assess whether these are increasing, stable, or declining, defined by the ratio of 2015 to 2010 results, specifically >1.1 increasing, 0.9 to 1.1 stable, or < 0.9, declining. There are a considerable number of springs where despite stabilization of nearby bore levels, flows continue to decline. This underscores the urgent need to (1) reinstitute detailed spring-by-spring and bore-by-bore monitoring and reporting of data; (2) enhance understanding of the conceptual model of each spring and its relationship to hydrological, geological, and other potential influences on flow; and (3) monitor for the long-term and consider and account for time lags in the response of springs to groundwater extraction (e.g., Currell, 2016).

6.4.3. Discussion: Protecting Springs

The primary compliance criteria for Wellfield A are drawdowns, namely a maximum of 5 m at Jackboot bore (northern most point of Wellfield A) and 4 m at GAB08 and HH2. There appears to have been no clear hydrogeological basis for developing these thresholds in the 1980s, such as pressure-flow relationships or an understanding of minimum pressures required to maintain ecological flows and values in springs. To our knowledge, there has been no further research or technical reviews or assessments since that time that could further underpin such an approach. Furthermore, the allowance of 4 m drawdown at GAB8 and HH2, which are both on the opposite side of the low permeability Norwest Fault Zone (Fig. 6.3) to the wellfield subbasin, implies a recognition of modest permeability across this zone such that drawdown would be measurably significant (supported by the subsequent monitoring data), and calls into question the ongoing use of Wellfield A. As such, the emphasis appears to have been maintaining the extraction rate for the Olympic Dam project, especially given the delay of the onset of Wellfield B extraction.

It is unclear whether and to what extent spring flows may recover in subsequent years in response to reductions in extraction, and if so, what the time lags and key hydrogeological controls are. As outlined in Currell (2016), a comprehensive approach to the monitoring and management of springs requires an analysis of both (1) changes in groundwater levels (and associated hydraulic gradients to key receptors) in response to extraction, and (2) analysis of the volumetric rate of capture induced by extraction. These two factors are clearly related, but the relationship between them is not straightforward and requires robust conceptual (and ideally numerical) modeling, which can quantify their interrelationships.

The GAB is managed by the state (or provincial) governments of Australia, which maintain the constitutional responsibility for water resources, through the nationally agreed Strategic Management Plan 2020 (GABCC, 2020) acknowledging earlier plans, complemented by the

Table 6.3 Synthesis of key hydrogeological impacts due to Wellfield A

Monitoring point	Reference values[a]	Adopted baseline Time period (no.)	Adopted baseline Value	Average peak impact Period	Average peak impact Peak (no.)	Averages 2010–2015	Comparative ratios 2015/2010	Comparative ratios 2015/baseline
Flows (L/s)								
New Years Gift bore	NR	All 1985 (26)	3.9	April-June 1994	0.18 (3)	NR	NR	NR
Venables bore	NR	All 1983 (8)	1.1	All 1993	0.03 (8)	NR	NR	NR
Beatrice bore HBS004	NR	All 1983 (6)	0.80	1995 to 1998	0.27 (7)	0.049 (12)	0.53, declining	0.050
Bopeechee bore HBO013	NR	All 1983 (7)	0.77	Oct.-Dec. 1995	0.14 (3)	0.39 (12)	1.08, stable	0.54
McLachlan Spring LMS004B	NR	All 1988 (3)	1.2	All 1998	0.74 (12)	0.93 (11)	0.96, stable	0.75
Fred Spring LFE001	NR	All 1983 to 1984 (9)	0.22	All 1994	0.042 (3)	0.096 (12)	0.83, declining	0.34
Gosse Spring LGS004	NR	All 1983 to 1984 (4)	1.9	1993 to 1994	1.3 (23)	1.5 (11)	1.03, stable	0.81
Emerald Spring LES001[b]	NR	All 1984 to 1985 (9)	1.8	1996 to 1997	1.3 (20)	1.6 (11)	1.39, increasing	1.25
Priscilla Spring LPS001	NR	All 1983 (5)	0.19	July to Sept. 1990	<0.005 (2)	NR	Extinct (1990)	Extinct (1990)
Hermit Hill Spring HHS170	NR	June 1984 to Oct 1985 (10)	0.54	1995 to 1999	0.34 (50)	0.30 (12)	0.56, declining	0.31
Old Finniss Spring HOF096	NR	June 1984 to Oct 1985 (12)	0.14	No change	–	0.22 (10)	1.13, increasing	1.57
Old Finniss Spring HOF081	NR	Dec. 1984 (1)	0.13	1987 to 1990	0.013 (38)	0.083 (11)	0.98, stable	0.64
Levels (m AHD)								
Bore GAB05A	27.7	all 1984 (4)	28.2	All 1996	9.7 (5)	21.4 (21)	0.91, stable	0.73
Bore GAB08	11.7	all 1984 (3)	11.9	All 1996	8.9 (12)	9.9 (20)	0.95, stable	0.81
Bore GAB11	20.7	all 1984 (4)	21.0	All 1996	13.2 (12)	18.3 (21)	0.95, stable	0.85
Bore GAB13A	30.4	Dec. 1986 to Nov. 1987 (9)	27.5	Jan. to Oct.1996	1.4 (9)	16.0 (22)	0.98, stable	0.58
Bore GAB24	39.2	all 1989 (3)	34.9	Mar. to Aug. 1996	6.3 (7)	33.5[c] (16)	0.95, stable	0.92
Bore HH1	11.1	all 1984 (4)	11.1	1993 to 1996	10.8 (46)	11.1[c] (18)	0.99, stable	1.01
Bore HH2	NR	all 1984 (4)	8.2	1995 to 1996	6.9 (25)	7.4 (20)	0.98, stable	0.89
Bore HH3	9.3	all 1984 (5)	9.2	No change	–	9.4 (20)	0.99, stable	1.02
Bore HH4	14.0	all 1986[d] (10)	13.9	1993 to 1996	13.6 (48)	13.5 (19)	0.93, stable	0.93

Note: All data should be considered approximate as it is digitized from the various annual reports; no. = number of data points used; NR = not reported.

[a] Reference (or compliance) values from AGC (var.), WMCR-ODC (var.), and BHP (var.).
[b] Emerald Spring is just outside the map in Figure 6.2, to the west of Kati Thanda South itself.
[c] Average over period 2016 to 2020.
[d] A slightly later period is adopted for HH4 to avoid rebound after the reduction in pastoral use on Finniss Springs station in the early 1980s.

specific plan for the adaptive management and protection of springs (Brake et al., 2020). While the springs plan acknowledges the critical role of artesian heads in driving spring flows, there appears to be little appetite for revisiting the compliance criteria for Wellfield A with respect to minimum artesian levels for individual springs and/or flux-based criteria, in addition to drawdown thresholds at specified bores.

The lack of recovery to baseline or prewellfield levels in the majority of bores and flows to spring highlights that the impacts from Wellfield A are ongoing and may continue to drive further spring degradation in the future (see Table 6.3, especially the comparative ratios). The concerns raised by the Arabana people and other community stakeholders over the potential ongoing loss of springs due to water extraction for the mine need to be heard and addressed through additional scientific studies that resolve these questions, and regulatory actions that ensure these unique and ancient springs and the cultural and ecological values they sustain are not lost.

6.5. SUMMARY AND CONCLUSIONS

This chapter has presented the case of Wellfield A, which supplies water to the Olympic Dam mining project in central South Australia, and the impacts and risks to Kati Thanda Springs. The springs are unique and irreplaceable cultural and ecological heritage but those in the vicinity of Wellfield A have clearly experienced significant impacts and some have gone extinct (Priscilla, Venables). This has been demonstrated through a detailed synthesis and assessment of monitoring data for Wellfield A, especially the lack of full recovery in levels or flows in many bores or springs following the reduction in extraction in the mid-1990s after the development of Wellfield B. The monitoring frequency has been substantially reduced since February 1999 for bores and 2002 for springs, while from 2015 reporting has reduced individual spring and bore monitoring data, making it impossible to assess in detail the state of individual springs or bores. A key finding of this chapter is the unique relationships between nearby artesian pressure levels and spring flows, arguably the first time such an analysis has been synthesized for the Kati Thanda Springs and confirming the intuitive expectation that small reductions in pressure can lead to significant reductions in spring flows. The special case study highlights the critical importance of aligning monitoring closely with the science and determining compliance criteria accordingly, something that clearly needs to be improved in this case. Given the ever-present potential for a third GAB wellfield for Olympic Dam, as well as an expansion of extraction from Wellfield B and even potentially Wellfield A, it is clear that threats to the cultural and ecological values of the Kati Thanda Springs remain real and significant.

ACKNOWLEDGMENTS

This chapter builds on previous independent research, including the early pioneering work of Daniel Keane and more recently students Felix Owusu and Toby Firth. Staff at the SA Department for Energy & Mining was extremely helpful in chasing down and digitizing the many historical wellfields reports for Olympic Dam, we sincerely thank them for their efforts, as it has certainly helped ensure a more comprehensive historical analysis than was otherwise possible. We would also like to acknowledge the many citizens who have contributed to public education on the state of and risks to the springs, especially Isla Marks, Eric Miller, Linda Marks, Jan White, Al Lad, and David Noonan (among many others). Finally, we would like to pay our respects to the Arabana elders and community on whose land the Kati Thanda Springs reside, paying particularly special tribute to Uncle Kevin Buzzacott and Reg Dodd who have inspired this revisiting of the state of the springs and the Olympic Dam wellfields. We also thank the anonymous reviewers for frank and comprehensive comments and PhD students Clint Hansen and Angus Campbell for technical reviews of this work.

REFERENCES

AAC (2020). *Submission to the inquiry into the destruction of 46,000 year old caves at the Juukan Gorge in the Pilbara Region of Western Australia*. Arabana Aboriginal Corporation (AAC), Canberra, Australia, Submission 92. https://www.aph.gov.au/DocumentStore.ashx?id=41e65957-4b2f-4245-bc55-f0198241c6ce&subId=690905

AGC (var.). *Wellfield A production history and aquifer response*. Prepared by Australian Groundwater Consultants Pty Ltd (AGC; later AGC Woodward Clyde Pty Ltd) for Roxby Management Services Pty Ltd, Adelaide, South Australia, Years 1986–87 to 1993–94, Adelaide, South Australia. https://map.sarig.sa.gov.au/

Arthington, A. H., Jackson, S. E., Tomlinson, M., Walton, C. S., Rossini, R. A., & Flook, S. C. (2020). Springs of the Great Artesian Basin: Oases of life in Australia's arid and semi-arid interior. *Proceedings of The Royal Society of Queensland*, 126, 1–10.

Bense, V. F., Gleeson, T., Loveless, S. E., Bour, O. & Scibek, J. (2013). Fault zone hydrogeology. *Earth-Science Reviews*, 127, 171–192.

BHP (var.). *Great Artesian Basin Wellfields report*. BHP Billiton Ltd / BHP Olympic Dam ('BHP'), Adelaide, South Australia, Years 2003 to 2019–20, Adelaide, South Australia. https://map.sarig.sa.gov.au/

BHPB (2009). *Olympic Dam expansion: Draft environmental impact statement*. BHP Billiton Ltd (BHPB), Adelaide, SA, May 2009Adelaide, SA.

Boyd, W. E. (1990). Mound Springs. In M. J. Tyler, C. R. Twidale, M. Davies, & C. B. Wells (Eds.), *Natural history of the north east deserts* (pp. 107–118). Adelaide, South Australia: Royal Society of South Australia Inc.

Brake, L., Harris, C., Jensen, A., Keppel, M., Lewis, M., & Lewis, S. (2020). *Great Artesian Basin Springs: A plan for the future, evidence-based methodologies for managing risks to spring values*. Prepared for the Australian Government Department of Agriculture, South Australian Department for Environment & Water, Queensland Department of Natural Resources, Mines & Energy, New South Wales Department of Planning, Industry & Environment, and the Northern Territory Department of Environment & Natural Resources, January 2020.

Crossey, L., Priestley, S., Shand, P., Karlstrom, K., Love, A., & Keppel, M. (2013). Source and origin of western GAB spring water. In *Allocating water and maintaining springs in the Great Artesian Basin*. Canberra, Australia: National Water Commission.

Currell, M. J. (2016). Drawdown "triggers": A misguided strategy for protecting groundwater-fed streams and springs. *Groundwater, 54*(5), 619–622.

Currell, M. J., Werner, A. D., McGrath, C., Webb, J. A., & Berkman, M. (2017). Problems with the application of hydrogeological science to regulation of Australian mining projects: Carmichael mine and Doongmabulla Springs. *Journal of Hydrology, 548*, 674–682.

Dafne, E. (2016). The Great Artesian Basin: Is it that great? *Hydrogeology Journal, 24*, 1329–1332.

Davis, J., Pavlova, A., Thompson, R. M., & Sunnucks, P. (2013). Evolutionary refugia and ecological refuges: Key concepts for conserving Australian arid zone freshwater biodiversity under climate change. *Global Change Biology, 19*(7), 1970–1984.

Endersbee, L. (2001). A new understanding of the groundwater resources of the Great Artesian Basin. *Engineering World*, June / July, 14–19.

Fatchen, T. J., & Fatchen, D. H. (1993). Dynamics of vegetation on mound springs in the Hermit Hill region, northern South Australia. Prepared by TJ Fatchen & Associates for WMC (Olympic Dam Operations) Pty Ltd, Mount Barker, South Australia, July 1993, Mount Barker, South Australia.

Fensham, R. J., & Fairfax, R. J. (2003). Spring wetlands of the Great Artesian Basin, Queensland, Australia. *Wetlands Ecology and Management, 11*(5), 343–362.

Fensham, R. J., Ponder, W. F., & Fairfax, R. J. (2010). *Recovery plan for the community of native species dependent on natural discharge of groundwater from the Great Artesian Basin*. Prepared for the Australian Department of the Environment, Water, Heritage & the Arts and Queensland Department of Environment & Resource Management, Brisbane, Australia.

Fensham, R. J., Silcock, J. L., Powell, O., & Habermehl, M. A. (2016). In search of lost springs: A protocol for locating active and inactive springs. *Groundwater, 54*(3), 374–383.

Flook, S., Fawcett, J., Cox, R., Pandey, S., Schoning, G., Khor, J., et al. (2020). A multidisciplinary approach to the hydrological conceptualisation of springs in the Surat Basin of the Great Artesian Basin (Australia). *Hydrogeology Journal, 28*, 219–236.

GABCC (2014). *Great Artesian Basin resource study 2014*. Great Artesian Basin Co-ordinating Committee (GABCC), Canberra, Australia.

GABCC (2020). *Great Artesian Basin strategic management plan*. Great Artesian Basin Co-ordinating Committee (GABCC), Canberra, Australia.

Gleeson, T., Befus, K. M., Jasechko, S., Luijendijk, E., & Bayani Cardenas, M. (2016). The global volume and distribution of modern groundwater. *Nature Geoscience, 9*, 161–167.

Habermehl, M. A. (1982). *Springs in the Great Artesian Basin, Australia: Their origin and nature*. Bureau of Mineral Resources, Canberra, Australia.

Habermehl, M. A. (2020a). Hydrogeological overview of springs in the Great Artesian Basin. *Proceedings of The Royal Society of Queensland, 126*, 29–46.

Habermehl, M. A. (2020b). Review: The evolving understanding of the Great Artesian Basin (Australia), From discovery to current hydrogeological interpretations. *Hydrogeology Journal, 28*, 13–36.

Habermehl, M. A., Devonshire, J., & Magee, J. (2009). *Sustainable groundwater allocations in the intake beds of the Great Artesian Basin in New South Wales (recharge to the New South Wales part of the Great Artesian Basin): Final report for the National Water Commission*. Bureau of Rural Sciences (BRS), Australian Government, Canberra, Australia.

Harrington, G. A., Smerdon, B. D., Gardner, P. W., Taylor, A. R., & Hendry, J. W. (2012). Diffuse discharge. In A. J. Love, P. Shand, L. J. Crossey, G. A. Harrington, & P. Rousseau-Gueutin (Eds.), *Allocating water and maintaining springs in the Great Artesian Basin, volume III: Groundwater discharge of the western Great Artesian Basin*. Canberra, Australia: National Water Commission.

Harris, C. R. (1980–1981). Oases in the desert: Mound springs of northern South Australia. *Proceedings of The Royal Geographical Society of Australia (South Australia Branch), 8*, 26–39.

Hercus, L. A. (1990). Aboriginal people. In M. J. Tyler, C. R. Twidale, M. Davies, & C. B. Wells (Eds.), *Natural history of the north east deserts* (pp. 149–160). Adelaide, South Australia: Royal Society of South Australia Inc.

Jenkin, T. J. R. (2001). *Place, image and environmental conflict: World heritage and the Lake Eyre Basin*. Royal Geographical Society of South Australia Inc, South Australian Geographical Papers, Adelaide, South Australia.

Jewett, C., & Garavan, M. (2019). Water is life: An Indigenous perspective from a standing rock water protector. *Community Development Journal, 54*, 42–58.

K-S (1984). *Olympic Dam Project: Supplementary environmental studies, Mound springs*. Prepared by the Kinhill-Stearns Pty Ltd (K-S) for Roxby Management Services Pty Ltd, August 1984, Adelaide, South Australia.

K-S (1997). *Olympic Dam Expansion Project: Environmental impact statement*. Prepared by the Kinhill-Stearns Pty Ltd (K-S) for Western Mining Corporation (WMC), Parkside, South Australia, May 1997.

Keane, D. (1997). *The sustainability of use of groundwater from the Great Artesian Basin with particular reference to the southwestern edge of the basin and impact on the mound springs*. (B

Env Eng (Hons), Final Year Environmental Engineering Investigation Thesis), RMIT University, Melbourne, Australia.

Kellett, J. R., Ransley, T. R., Coram, J., Jaycock, J., Barclay, D. F., McMahon, G. A., et al. (2003). *Groundwater recharge in the Great Artesian Basin intake beds, Queensland*. Final Report for the NHT Project #982713 Sustainable Groundwater Use in the GAB Intake Beds, Queensland. Bureau of Rural Sciences (Australian Government), Canberra, Australia and Department of Natural Resources & Mines (Queensland Government), Brisbane, Queensland.

Keppel, G., Van Niel, K. P., Wardell-Johnson, G. W., Yates, C. J., Byrne, M., Mucina, L., et al. (2012). Refugia: Identifying and understanding safe havens for biodiversity under climate change. *Global Ecology and Biogeography, 21*(4), 393–404.

Keppel, M., Karlstrom, K., Love, A. J., Priestley, S., Wohling, D., & De Ritter, S. (Eds.) (2013). *Hydrogeological framework of the western Great Artesian Basin*. Canberra, Australia: National Water Commission.

Keppel, M. A., Clarke, J. D. A., Halihan, T., Love, A. J., & Werner, A. D. (2011). Mound springs in the arid Lake Eyre south region of South Australia: A new depositional tufa model and its controls. *Sedimentary Geology, 240*, 55–70.

Keppel, M. N., Karlstrom, K., Crossey, L., Love, A. J., & Priestley, S. (2020). Evidence for intra-plate seismicity from spring-carbonate mound springs in the Kati Thanda–Lake Eyre region, South Australia: Implications for groundwater discharge from the Great Artesian Basin. *Hydrogeology Journal, 28*, 297–311.

Konikow, L. F., & Leake, S. A. (2014). Depletion and capture: Revisiting "The source of water derived from wells." *Groundwater, 52*(1), 100–111.

KSR (1982). *Olympic Dam Project: Draft environmental impact statement*. Prepared by the Kinhill-Stearns Roger (KSR) Joint Venture for Roxby Management Services Pty Ltd, Adelaide, South Australia, October 1982.

Lewis, S., & Harris, C. R. (2020). Improving conservation outcomes for Great Artesian Basin springs in South Australia. *Proceedings of The Royal Society of Queensland, 126*, 271–287.

Marshall, V. (2017). *Overturning aqua nullius: Securing Aboriginal water rights*. Canberra, Australia: Aboriginal Studies Press.

Mazor, E. (1995). Stagnant aquifer concept part 1: Large-scale artesian systems, Great Artesian Basin, Australia. *Journal of Hydrology, 173*(1–4), 219–240.

Miraldo Ordens, C., McIntyre, N., Underschultz, J., Ransley, T., Moore, C., & Mallants, D. (2020). Preface: Advances in hydrogeologic understanding of Australia's Great Artesian Basin. *Hydrogeology Journal, 28*, 1–11.

Moggridge, B. J. (2020). Aboriginal people and groundwater. *Proceedings of The Royal Society of Queensland, 126*, 11–27.

Moya, C. (2015). *Hydrostratigraphic and hydrochemical characterisation of aquifers, aquitards and coal seams in the Galilee and Eromanga basins, central Queensland, Australia*. PhD Thesis, Queensland University of Technology, Brisbane, Australia.

Mudd, G. M. (2000). Mound springs of the Great Artesian Basin in South Australia: A case study from Olympic Dam. *Environmental Geology, 39*, 463–476.

Murphy, N. P., Guzik, M. T., Cooper, S. J. B., & Austin, A. D. (2015). Desert spring refugia: Museums of diversity or evolutionary cradles? *Zoologica Scripta, 44*(6), 693–701.

Oki, T., & Kanae, S. (2006). Global hydrological cycles and world water resources. *Science, 313*(5790), 1068–1072. https://doi.org/10.1126/science.1128845

Priestley, S., Shand, P., Love, A. J., Crossey, L. J., Karlstrom, K. E., Keppel, M. N., et al. (2020). Hydrochemical variations of groundwater and spring discharge of the western Great Artesian Basin, Australia: implications for regional groundwater flow. *Hydrogeology Journal, 28*, 263–278.

Radke, B., & Ransley, T. (2020). Connectivity between Australia's Great Artesian Basin, underlying basins and the Cenozoic cover. *Hydrogeology Journal, 28*, 43–56.

Ransley, T. R., & Smerdon, B. D. (Eds.) (2012). *Hydrostratigraphy, hydrogeology and system conceptualization of the Great Artesian Basin: A technical report to the Australian government from the CSIRO Great Artesian Basin Water Resource Assessment*. Canberra, Australia: CSIRO Water for a Healthy Country Flagship.

Rasmussen, T. C., & Crawford, L. A. (1997). Identifying and removing barometric pressure effects in confined and unconfined aquifers. *Groundwater, 35*(3), 502–511.

Read, J. L. (2003). *Red sand, green heart: Ecological adventures in the outback*. South Melbourne, Australia: Lothian Books.

Rossini, R. A., Fensham, R. J., Stewart-Koster, B., Gotch, T. B., & Kennard, M. J. (2018). Biogeographical patterns of endemic diversity and its conservation in Australia's artesian desert springs. *Biodiversity Research, 24*(9), 1199–1216.

Smerdon, B. D., Ransley, T. R., Radke, B. M., & Kellett, J. R. (2012). *Water resource assessment for the Great Artesian Basin: A report to the Australian government*. CSIRO Water for a Healthy Country Flagship, Canberra, Australia, December 2012.

Theis, C. V. (1940). The source of water derived from wells: Essential factors controlling the response of an aquifer to development. *Civil Engineering, 10*, 277–280.

Welsh, W. D., Moore, C., Turnadge, C., Smith, A. J., & Barr, T. M. (2012). *Modelling of climate and groundwater development: A technical report to the Australian government from the CSIRO Great Artesian Basin Water Resource Assessment*. CSIRO Water for a Healthy Country Flagship, Canberra, Australia, December 2012.

White, D. C., & Lewis, M. M. (2011). A new approach to monitoring spatial distribution and dynamics of wetlands and associated flows of Australian Great Artesian Basin springs using QuickBird satellite imagery. *Journal of Hydrology, 408*, 140–152.

Williams, A. F., & Holmes, J. W. (1978). A novel method of estimating the discharge of water from mound springs of the Great Artesian Basin, central Australia. *Journal of Hydrology, 38*(3), 263–272.

WMCR-ODC (var.). *Great Artesian Basin wellfields: Abstraction history history and aquifer response*. WMC Resources (Olympic Dam Corporation) Pty Ltd (WMCR-ODC), Roxby Downs, South Australia, Years 1988–90 to 2002. https://map.sarig.sa.gov.au/

Part II
Methods, Tools, and Techniques to Understand Spring Hydrogeology

7
Environmental Tracers to Study the Origin and Timescales of Spring Waters

Axel Suckow and Christoph Gerber

ABSTRACT

Environmental tracers can add valuable insights to the assessment of spring vulnerability that are not available with any other method. They can identify the origin of contaminants, help to specify from which aquifer formation(s) the water originates, give clues on the infiltration conditions of the water such as temperature and altitude to identify recharge locations and processes, and define the timescales of groundwater flow and, in ideal cases, deconvolute the age distribution. Environmental tracers will rarely directly define the vulnerability of springs toward changes in hydraulic gradient, but they may give invaluable contributions to a conceptualization that is the basis for assessing the impact of a change in hydraulic gradients. The available tracer tools are discussed in this chapter and brought into the context of other investigations such as conceptual hydrogeological models, geophysics, and hydrochemistry. A detailed case study demonstrates the capabilities under ideal conditions.

7.1. INTRODUCTION

Next to freshwater lakes and rivers, springs through all times have been the most important source of drinking water for all mammals, including humans. The origin of this water emerging from Earth has been connected to myths and legends such as the rainbow serpent in Australia (Konishi, 2021). While modern hydrogeology demystifies the origins of water emerging in springs, they remain important for water supply, as the origin of rivers, or in supporting groundwater dependent ecosystems. Nonetheless, and despite the advances in geophysical technologies, looking into the ground without drilling, and capable to refine conceptual geologic models in great detail, the dynamics of groundwater flow are still difficult to assess, and this is where tracers can be invaluable help. For some springs, it remains one of the most difficult tasks to pinpoint the origin of its waters. This encompasses the groundwater system it is dependent on, delineating its recharge area, and quantifying the underground travel times of the original rainwater that contributes to this spring (Flook et al., 2020). All these questions are inherently connected, and all have to be answered jointly to assess the vulnerability of a spring to human-induced changes of groundwater level, contamination, or climate change.

During the last decades a new class of tools has emerged, called "environmental tracers." Similar to a trace of footprints in the sand demonstrating the former presence of an animal, these tracers allow scientists to quantify groundwater travel time and identify where the water has been flowing. Similarly, environmental tracers often are an isotopic signature marking the dissolved species in water, which pinpoints the origin of this dissolved species. These chemical and isotopic signatures can identify host rocks, which the groundwater has been in contact with along its flow path. They can also reveal

CSIRO Land and Water, Adelaide, Australia

Threats to Springs in a Changing World: Science and Policies for Protection, Geophysical Monograph 275, First Edition.
Edited by Matthew J. Currell and Brian G. Katz.
© 2023 American Geophysical Union. Published 2023 by John Wiley & Sons, Inc.
DOI:10.1002/9781119818625.ch07

whether a contamination (such as nitrate) is of natural or human origin. Furthermore, tracers can elucidate the conditions of recharge, such as temperature, altitude, and salinity of the water reaching the groundwater table and whether large fluctuations of the water table were present or a deep unsaturated zone. Environmental tracers are most widely known as a technology to determine the "age" of water, a concept that needs some detailed discussion in this chapter. While there are various textbooks on environmental tracer applications in groundwater, in surface water catchments, or in the oceans, the focus of this chapter is specific problems for springs: it discusses what to consider when planning to use tracers, introduces the most important of these tracers, and illustrates their use to assess the origin of spring water and the vulnerability of the springs.

7.2. BEFORE TRACERS ARE APPLIED

To better understand the value tracers provide, some conceptual ideas need to be introduced first. Also, while the topic of this chapter is environmental tracers and how they can contribute to a deeper understanding of spring water origin and its vulnerability, environmental tracers are never the only method to be applied. They are most powerful tools when embedded in a holistic approach that at least also encompasses hydraulic heads, as many surrounding wells as possible, geophysical tools, and, of course, basic hydrogeology and hydrochemistry.

7.2.1. Propagation of Pressure Versus Transport of Water

Vulnerability of a spring and the ecosystems depending on it can be related either to the propagation of pressure or to water mass transport, both of which are under human influence and under the influence of climate change. A simple comparison to a filled garden hose laying in the sun can explain the difference: When the tap is turned on, water starts flowing very quickly, which characterizes the pressure propagation. But the water flowing out of the hose is warm, although the water from the tap is colder, and it will take a while until the warm water in the hose is flushed out. This is the timescale of water mass transport.

Changing groundwater levels, whether from human induced pumping or from seasonal variations in recharge or from climate change, changes the pressure gradient that causes the spring to flow in the first place, leading to a change in discharge or even to the spring falling dry. For example, it is known that some springs flow only in the wet season and fall dry or have much less discharge during the dry season (Blavoux et al., 1992; Fiorillo et al., 2021; Fiorillo et al., Chapter 9 this volume) due to the propagation of hydraulic pressure. In the wet season, when recharge into the groundwater catchment is high, groundwater levels are high and drive the high discharge of the spring. In the dry season, no more recharge happens, deep-rooted plants may even transpire groundwater, and thus the groundwater level drops, resulting in reduced or no discharge of the spring. This phenomenon is most pronounced in karst springs (Stevanović, 2015) since karst propagates hydraulic pressure very quickly and the difference can be directly seen in experiments (Luhmann et al., 2012). ("Karst" is a geological phenomenon occurring in soluble rocks such as salt, gypsum or limestone, where the dissolution of the host rock creates large open cavities such as sinkholes and caves in which water can flow freely.) In several prominent springs, simple or detailed models such as bucket-like reservoirs that overflow to feed the spring have been very successful to describe both flow and the vulnerability of the spring (Blavoux et al., 1992; Emblanch et al., 2003; Fleury et al., 2007; Ollivier, 2019; Ollivier et al., 2019; Ollivier et al., 2020). Hydraulic head and discharge time series are the most common data used to investigate pressure-related processes, and environmental tracers are of limited use here.

Environmental tracers are transported with the water and therefore always lag behind any propagation of pressure. In the garden hose example above, they correspond to the timescale of warm water being flushed out of the hose. The transport timescale, on the other hand, relates to questions of pollution. Environmental and artificial, or active, tracers (see section 7.3.2 for the difference) always assess the transport timescale, not the pressure propagation timescale. Environmental tracers are therefore excellent tools to conceptually understand the aquifer systems and to conclude about the vulnerability of a spring from chemical components such as nitrate from fertilizers or anthropogenic contaminants (Wilske et al., 2020), but bad predictors to the effect of changes of hydraulic gradient.

The difference between these two timescales is important: a karst spring can clearly react to a seasonal pressure timescale and show seasonal variations of a factor of 3 in flow, while a detailed evaluation using environmental tracers finds a transport timescale of a century (Ozyurt, 2008). As we have seen from earlier sections in this volume (Mudd & Currell, Chapter 6 this volume), a spring can discharge water that is more than a million years old, but still fall dry due to human influence within a decade if nearby wells compromise the hydraulic groundwater gradient causing the discharge. The present chapter will focus on environmental tracers. It will deal only with vulnerability of springs to hydraulic pressure changes insofar as tracers improve the conceptualization and clarify pressure and climate change influences.

7.2.2. Springs Versus Wells

When constructing a well, there is a high level of control over which part of an aquifer system is tapped. A well can be drilled into a certain aquifer and sealed to the overlying and underlying layers to make sure it produces only water from a certain layer. Multilevel wells or well nests provide access to different depths within one aquifer or different aquifers at the same location. This way it is possible to obtain chemical gradients with depth (Böttcher et al., 1992; Frind et al., 1990) or even investigate how groundwater age increases with depth (Schlosser et al., 1989; Schlosser et al., 1988). Up to a certain depth, water samples can be taken even while drilling (Hofmann et al., 2010; Houben et al., 2018; Taylor et al., 2018), allowing us to take groundwater samples that are as free as possible from any human influence before we sample them.

All this is not possible the same way when sampling a spring. A spring is a surficial effluent of groundwater from an a priori unknown origin. We observe water coming out of the ground and have no choice where to take a sample, often even how to take a sample is restricted. It is not always clear from what layer the water originates and additional information of the surrounding geology and hydrogeology is needed to identify from which aquifer the waters originate. Only if we have a regional understanding of this hydrogeology and know all the outcrop locations of those layers that may contribute recharge water to the spring may we conclude where the water recharged. But more often than not, a spring will integrate over many flow paths, possibly over several aquifers, nearly always over several timescales of flow. To further complicate things, a spring can have variable discharge with time, resulting in a time-variable mixture of different groundwaters. Due to this lack of control, it is necessary to not only investigate the spring itself but to also take samples from wells in potential source aquifers, ideally upstream of the spring. Environmental tracer analyses from such wells can be pivotal to constrain the origin and flow timescales of water flowing to the spring (Flook et al., 2020).

7.2.3. Hydrochemistry and Geophysics

Investigating the origin of spring waters to assess the vulnerability of a spring is by no means an easy task. Therefore, all possible lines of evidence have to be integrated to properly assess a spring. Logical partners to the toolbox of environmental tracers are any tools of the hydrochemical kind. For instance an accurate interpretation of radiocarbon (^{14}C) or the stable carbon isotope (^{13}C) is not feasible without accounting for the underground chemical processes controlling the alkalinity in groundwater. Similarly, an age interpretation using ^{36}Cl would be impossible without detailed knowledge of the evolution of chlorine in the groundwater upstream of this spring.

But besides the specific need of hydrochemistry to interpret tracers, hydrochemistry itself is normally the first step to assess springs, and environmental tracers are second. The reason is not only the mere price difference between a full analysis of major and trace ions compared with a full suite of tracer analyses (tracers are typically a factor of 10 to 50 more expensive). At least as important is that the hydrochemical interpretation is still more commonplace than the more specialist knowledge to sample for and interpret environmental tracers. It is therefore important that the interpretation of tracer results take the findings of hydrochemistry into account. No one would easily believe the tracer interpretation if it contradicts the findings of major ions, but existing hydrochemistry is often used to guide the collection of tracer data and their interpretation.

From the previous sections, it is also clear that it is necessary to gain as much structural geological information as possible to assess a spring. Geophysical methods such as seismics, ground penetrating radar, and ground-based or airborne electromagnetics can provide valuable information to refine an initial conceptual model of aquifer geometry. Seismic and electromagnetic methods can show the layering of sediments upstream of the springs in a much more complete way than geological investigations at the surface or drilling cores can give. Electromagnetic methods can also elucidate whether the layers have higher or lower electric conductivity, giving hints to clay layers or the salinity of the groundwater. If applied as airborne electromagnetics, such surveys can derive a detailed picture of the underground layers in a short amount of time. Gabriel et al. (2003) give a good example of investigating sediment layers and buried valleys leading to submarine groundwater discharge with a diverse combination of geophysical methods (AEM, seismics, and ground truthing) in northern Germany, and Rutherford et al. (2021) use an airborne electromagnetic survey in a similar setting to investigate the origin of spring water in northern Australia. Often these methods can clarify whether or not a spring is sitting on a fault system that created the pathway for the groundwater to emerge to the surface (Banks et al., 2019). However, all such geophysical methods need detailed ground truthing and are in turn more expensive than environmental tracer studies in a similar area.

The final goal is a unified conceptual model integrating all different lines of evidence: hydrochemistry, geophysical methods, hydraulic heads, discharge time series, active and environmental tracers, and numerical modeling of groundwater flow and transport have to agree in their description of the groundwater system. Only such a thorough conceptualization may possibly answer questions

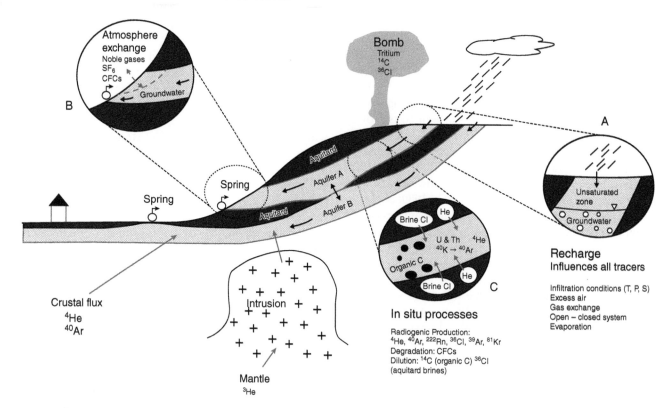

Figure 7.1 Conceptualization of groundwater flow from recharge to the spring, and illustration of different processes discussed in this chapter, influencing environmental tracers (design: Cornelia Wilske, CSIRO).

on the vulnerability of a spring under future scenarios of climate change or human-induced changes such as groundwater abstraction, changes in hydrological setting (e.g. dams or mines), or groundwater contamination. Figure 7.1 gives a very simplified version of such a conceptualization and this chapter will discuss the available tracers and how these are influenced by, and therefore allow conclusions on, processes during recharge and underground flow in the aquifer.

7.3. ENVIRONMENTAL TRACERS: KINDS, THE CONCEPT OF AGE, AND SIMPLE MODELING

Environmental tracers are substances that are existing in the environment already and are transported with the water. They can provide information on a certain aspect of the water movement and its origin, such as transport (advection, diffusion, and dispersion). Environmental tracers fill several textbooks (Cook & Herczeg, 2000; IAEA, 2006, 2013; Kazemi et al., 2006; Kendall & McDonnell, 1998) and are part of several university curriculums around the world. This chapter cannot cover the details but tries to give an eagle's eye overview covering the most important properties and applications.

Environmental tracers cover three different purposes, generally speaking. The most common and known application is to determine what sometimes is called the age of groundwater (or fluids in general, since similar techniques can be applied to lake water, ocean water, oil, and gas reservoirs or even fluid inclusions entrapped in minerals). This application often requires additional parameters and assumptions and some modeling to convert the measured tracer concentrations into time information, which we will cover later in this chapter. First, we will introduce a class of environmental tracers that can shed light on infiltration conditions such as soil temperature, and second environmental tracers that can elucidate the origin of contaminants and characterize rock-water interaction.

Tracers are often classified according to the degree their transport properties deviate from the transport properties of the water itself, and throughout this chapter we will follow this classification. The most ideal tracers are part of the water molecule itself, and indeed the isotopes of hydrogen (^{2}H or deuterium and ^{3}H or tritium) and oxygen (^{18}O) are among the first that have been historically used (Craig, 1961; Eriksson, 1958; Kaufman & Libby, 1954), and often the easiest applied today. Second best would be a class of substances that does not show any chemical reactions: the chemically inert noble gases. Since they do not undergo any chemical or biological reactions, they are only influenced by advection, diffusion, dispersion, and mixing, which are physical processes only and are thus relatively well understood and quantifiable. Another very well quantifiable physical process is radioactive decay, which leads to decreasing concentration at a constant and very well-known rate. Some noble gas isotopes are radioactive (e.g., ^{85}Kr, ^{39}Ar, ^{81}Kr) and provide the most reliable tracers for transport times. Other noble gas isotopes are radiogenic, that is, being produced by radioactive decay, for example, from the aquifer rocks, or primordial, that is, are remnants from the early formation of Earth from the solar cloud (see Ballentine & O'nions, 1944, and Pinti & Marty, 2000, for the characteristic noble gas isotope patterns from different terrestrial sources). Next in hierarchy are tracers that are quite inert, but still show some chemical reactions, or that are even part of a chemical solute. The most-used species here are ^{14}C and ^{36}Cl, and some processes influencing these nonideal tracers are discussed later in this chapter. It is easily understandable that the difficulty in interpreting the different classes of environmental tracers increases with the complexity of the processes they may undergo.

7.3.1. Tracers for Groundwater or Solute Origin and Infiltration Conditions

Information on recharge temperature and altitude can be very useful for identifying recharge areas for springs. Tracers that can provide such information are stable isotopes of water (^{2}H and ^{18}O) and dissolved noble gases. We will mainly focus on noble gases here and only briefly touch on ^{2}H and ^{18}O, because their application for this purpose has been discussed in detail in textbooks available free of charge for download (Gat et al., 2001; Geyh, 2001; Mook, 2001). After that we will briefly cover tracers that can be used to investigate the origin of solutes.

Infiltration Conditions: Temperature, Altitude

The different noble gases have different solubility and, even more important, the solubility depends on temperature differently for each noble gas. The mixing ratio of noble gases in the atmosphere is globally constant. However, their dissolved amount per gram of groundwater depends on the ambient pressure, which is dependent on atmospheric pressure and decreasing with altitude. This motivated Emmanuel Mazor to use them to derive infiltration conditions for some springs in Israel (Mazor, 1972). Later it also became evident that there is a surplus of noble gas concentration above the simple solubility equilibrium, which was called "excess air" (Heaton & Vogel, 1981). In the decades following, these two components were investigated in great detail and allowed us to derive the processes that lead to excess air (Holocher et al., 2002; Ingram et al., 2007; Klump et al., 2008). It was also possible to derive paleoclimate records from groundwater, which allowed the most reliable and precise estimates of soil temperature during the last glacial maximum. Most of these paleoclimate studies have recently been reviewed by Seltzer et al. (2021b). A very recent application of noble gases is that high-precision isotopic measurements of dissolved krypton and xenon in groundwater also allow the reconstruction of past thicknesses of the unsaturated zone and this provides a new tool for paleoclimate studies especially in arid regions (Seltzer et al., 2019).

Both the infiltration temperature and excess air can give clues on the source of spring water. Excess air will be higher if groundwater levels vary a lot in the infiltration area, which for instance is the case in Savanna regions with monsoon recharge patterns. Or excess air will be higher if an ephemeral stream infiltrates into a dry riverbed, enclosing lots of air bubbles during a flash flood, in contrast to a perennial loosing river infiltrating through an all-year saturated zone. In a mountain region, mean annual temperature will decrease with altitude and therefore result in higher concentrations if the infiltration area is higher. Unfortunately, atmospheric pressure also decreases with altitude, which will cause lower noble gas concentrations in the spring waters, however, with a different pattern of elemental noble gas ratios. It takes a bit of reverse fitting and modeling to distinguish the patterns, but this can be done with public domain software (Jung & Aeschbach, 2018). All these results can help identify or constrain possible source regions of spring water.

Sampling for noble gases is done establishing a gas-tight connection with the water body and enclosing the water in copper tubes sealed with special pinch-off clamps. These copper tube containers are supplied by the analyzing laboratory. This necessity to avoid any gas exchange or gas loss excludes the application of noble gases at springs with very low discharge rates, although the amount of water needed is comparably small (10–30 ml). Sampling from a spring pond with a pump or a spring tapping will give good results but a possible gas exchange of the groundwater with atmosphere prior to emerging in the spring always must be considered for any gas tracer, depending on the aquifer geometry (Fig. 7.1 B).

The stable isotopes of water, ^{18}O and ^{2}H, can also provide indirect proxy information on temperature and altitude. Both are a natural part of the water molecule, in which either the more abundant oxygen atom (^{16}O) is replaced by ^{18}O, or one of the more abundant hydrogen atoms (^{1}H) is replaced by ^{2}H. The stable isotope composition of both precipitation and groundwater has been the subject of decades of scientific research, and for many regions of the world, detailed records of the isotopic composition are available in public databases such as GNIP (IAEA/WMO, 2016). Also, every developed country today has typically several laboratories able to measure stable isotopes in water with a throughput of hundreds of samples per year. Therefore, it is well established that also the stable isotope composition shows an altitude effect and a temperature effect, the latter leading also to seasonal variation of the isotopic composition of precipitation. Free textbooks by UNESCO, WMO, and IAEA provide the detailed established scientific background for the interested reader (Gat et al., 2001; Geyh, 2001; IHP, 2000; Mook, 2001; Rozanski et al., 2001; Seiler, 2001; Yurtsever et al., 2001). Both altitude and temperature effects can be used to reconstruct the origin of springwater. However, in contrast to the globally uniform concentrations of noble gases in the atmosphere, the isotopes in precipitation vary regionally very much and even the altitude effect varies from place to place. The ease and low price of sampling and measurement applying stable isotopes is therefore counterbalanced with the need to establish time-consuming local background studies involving samples from different altitudes and seasons. Stable isotopes of the water and noble gases are also complementary, because stable isotopes of water record the altitude at which the precipitation falls, whereas noble gases record the altitude and temperature at which groundwater recharge takes place (Manning & Solomon, 2003).

Solute Origin

The origin of nitrate can be elucidated by investigating its isotopic composition: NO_3 from manure has a different isotopic composition than artificial fertilizers, and termites again create their own signature (Clark, 2015). Nitrate reduction will shift both isotopes (^{18}O and ^{15}N) in a characteristic way (Böttcher et al., 1990). This can be used to clarify the source of nitrate in a spring and whether changes in nitrate concentrations are driven by denitrification or mixing.

Also the Sr isotope ratio ($^{87}Sr/^{86}Sr$) can have a signature characteristic of an aquifer host rock of a spring. The reason for this is that older rocks had more time to have ^{87}Rb (half-life 49.23 billion years) decaying to ^{87}Sr, which provides the radiogenic component of this isotope ratio. Since Rb is also enriched in crustal rocks rather than magmatic melts, the $^{87}Sr/^{87}Sr$ ratio differs usefully between different aquifer host rocks, which can be favorably used to conclude on the origin of spring waters (Keppel et al., 2020; Lee et al., 2011; Lyons et al., 1995)

Isotope analyses of Sr as well as NO_3 and SO_4 are established methods requiring normally less than a liter of water, which in case of Sr just needs to be filtrated or in the case of NO_3 and SO_4 has to undergo varying preservation methods, which have to be queried from the analyzing laboratory. However, technology continues to evolve, and not only for strontium but also for many other elements such as lithium (Meier et al., 2017) and boron (Gäbler & Bahr, 1999; Vengosh & Spivak, 2000; Vengosh et al., 2005) isotopic analyses can separate natural and anthropogenic influences on groundwaters and springs. Most favorably, these stable isotope techniques are combined all together to conclude on anthropogenic influence, weathering processes, and isotopic signatures of the aquifers with which the water was in contact (Dotsika et al., 2010; Meredith et al., 2013; Millot et al., 2011). Such techniques are now available or emerging thanks to progress in multicollector inductively coupled plasma mass spectrometric methods (MC-ICP-MS). It is useful to keep an eye open in this field because new applications and isotope systems are established continuously.

7.3.2. Tracers for the Timescales of Water Movement

Several environmental tracers can be used to get an idea on what is commonly called the "age" of groundwater (or travel time for those more familiar with catchment hydrology). This term is highly misleading because we humans have a very clear understanding of what age is, and this understanding cannot simply be transferred to water (Suckow, 2014a), despite the fact that textbooks use the term in the title (Kazemi et al., 2006). Water inherently mixes, and an equivalent to the human individual is not existing for water. The mixing is especially important in the case of a spring, where we cannot select a specific part of the aquifer contributing to the water sample (see section 7.2.2). For instance, in Figure 7.1, both springs could discharge (time varying amounts of) water from aquifer A and aquifer B depending on recharge and hydraulic pressure variations in these two aquifers. Spring water will always consist of an unknown mixture of different ages. As a consequence, environmental tracers can at best give an idea on the distribution of this age mixture, and indeed we will discuss a very successful example in the case studies. But first we have to introduce what tracers are available on which timescale of groundwater flow.

Available Timescales

Each different timescale of water movement needs a different environmental tracer (Fig. 7.2). Unfortunately, for some timescales only one tracer is applicable, whereas for others several are available. Generally, one can distinguish tracers that are radioactive (e.g., ^{222}Rn, ^{85}Kr, ^{3}H, ^{39}Ar, ^{14}C, ^{81}Kr, ^{36}Cl), those that are assumed to be stable but have an input function that varies with time, usable to infer time (SF_6, CFCs, ^{18}O, ^{2}H), and those that are radiogenic, that is, are produced by radioactive decay such as ^{4}He and ^{40}Ar. Combinations are possible, for instance, ^{3}H has a time-variable input function and ^{222}Rn, ^{39}Ar, and ^{36}Cl are radioactive as well as produced in the aquifer by

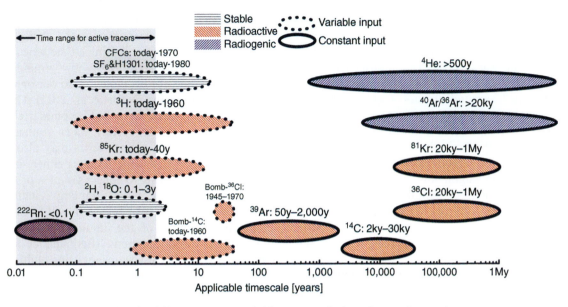

Figure 7.2 Timescales for the different tracers available to quantify the velocity of groundwater movement.

radioactive decay. As a rule of thumb, radioactive tracers can be used in groundwater from one-fourth up to five to seven half-lives, which is why these tracers have a time limit in application. Before this time, the change in concentration is too small to be detectable; after this time, it is difficult or impossible to detect the tracer.

For environmental tracers, we have no control on the input function. To understand this, it is useful to first discuss the difference between an environmental tracer and a purposefully applied artificial tracer.

Environmental Tracers Versus Active Tracers

Dye or artificial tracer experiments, which in the following we will call "active tracers" since they are actively applied during the experiment, have a long and useful tradition, especially to study the origin of water in karst springs and to demonstrate connections between sinkholes and springs. Leibundgut et al. (2009) is a good textbook focusing on the application and interpretation of such experiments. Dye tracers have also been successfully used in sedimentary aquifers, most prominently in the Borden Aquifer in Ontario in 1983 and 1986 (see Rivett et al., 1994, and citations therein) and at the Krauthausen site in Germany (Vereecken et al., 2000), and also in surface waters such as lakes and streams (Clark et al., 1996; Maiss et al., 1994a; Maiss et al., 1994b). In soil science they are routine techniques (Flury & Wai, 2003; Gerke et al., 2013; Gupta & Sharma, 1984; Zimmermann et al., 1967). Substances used in such tracer tests include dyes, salts, isotope markers, and pollen or microplastic. Adequate time resolution is always necessary to obtain the breakthrough curve of the tracer as completely as possible and to assess whether tracer was lost on another pathway that was not sampled. Also, with active tracers, additional geophysical methods (as discussed in section 7.2.3) can support the results, and breakthrough curves are measured at multilevel wells with much better spatial resolution (Müller et al., 2010). The most obvious result of a successful dye tracer test in groundwater is that a flow line between two points in an aquifer exists at all, and how much time the tracer needs to travel this distance, which is a direct measure of groundwater velocity. Beyond that, the shape of the breakthrough curve gives clues about dispersion on the flow path and if several tracers are applied also on adsorption of single tracers. Different tracers in the same system will not give the same breakthrough curve, depending both on properties of the tracer such as diffusivity, adsorption coefficients, and properties of the aquifer matrix such as dual porosity and stagnant pore volumes.

The one big advantage of applying an active tracer is complete control over the input of the tracer. It can be continuously applied or transferred into the aquifer as a one-time spike. However, with this advantage comes a big disadvantage: the restriction to a comparably short timescale. In many cases, the experiment is stopped after a few months, and even if a PhD student has more endurance, no active tracer test will last much longer than a year (Fig. 7.2). If we want to study timescales of many years up to a million years, we have no other choice than to work with substances that already are present in the system, which is the idea behind an environmental tracer. This may include natural substances such as natural radioactive isotopes, for instance radiocarbon (^{14}C) or ^{39}Ar. Or these substances can originate from a long anthropogenic input into the natural system in question, for instance the CFCs or tritium. A special case is the "bomb" pulse resulting from atmospheric testing of nuclear devices in several regions of the globe (mainly Northern Hemisphere) in the 1960s. This was a short-term anthropogenic input of tritium (^{3}H), ^{14}C, and ^{36}Cl, discernible as peak in the curves in Figure 7.3 around 60 years of age, which is used extensively in various age-dating contexts, including in groundwater. The limitation of the timescale of active tracer experiments also limits the length scale. The experiments in Borden and Krauthausen were done on fields of little above 100 m length; only in karst systems where groundwater flows tens or hundreds of meters a day, distances of kilometers can be observed with active tracer experiments. If the aquifer system upstream of the spring in question spans many tens or hundreds of kilometers, only environmental tracers allow understanding the system as a whole. Active tracers and environmental tracers can also be applied together in the same study and then provide information on different times scales present in the spring water (Leblanc et al., 2015).

Snapshots Versus Time Series

When using tracers to gain information of the age distribution of groundwater, two distinct ways of application are possible: A single snapshot typically using multiple tracers operating on different timescales to better understand the distribution of ages; and collecting a number of samples over a period of time at regular or irregular intervals, typically only of one or a few tracers. Both approaches have their advantages and disadvantages. The main advantage of collecting a time series of data is that it enables investigations of how groundwater age changes over time, for example due to seasonal changes or precipitation events (Duvert et al., 2015). This is most relevant at shorter timescales and becomes irrelevant if a spring does not discharge any water younger than a few years. Collecting a time series may also be more demanding logistically, as it requires either someone to regularly visit the spring to collect samples, or the implementation of an automated sampling system (Sahraei et al., 2020). Hydrograph separation will further require automated sampling of the input water such as time-resolving rainwater sampling (Berglund et al., 2019). The main advantage of a snapshot using multiple tracers is that tracers can be combined that span a very wide age range (e.g., tritium, radiocarbon, and ^{81}Kr), which is particularly valuable in springs where groundwater of very different ages mixes. Using a time series of, for example, ^{18}O alone in such a system will only reveal age information on the young fraction of groundwater, but will be blind to the older groundwater (Kirchner, 2016).

Tracer Properties

This section goes briefly through the tracers in Figure 7.2 and characterizes some of their properties as far as they are important for the assessment of springs.

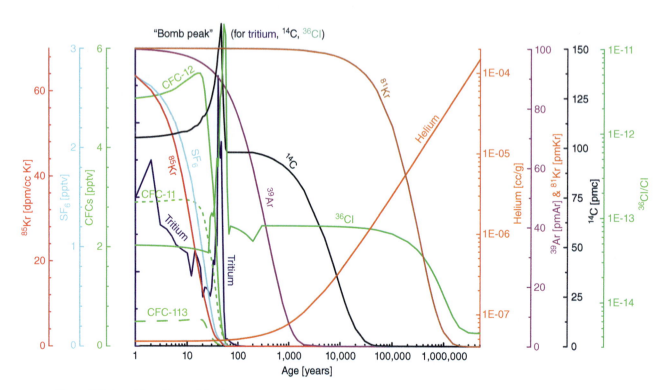

Figure 7.3 Input functions of the tracers.

Of course, this short overview cannot go into all details and therefore often refers to the according literature. Input functions of the tracers discussed are given in Figure 7.3. Each tracer is most useful in the area where its corresponding curve of the input function varies most. Again, we will follow the classification according to how much the transport properties of a tracer deviate from the transport properties of the water itself.

The Most Ideal Ones, Part of the Water Molecule

As discussed earlier in this chapter, the most ideal tracer is the water molecule itself. This is possible since normal water (H_2O) consists of the stable and most abundant isotopes ^{16}O and ^{1}H, with small amounts of two stable isotopes of hydrogen and oxygen (^{2}H and ^{18}O) and a tiny amount of a radioactive isotope of hydrogen (^{3}H). There is a third stable isotope of oxygen (^{17}O), which to our knowledge has not been applied in spring studies, and with further analytical developments such applications may emerge in future. The stable isotopes are natural and also show seasonal natural variations, which can be used favorably to determine timescales of groundwater movement from recharge to discharge in a spring of up to a few years (Fig. 7.2). Radioactive hydrogen (^{3}H or tritium) also occurs naturally, but its natural occurrence was overprinted by up to a factor of 500 during the atmospheric tests of hydrogen bombs in the 1960s (Figs. 7.1 and 7.3). With a half-life of 12.3 years (4,500 days; Unterweger & Lucas, 2000), tritium is good to learn about water flow times up to a few decades.

The Second Best: Noble Gases

Second best are substances that show no chemical reactions and are transported the same way as water itself. These second-best tracers are the noble gases. Being chemically completely inert they are not influenced by any chemical rock-water interaction or by biological processes. Again, we can distinguish subclasses: the elemental concentrations of noble gases, their stable isotope composition, and the radioactive noble gas species. We discussed how to use the elemental composition to determine recharge conditions in section 7.3.1. A special candidate here is helium, which is produced naturally by the radioactive decay from uranium and thorium in the rock matrix (Fig. 7.1C). This natural underground production doubles the concentration relative to the solubility equilibrium with air after a few hundred to thousand years (depending on rock porosity and U and Th content). Aquitards tend to have a higher helium production due to smaller grain sizes (higher helium release) and often higher U and Th content of the clays (Fig. 7.1C). Additional fluxes to the aquifer from below can easily create dissolved helium concentrations a factor of 10,000 above solubility equilibrium (Torgersen & Ivey, 1985; Torgersen & Clarke, 1985, 1987); see "crustal flux" in Figure 7.1. The elemental concentration of helium is therefore an excellent and very sensitive indicator for the admixture of old water (Torgersen & Stute, 2013).

Helium also stands out regarding the stable isotope composition of noble gases. Its natural isotopic composition varies by 5 orders of magnitude: deep groundwaters with purely radiogenic helium have $^{3}He/^{4}He$ isotopic compositions in the range of $1 \cdot 10^{-8}$, air has a composition around $1 \cdot 10^{-6}$, and the mantle is in the range of $1 \cdot 10^{-4}$, which altogether is a variation of a factor of 10,000, a higher variation in stable isotopic composition than any other element shows on Earth. (The element with the second largest isotopic variation is hydrogen, which in natural water has a variation of $^{2}H/^{1}H$ of only a factor of 2, between Antarctic ice and highly evaporated lakes such as Lake Eyre.) This means that the helium isotopic composition can be an excellent indicator demonstrating that the spring water is connected to a deep volcanic system (elevated $^{3}He/^{4}He$ ratios) or originating only from sedimentary rocks ($^{3}He/^{4}He$ in the range of 10^{-8}). In the Northern Hemisphere, another important application of helium isotopic composition is the combination of ^{3}He with tritium (^{3}H). Since ^{3}H decays to ^{3}He, the two combined can give a radioactive mother-daughter relation similar to potassium-argon dating (Schlosser et al., 1989; Schlosser et al., 1988; Schlosser et al., 1998; Solomon & Sudicky, 1991). Unfortunately, the factor 50–100 smaller tritium input in the Southern Hemisphere means that the signal of ^{3}He from tritium decay (tritiugenic helium or $^{3}He_{trit}$) is hardly detectable in groundwater south of the equator.

The next noble gas with useful variation in its stable isotopic composition is argon. ^{40}Ar is produced by radioactive decay of ^{40}K, a very abundant element in aquifer rocks (Fig. 7.1C). But the atmospheric argon concentration is nearly 1% (compared with 5.24 ppm for helium) and its ratio to the next lighter stable isotope ($^{40}Ar/^{36}Ar$) is nearly 300. This means the underground production of ^{40}Ar is not as easy to detect. Even with high-precision measurements, the signal is detectable only after several thousand years (Seltzer et al., 2021a), with routine measurements only after more than 100,000 years (Beyerle et al., 2000; Stanley et al., 2009; Suckow et al., 2019).

Of the radioactive noble gas isotopes, radon is a very commonly applied element. ^{222}Rn has the longest half-life of all radon isotopes, but this is still short with only 3.8 days. It is produced by the decay of radium (^{226}Ra) in the aquifer (Fig. 7.1C) and reaches equilibrium between production and decay within 4 weeks (Froehlich, 2013). It is very useful to detect submarine springs (Chen et al., 2020; Oliveira et al., 2003; Santos et al., 2011; Wang et al., 2017) and groundwater inflows to rivers (Banks et al., 2019; Lamontagne & Cook, 2007; Lamontagne et al., 2015) or, reversely, bank infiltration (Frei & Gilfedder, 2021).

With its short half-life, ^{222}Rn often only represents the last few meters of the flow path to a spring.

There are three more radioactive noble gas isotopes in Figures 7.2 and 7.3, which operate on much longer timescales: ^{85}Kr is useful on the scale of decades, ^{39}Ar for centuries, and ^{81}Kr for up to one million years. ^{85}Kr is anthropogenic, produced in nuclear power plants similar to ash in a normal fire and released when the nuclear fuel is reprocessed. ^{39}Ar and ^{81}Kr are naturally produced in the upper atmosphere by cosmic radiation. Natural concentrations in air as well as dissolved in water are tiny, which implies that all three tracer methods presently still need large amounts of water (on the order of 1,000 L extracted in the field; Burk et al., 2013; Matsumoto et al., 2013; Ohta et al., 2009) or of in situ extracted gas (Musy et al., 2021), and that they can be measured only by very few (less than a handful) laboratories around the globe (Purtschert et al., 2013). But the ATTA technology (Atom Trap Trace Analysis) will be a game changer in this respect, reduce the necessary amount of water and increase the sample throughput drastically (Lu, 2016; Lu & Mueller, 2010; Lu et al., 2014). All three tracer methods have the advantage that only the isotopic ratio is used in the time calculations, not the absolute concentration. This means that gas exchange is still a potential problem (Fig. 7.1C), but gas loss (e.g., due to emanating bubbles in the spring pond or in thermal springs) is of minor importance. Applications in springs are emerging (Delbart et al., 2014; Momoshima et al., 2011; Sidle, 2006, 2008; Sturchio et al., 2004) and the prospect of having a technology on the Southern Hemisphere for young water that corresponds in age range to the ^{3}H/^{3}He technology in the Northern Hemisphere and is even more robust in its application is exciting.

Tracers That Are Least Certain

Finally, there are the least ideal dating tracers that nevertheless have a wide range of applications, which we will discuss from the longest timescale to the shortest (or from right to left in Fig. 7.2). The most ideal in this group is probably ^{36}Cl, since the chloride ion does not really show many reactions in groundwater. ^{36}Cl is produced naturally in the upper atmosphere by cosmic rays and the half-life of ^{36}Cl is 301 ky, which means it has a dating range of up to one million years. To complicate things, ^{36}Cl was also produced in the atmospheric bomb tests (Fig. 7.1A) and this peak can be successfully used to determine timescales, although the peak varies considerably from place to place (Bouchez et al., 2019; Corcho Alvarado et al., 2005; Wilske et al., 2020). There are three main difficulties involved with using the ^{36}Cl/Cl ratio in groundwater. First, the natural input function is variable with latitude (due to cosmic ray production being dependent on the Earth's magnetic field) and distance from the coast (because sea spray dilutes the ^{36}Cl/Cl signal) (Phillips, 2013). While for tritium and ^{2}H and ^{18}O, there is a plethora of data available from the Global Network of Isotopes in Precipitation (GNIP; IAEA/WMO, 2016) to deduce a local input function with some certainty, such data are lacking for ^{36}Cl. Therefore, each application of ^{36}Cl needs data in the recharge area, and, even then, the value measured in the recharge area today may differ from the value 100 ky ago. Second, ^{36}Cl is also produced in the underground (Andrews et al., 1989), which means that the final value for water beyond the dating range is unknown (Fig. 7.1C). And third, aquitards may leak "dead" (that is: ^{36}Cl-free) chloride into the groundwater, which again disturbs the ^{36}Cl/Cl ratio such that a decrease may be due not only to radioactive decay. Nevertheless, ^{36}Cl is a very useful tracer for old groundwater systems. For instance, the mount springs at the end of the flow path of the Great Artesian Basin are close to background in ^{36}Cl, indicating the water emerging here is older and contains only admixtures of younger water (Keppel et al., 2016).

One of the earliest and most commonly applied tracers in groundwater is radiocarbon or ^{14}C. Radiocarbon is produced in the upper atmosphere by cosmic radiation, quickly oxidized to CO_2 and incorporated in the whole cycle of life on Earth. Its half-life of 5,730 years (Godwin, 1962) and wide application during the 1950s and 1960s in archeology and the geosciences soon also enabled the assessment of groundwater velocities (Münnich, 1957). This $^{14}CO_2$ dissolves in groundwater and becomes part of the carbonate system (Fig. 7.1A), which is where the actual dating problem starts: charge balance requires a certain amount of carbonates to be dissolved, which originate from the aquifer rocks and have an unknown ^{14}C content that in most cases is assumed to be zero (incorrectly; see Geyh, 1970; Meredith et al., 2016). Also, the assumption that soil CO_2 always has atmospheric ^{14}C values has been demonstrated to be wrong (Gillon et al., 2012; Wood et al., 2014). Various theoretical models exist trying to correct the initial radiocarbon concentration for these geochemical processes under open and closed system conditions, and they are discussed in detail in Plummer and Glynn (2013).Other researchers applied a more practical approach investigating the geochemical processes directly with isotope methods (Gillon et al., 2009; Meredith et al., 2016; Wood et al., 2014).

Unfortunately, it gets even more complex for radiocarbon because also in the underground passage of groundwater further fossil carbon can be taken up (Fig. 7.1C) if organic material in the aquifer is oxidized, either by dissolved oxygen, dissolved nitrate, or dissolved sulfate (Plummer & Glynn, 2013). All this together means that the ^{14}C/C ratio prior to radioactive decay is badly constrained and that the straightforward calculated "conventional radiocarbon age" delivered by many laboratories, albeit very useful for archeologists, is quite meaningless in groundwater (Mook & Plicht, 1999). Nonetheless, interpreted with the necessary caution and supported by the other tracers discussed here, ^{14}C is a very valuable tool in the assessment of groundwater resources and the vulnerability of springs. Luckily, ^{12}C carbon has another stable isotope, ^{13}C, which is in similar ways influenced by the carbonate processes mentioned above, and both ^{13}C and ^{14}C together allow some conclusion on the geochemical processes on the way.

The final group of tracers is anthropogenic trace gases such as the chlorofluorocarbons (or CFCs; in use in groundwater are the species CFC-11, CFC-12, and CFC-113), sulfur hexafluoride (SF_6), and more recently a halon species (H1301). All these species have known and variable time series of atmospheric concentrations (Figs. 7.1A and 7.3), and if they were inert and all infiltration parameters (such as temperature, altitude, and excess air discussed in section 7.3.1) were known, the infiltration year of the water could be directly calculated. The principles and applications of CFCs and SF_6 in groundwater are the topic of a standard and freely available public domain textbook (IAEA, 2006) and can be directly transferred to H1301 (Beyer et al., 2014; Beyer et al., 2015; Beyer et al., 2017). CFCs became famous for destroying the ozone layer and their production was phased out in the eighties, which means their atmospheric concentration currently decreases. This creates the first problem in application if the species are applied singly: since the atmospheric concentration curve decreased during the last four decades, one measured concentration always corresponds to two infiltration times. SF_6 and H1301 do not yet have this disadvantage, since their atmospheric concentration steadily increases. The combination of all five species together in turn allows resolving the age distribution of the last decades (see Figure 7.3). The CFCs and H1301 have a strong dependency of dissolved concentrations on temperature, whereas the solubility of SF_6 and H1301 is so low that they are extremely influenced by the amount of excess air (see section 7.3.1; Fig. 7.1A). A final complication in the application of CFCs is degradation by microbes in anaerobic groundwater (Fig. 7.1A), whereas for SF_6, underground production is discussed (Deeds et al., 2008; Friedrich et al., 2013; Rohden et al., 2010). While the degradation of CFCs happens beyond doubt (IAEA, 2006), the underground production of SF_6 is in contradiction to the general assumption that SF_6 is completely stable in groundwater and widespread evidence that old groundwater is free of SF_6. Our opinion is rather that the details of the excess air creation and reequilibration is badly understood for SF_6 (Gerber et al., in preparation).

While this last section focused on problematic properties of the "dating" tracers and therefore may sound a bit disheartening, the usefulness of environmental tracers to assess groundwater flow and transport is beyond any doubt, if the interpretation is done with the necessary care. The next section will therefore go a little bit more into detail about how this interpretation is done.

Converting Tracer Concentrations to Time Information

As discussed in section 7.2.2, groundwater has an inherent property to mix, already during infiltration but even more during underground flow. This means that no well will ever deliver an unmixed sample, but the well construction can restrict the mixture (e.g., with short screens). For a spring, such a restriction of flow paths is completely beyond human control, and we have to take what we get when sampling a spring. In assessing springs, the task is therefore rather to answer the question "Which age is present in which proportion in the spring water?" A very useful approach in which tracers can give a direct and nondebatable answer is to reverse this question thus: "Which ages are NOT in the spring?" If a 2 year monthly time series of stable isotopes (^{2}H and ^{18}O) in a spring shows no variation of the results beyond the uncertainty of the measurement, the conclusion will be the water is older than a few years, because otherwise a seasonal variation would be visible. Similarly, if a spring water contains no tritium at all, then we can exclude the presence of any significant amount of water infiltrated later than the 1960s. Similar and very robust statements can be made for the other tracers: (1) no radiocarbon, meaning no water younger than a few thousand years; and (2) no ^{36}Cl, meaning no water younger than a few hundred thousand years, although such a result is very unlikely for a spring (Keppel et al., 2016).

When applying this exclusion principle, it is important to be aware of the detection limit and accuracy of any measurement. For instance, if the seasonal amplitude of rain is 50‰ in ^{2}H and the measurement accuracy is 5‰ (that is: very bad) then the "older than a few years" mentioned above corresponds to "older than 1.6 years," whereas the typical accuracy of modern laboratories of 1‰ for ^{2}H would result in 8 years. See Małoszewski et al. (1983) to learn how these limits were calculated. Similarly, in Australia, modern rain (2020) has tritium values around 0.5–2 TU (Tadros et al., 2014). If a laboratory delivers a tritium detection limit of 1 TU for a spring sample, then the statement "contains no tritium" is meaningless (since the lab would not be able to detect tritium in modern rain anyway), whereas if the lab delivers the state-of-the-art detection limit of 0.05 TU, then the measurement of "no tritium" (that is, <0.05TU) can be translated in Australia into "the amount of modern rain in this spring is smaller than 10%."

On the other hand, a radiogenic tracer allows the application of an *inclusion* principle. For instance, if large amounts of helium are present in a spring, then this is evidence that a significant part of the spring water is old, although it will be hard to quantify both the fraction of old water and its age if no detailed measurements are available from deeper strata.

If a tracer was measured in the spring discharge, it is evident that there is something "younger than" present; but how to quantify the contributions of different ages? To move beyond simple exclusion/inclusion principles in this case, a more rigorous mathematical approach is needed to quantify the distribution of different groundwater ages. This is where it becomes absolutely necessary either to have a time series of environmental tracer measurements or to measure at least as many tracers as possible on a spring in a snapshot in time (ideally both).

The mathematical formulation transforming a time series of environmental tracer measurements in the input $C_{in}(t)$ into an output concentration $C_{out}(t)$ while at the same time deriving some parameters of an age distribution is called a *convolution integral* (Maloszewski & Zuber, 1982; Małoszewski et al., 1983):

$$C_{out}(t) = \int_{-\infty}^{t} C_{in}(t') \cdot g(t-t') \cdot e^{-\lambda(t-t')} dt'. \quad (7.1)$$

Here $g(t-t')$ is the age distribution of the mixture, and the exponential term accounts for radioactive decay with a decay constant λ if the tracer is radioactive (otherwise $\lambda = 0$ and the exponential term converts into 1). Characterizing the mathematical formalism in words, this is a weighted mixture of waters of different ages. The sum of all contributions over all ages (that is, the integral over $g(t-t')$ alone) equals one, and the different contributions of a certain age $(t-t')$ are added up multiplying with the percentage with which this age is present (which is $g(t-t')$). This age distribution cannot be directly measured, but using a known input function (that is, the tracer concentration $C_{in}(t)$ over time at the recharge region) and several tracer measurements at the spring ($C_{out}(t)$), one can deduce important parameters of this age distribution, such as the mean (called Mean Residence Time or MRT) or the ratio between dispersion and advection (the Peclet number). To apply this formula, it is therefore necessary to have a time series of input measurements ($C_{in}(t)$), which is available from public sources for all tracers discussed here, and the time development of the tracers is also displayed in Figure 7.3. Further, one needs at least one or, most usefully, several measurements (different tracers or time series or both) of the spring ($C_{out}(t)$). These measurements are compared with the numerical output of the convolution integral, usually using public domain computer codes that we show later.

Historically, due to limited data availability and computational resources, the assumption has usually been made that the system is in a dynamic equilibrium and that the age distribution does not change over time. Even a time-invariant age distribution can in principle still have an arbitrary shape, which is generally impossible to determine with only a few measurements. Therefore, the problem is often further simplified by parameterizing the shape of the age distribution with specific functions representing analytical descriptions of idealized flow systems. These are generally known as lumped parameter models, or box models.

The most frequently used models are (1) the piston flow model (PM), which assumes that the sample consists of only one age and underwent no mixture at all during flow; (2) the exponential model (EM), which mathematically corresponds simultaneously to a well-mixed chemical reactor (Eriksson, 1958) and a complete screened well in a very idealized homogenous aquifer of finite depth and homogenous recharge first described in Vogel (1967); and (3) the dispersion model (DM), which corresponds to a transport process involving only advection and dispersion. There are other models in use, such as the gamma model for which a physical reasoning is not really clear, or variations of the exponential model that cut out parts of the idealized Vogel aquifer and have up to three parameters. Maloszewski and Zuber (1982) as well as Solomon et al. (2006) and Suckow (2014b) provide good explanations of which model describes which idealized flow system and under which conditions these are applicable.

In practice, this means that a known tracer input function is introduced into a computer code describing the convolution integral, and the output is compared (or fitted) with only one or two, rarely three, free parameters to the measured output concentration. One of these free parameters is always the flow-weighted mean age, called the mean residence time (MRT), and another parameter may describe the degree of mixing occurring on the flow path in relation to flow time, such as the Peclet number. Several such software packages are available in the public domain either as Excel spreadsheets or as stand-alone code (Bayari, 2002; Dávila et al., 2013; Jurgens et al., 2012; Ozyurt & Bayari, 2003; Suckow, 2012, 2014b; Zoellmann et al., 2002), the more sophisticated ones allowing to calculate two convolution integrals (two box models) in parallel that are mixed and include the necessary input functions and solubilities for all tracers discussed here. The number of free parameters in this modeling process must be smaller than the number of available measurements. It is also a good idea to compare modeling results directly against measured values: some modeling codes, for instance, do not compare against the measured CFC and SF_6 concentrations but against modeled atmospheric concentrations. This may be numerically simpler and faster, but it is easy to lose track of which number is a hard measured value and which one is interpreted under a certain assumption. Even with many measured tracers, there will be a high degree of nonuniqueness in the result (McCallum et al., 2015; McCallum et al., 2014a; McCallum et al., 2014b). Nevertheless, in rare cases, a model approach with two parallel and mixed convolution integrals may be appropriate, which gives a total of more than five fit parameters. One such case will be discussed in the next section.

7.4. CASE STUDY: THE FISCHA-DAGNITZ SPRING, AUSTRIA

We will present only one case study here, the Fischa-Dagnitz Spring in Austria, which admittedly is a very ideal case in which a very long time series of tritium (several decades from 1963 until 2016, most of them sampled by one person, Dieter Rank) and a few other environmental tracer measurements are available in a well-described sedimentological context. The long time series of tritium measurements attracted further activities from the International Atomic Energy Agency (IAEA) performing a gas tracer test in the stream emerging from this spring. It also triggered a cooperation between the Austrian Environment Agency (Umweltbundesamt) and Bern University, during which also CFCs, SF_6, ^3He, and ^{85}Kr were measured. These measurements allowed detailed conclusions on the age distribution and underground processes influencing some tracer results. It must be stated that this is a very ideal case. The only other case study known to us with a similar long time series of tritium measurements on a spring is given in Blavoux et al. (2013). More realistic in a new and a priori unknown case is the application of several environmental and active tracer methodologies as presented in Leblanc et al. (2015).

7.4.1. Hydrogeological Setting

The Fischa-Dagnitz Spring is located approximately 35 km south of Vienna, Austria, within the Southern Vienna Basin (SVB). The basin contains up to 175 m of quaternary fluviatile channel and basin structures filled with coarse sediments, and constitutes one of Europe's largest groundwater reservoirs (Kupfersberger et al., 2020; Salcher & Wagreich, 2010; Salcher et al., 2012) (see Fig. 7.4a). In the southern part of the SVB aquifer, groundwater level fluctuations of more than 10 m and a steep groundwater gradient toward the northeast are

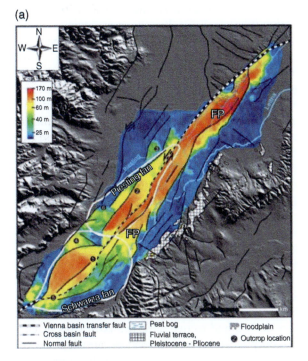
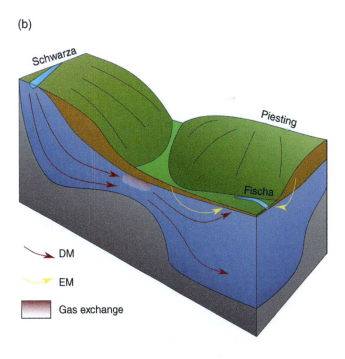

Figure 7.4 Conceptualization of the Fischa flow system. The underlying tectonic system and sedimentary thickness is displayed in (a) (Salcher & Wagreich, 2010/with permission of Elsevier). The Schwarza River infiltrates in the southeast and contributes the main flow to Fischa Spring. Additional local recharge happens into the Piesting sedimentary fan which is part of a smaller local groundwater system. These two flow systems are displayed as a simplified conceptualization in (b): the Schwarza infiltration and flow as dispersion model (DM), the local infiltration as exponential model (EM). It is assumed that some gas exchange happens at the transition between the two sedimentary fans, partly resetting the gas tracer clocks.

observed, whereas the northern part includes several gaining streams (e.g., Fischa, Kalter Gang, Leitha) resulting in reduced groundwater level dynamics (Kupfersberger et al., 2020). Previous authors have suggested recharge to the SVB is primarily driven by infiltration from losing streams in the south (e.g., Schwarza, Leitha) and, to a much lesser degree, by diffuse infiltration throughout the SVB (Kupfersberger et al., 2020; Rank & Papesch, 2003). However, this conclusion was drawn on the basis of only a minor change in ^{18}O signature between the infiltration water of Schwarza River (-11.1‰) and Fischa Spring (-10.8‰), where local rain has an average value of -9.5‰. This does not exclude contributions of local infiltration of 10%–20% (Rank et al., 2016). Understanding timescales of groundwater flow and spring discharge is of particular pertinence because a plume of chlorinated hydrocarbons from industrial contamination in the southernmost part of the SVB has been moving toward the Danube for many years (Rank & Papesch, 2003).

The Fischa-Dagnitz Spring itself consists of three distinct springs, located within about 200 m of each other, and the first-order stream that begins at the springs. Stolp et al. (2010) note that there is no existing stream channel upgradient of the springs and no perennial or obvious ephemeral stream tributaries along the first two kilometers of the stream. Stream discharge increases from a few tens of liters per second at the main spring to 400–700 L/s, 2 km downstream (Gerber, 2012; Stolp et al., 2010).

7.4.2. Science in Several Steps: The History of Environmental Tracers at Fischa-Dagnitz

The spring of Fischa-Dagnitz was a topic for environmental tracer measurements as early as 1966 when the first tritium measurements were reported on an international conference at the nearby International Atomic Energy Agency (IAEA) in Vienna (Davis et al., 1967). This presentation addressed several wells and springs in the Vienna Basin, presenting also monthly measurements of the Fischa-Dagnitz Spring between 1963 and 1966. Unfortunately, the measurements were reported only in paper format since electronic data storage at that time was unusual, and therefore they were initially forgotten in later studies.

From 1968 on, Dieter Rank, a member of Arsenal, the Austrian national laboratory engaged in tritium measurements, took randomly but frequently samples from the Fischa-Dagnitz Spring and thus established a time series of tritium measurements of several decades, until 2016 (Rank et al., 2016). A separate initiative of the IAEA was driven by work of Kip Solomon who, at that time, started to use springs as a naturally flow-weighted and integrated source for waters and to obtain integrated numbers for recharge over a catchment (Solomon et al., 2015; Solomon et al., 2007). The idea was to use not only tritium but also its decay product ^{3}He$_{trit}$ as a tracer to conclude on the flow-weighted mean residence time in the spring. The use of such a volatile gas tracer made it necessary to quantify the gas loss, which occurs when the spring water emerges and flows into a stream. Therefore, an attempt was made to quantify this gas loss and an active tracer experiment was performed that used a steady injection of helium and krypton to quantify gas exchange and therefore gas loss of ^{3}He. Also, the chlorofluorocarbons CFC-11, CFC12, and CFC-113 were measured by the IAEA lab on samples taken simultaneously, but these turned out to be contaminated by a military airbase nearby and not usable for quantification of the MRT. It can be stated that the age results from ^{3}He generally agreed with tritium insofar as the bulk of the waters were younger than a few decades, but in detail it was not possible to reconcile them completely. Therefore, Stolp et al. (2010) tried to compromise in their modeling between the two tracer results. This in turn attracted the attention of Martin Kralik from the Austrian Environmental Agency who invited the group of Roland Purtschert in Bern to perform also ^{85}Kr measurements on the waters from Fischa-Dagnitz and repeat some of the other tracer measurements. This allowed a similar gas tracer test as in Stolp et al. (2010), without the need of an active tracer application (Gerber, 2012). But the main achievement of Martin was to dig out the older original data from Davis et al. (1967), which in the comparison with the tritium data of Stolp et al. (2010) demonstrated that the discharge showed two distinct tritium peaks instead of only one, allowing more detailed lumped parameter modeling. This more detailed modeling in turn demonstrated that gas exchange happened also in the aquifer before the waters emerged at the Fischa-Dagnitz Spring.

7.4.3. What Is the Age of the Water at the Fischa-Dagnitz Spring?

As mentioned before, the tritium record in Fischa-Dagnitz shows two different peaks (Fig. 7.5), a sharp one of 300–350 TU emerging in 1965, followed by a dip going back to nearly 150 TU and a broader peak in 1970 with maximum around 240 TU. Both these peaks correspond to the peak input in tritium in precipitation in 1963 and therefore correspond to two different flow paths. Such a bimodal output cannot be described by one single convolution integral with any unimodal age distribution such as EM, PM, or DM, since one convolution integral as used in Stolp et al. (2010) and an input function with only one input peak can create only one output peak. Therefore, two parallel lumped parameter models had to be used (Suckow et al., 2013), which creates a bimodal age distribution.

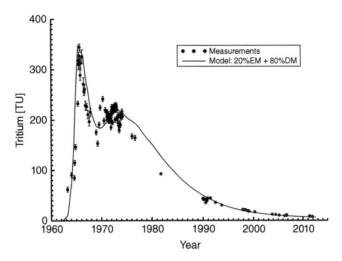

Figure 7.5 Comparison of the measured concentrations in the tritium time series of five decades with the output derived from two parallel box models (20% EM and 80% DM).

It was possible to describe the whole tritium time series reasonably well using an exponential model having a MRT of 2.6 years where the tritium spends 1.3 years in the unsaturated zone (representing local recharge) and a second dispersion model with a MRT of 15.5 years and a Peclet number of 6.5 (describing the infiltration from the Schwarza and subsequent flow in the Southern Vienna Basin; see Fig. 7.4). These two models were then mixed with a flow percentage of 20% for the EM and 80% for the DM. The resulting fit to the tritium time series is displayed in Figure 7.5. Up to 1976, tritium was measured without any electrolytic enrichment, which explains the larger uncertainties (error bars); after that the uncertainties are smaller than the symbol.

While this model description achieves an excellent fit to the tritium data, it does not provide a reasonable fit to any of the gas tracers SF_6, $^3He_{trit}$, and ^{85}Kr. Remembering section 7.3.2, tritium is probably the most ideal tracer available for water movement since it is part of the water molecule itself, whereas the other three tracers are dissolved and transported with the water. For all gas tracers, the deviation is systematic to the *younger* side, or, in other words, can be described by gas exchange happening on the flow path *upstream* of the Fischa-Dagnitz Spring. This at least partly resets the gas tracer clocks, releasing part of the tritiugenic ^3He that had built up and re-equilibrating ^{85}Kr and SF_6 with younger (higher atmospheric) concentrations (Fig. 7.6). A location of such a gas exchange is plausible at the transition between the two sedimentary fans of Schwarza and Piesting, through which the flow path described by the DM must pass (Fig. 7.4b).

The question now is, how old is the water at Fischa-Dagnitz? The model-derived age distribution is shown in Figure 7.7. It displays that in this model representation, the water contains no measurable contribution younger than 1.3 years. This result is due to the very first two measurements showing a sharp increase in tritium concentrations in 1964–65 in Figure 7.5. This implies that any contamination in the local area (e.g., pesticides from agriculture or a surface contaminant spill) would show up in the modeled spring earliest after 1.3 years, and this corresponds to the time the water spends in the unsaturated zone of the local system. However, the age distribution presented in this model uses a mathematically sharp increase at 1.3 years. This is probably not realistic since also the recharge in the unsaturated zone has preferential flow paths, but processes in the unsaturated zone were not the target of this study. Note that the mean residence time of the exponential model is 2.6 years, so the first occurrence of a contamination is not the same as the MRT. However, any local contaminant would observe a dilution down to 20% of whatever reaches the groundwater table, since most of the water in this model representation (80% of the flow discharging) is water that recharged much farther upstream in the infiltration section of the Schwarza River. And this part of the water has a mean residence time of 15.5 years. One can see in Figure 7.7 that any contamination that Schwarza carries would infiltrate and show up in Spring Fischa after a few years of delay with a peak after roughly 10 years and still tailing out after 30–40 years. It is also notable that 3% of the water is older than 40 years, and with the tracers used here, it is impossible to quantify how much older. We can only exclude that it is many millennia old, since no radiogenic ^4He was found in the spring nor was any other deep noble gas signature found that is known to exist in the underground of the Vienna basin (Ballentine & O'Nions, 1992).

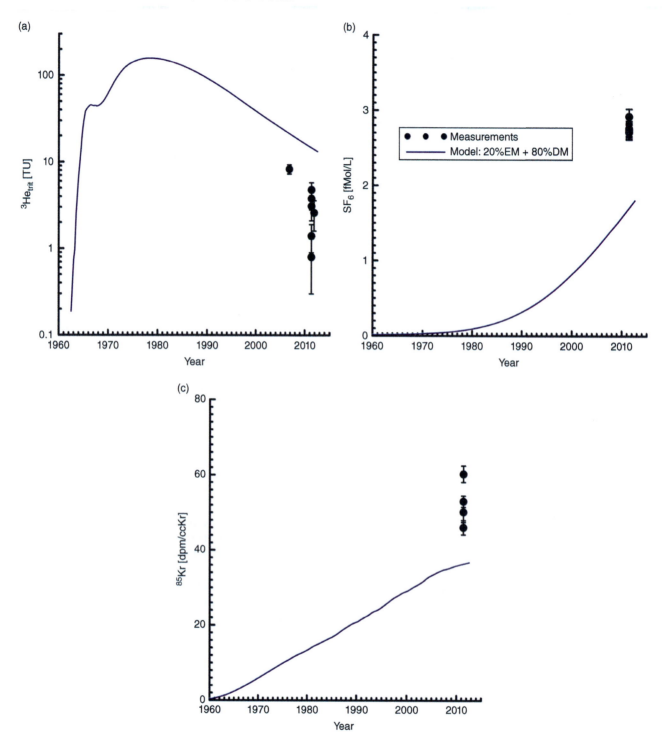

Figure 7.6 Comparison of the measured concentrations of (a): $^3\text{He}_{trit}$ (Gerber, 2012; Gerber et al., 2012; Stolp et al., 2010); (b): SF_6; and (c): ^{85}Kr (Gerber, 2012; Gerber et al., 2012) with the output derived with the same model combination as for the tritium time series in Figure 7.5.

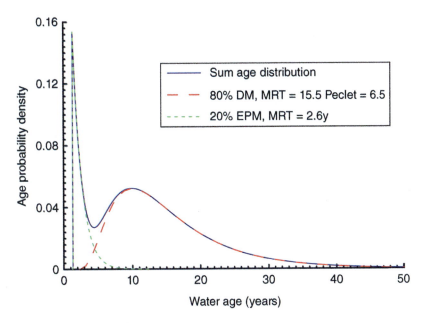

Figure 7.7 Age distribution of the Fischa-Dagnitz Spring.

7.5. SUMMARY

This chapter demonstrated the unique advantages that environmental tracer applications can bring to the investigation of springs. It also discussed many of the problems that each of the environmental tracer methods has, which should not dishearten the reader. Any scientific method has limits, but the method is still very valuable if the interpretation keeps these limits in mind. One of the clear outcomes of the methodological descriptions as well as the detailed case study at Fischa-Dagnitz is that a spring has no unique age. A spring always contains mixtures of different flow paths and different travel times. Assessing the vulnerability of a spring toward contamination therefore always is only as good or bad as our success in deconvoluting this age distribution. Single measurements of one tracer on a spring can simply not give enough information for this task. A good conceptual understanding of the surrounding hydrogeology is as important as an integrated interpretation of hydrochemistry, geophysics, and environmental tracers. Ideally, a spring needs to be addressed with a time series of measurements and several environmental tracers covering different timescales of groundwater movement in the spring itself and several wells upstream of the spring in the potential source aquifers.

ACKNOWLEDGMENTS

The authors wish to thank Cornelia Wilske for detailed discussions on didactics of the text and insisting that we add Figure 7.1, which she also designed. Comments by Matthew Currell and reviews by Florent Barbecot and an anonymous reviewer significantly improved the manuscript.

AVAILABILITY STATEMENT

Lumped Parameter model calculations were done using the public domain code "Lumpy," which together with the used Fischa data set is available from the authors (AS).

REFERENCES

Andrews, J. N., Davis, S., Fabryka-Martin, J., Fontes, J. C., Lehmann, B. E., Loosli, H. H., et al. (1989). The in-situ production of radioisotopes in rock matrices with particular reference to the Stripa granite. *Geochimica et Cosmochimica Acta*, *53*(8), 1803–1815.

Ballentine, C. J., & O'Nions, R. K. (1992). The nature of mantle neon contributions to Vienna Basin hydrocarbon reservoirs. *Earth and Planetary Science Letters*, *113*(4), 553–567.

Ballentine, C. J., & O'Nions, R. K. (1994). The use of natural He, Ne and Ar isotopes to study hydrocarbon-related fluid provenance, migration and mass balance in sedimentary basins. *Geological Society, London, Special Publications*, *78*(1), 347–361.

Banks, E. W., Hatch, M., Smith, S., Underschultz, J., Lamontagne, S., Suckow, A., et al. (2019). Multi-tracer and hydrogeophysical investigation of the hydraulic connectivity between coal seam gas formations, shallow groundwater and stream network in a faulted sedimentary basin. *Journal of Hydrology*, *578*, 124132.

Bayari, S. (2002). TRACER: An EXCEL workbook to calculate mean residence time in groundwater by use of tracers CFC-11, CFC-12 and tritium. *Computers & Geosciences*, *28*, 621-630.

Berglund, J. L., Toran, L., & Herman, E. K. (2019). Deducing flow path mixing by storm–induced bulk chemistry and REE variations in two karst springs: With trends like these who needs anomalies? *Journal of Hydrology*, *571*, 349–364.

Beyer, M., Morgenstern, U., van der Raaij, R., & Martindale, H. (2017). Halon-1301: Further evidence of its performance as an age tracer in New Zealand groundwater. *Hydrology and Earth System Sciences*, *21*(8), 4213–4231.

Beyer, M., van der Raaij, R., Morgenstern, U., & Jackson, B. (2014). Potential groundwater age tracer found: Halon-1301 (CF3Br), as previously identified as CFC-13 (CF3Cl). *Water Resources Research*, *50*(9), 7318–7331.

Beyer, M., van der Raaij, R., Morgenstern, U., & Jackson, B. (2015). Assessment of Halon-1301 as a groundwater age tracer. *Hydrology and Earth System Sciences*, *19*(6), 2775–2789.

Beyerle, U., Aeschbach-Hertig, W., Imboden, D. M., Baur, H., Graf, T., & Kipfer, R. (2000). A mass spectrometric system for the analysis of Noble gases and tritium from water samples. *Environmental Science & Technology*, *34*(10), 2042–2050.

Blavoux, B., Lachassagne, P., Henriot, A., Ladouche, B., Marc, V., Beley, J.-J., et al. (2013). A fifty-year chronicle of tritium data for characterising the functioning of the Evian and Thonon (France) glacial aquifers. *Journal of Hydrology*, *494*, 116–133.

Blavoux, B., Mudry, J., & Puig, J.-M. (1992). The karst system of the Fontaine de Vaucluse (southeastern France). *Environmental Geology and Water Sciences*, *19*(3), 215–225.

Böttcher, J., Strebel, O., & Kölle, W. (1992). Redox conditions and microbial sulphur reduction in the Fuhrberger Feld sandy aquifer. In G. Matthess et al. (Eds.), *Progress in hydrogeochemistry* (pp. 219–226). Berlin: Springer-Verlag.

Böttcher, J., Strebel, O., Voerkelius, S., & Schmidt, H.-L. (1990). Using isotope fractionation of nitrate nitrogen and nitrate oxygen for evaluation of denitrification in a sandy aquifer. *Journal of Hydrology*, *114*, 413–424.

Bouchez, C., Deschamps, P., Goncalves, J., Hamelin, B., Mahamat Nour, A., Vallet-Coulomb, C., et al. (2019). Water transit time and active recharge in the Sahel inferred by bomb-produced 36Cl. *Scientific Reports*, *9*(1), 7465.

Burk, L., Suckow, A., Cook, P. G., & Mathouchanh, E. (2013). *A system for field gas-extraction of 85Kr, 39Ar and 81Kr using SuperPhobic membrane contactors*. Paper presented at AGU Fall Meeting 2013, American Geophysical Union, San Francisco.

Chen, X., Cukrov, N., Santos, I. R., Rodellas, V., Cukrov, N., & Du, J. (2020). Karstic submarine groundwater discharge into the Mediterranean: Radon-based nutrient fluxes in an anchialine cave and a basin-wide upscaling. *Geochimica et Cosmochimica Acta*, *268*, 467–484.

Clark, I. D. (2015). *Groundwater geochemistry and isotopes*. Boca Raton: CRC Press.

Clark, I. D., & Fritz, P. (1997). *Environmental isotopes in hydrogeology*. Berlin: Springer Verlag.

Clark, J. F., Schlosser, P., Stute, M., & Simpson, H. J. (1996). SF6-3He tracer release experiment: A new method of determining longitudinal dispersion coefficients in large rivers. *Environmental Science and Technology*, *30*(5), 1527–1532.

Cook, P. G., & Herczeg, A. L. (Eds.) (2000). *Environmental tracers in subsurface hydrology*. Boston, MA: Kluwer Academic Press.

Corcho Alvarado, J. A., Purtschert, R., Hinsby, K., Troldborg, L., Hofer, M., Kipfer, R., et al. (2005). 36Cl in modern groundwater dated by a multi-tracer approach (H-3/He-3, SF6, CFC-12 and Kr-85): A case study in quaternary sand aquifers in the Odense Pilot River Basin, Denmark. *Applied Geochemistry*, *20*(3), 599–609.

Craig, H. (1961). Isotopic variations in meteoric waters. *Science*, *133*, 1702–1703.

Dávila, P. F., Külls, C., & Weiler, M. (2013). A toolkit for groundwater mean residence time interpretation with gaseous tracers. *Computers & Geosciences*, *61*(0), 116–125.

Davis, G. H., Payne, B. R., Dincer, T., Florkowski, T., & Gattinger, T. (1967). Seasonal variations in the tritium content of groundwaters of the Vienna Basin, Austria. In *Isotope Hydrology 1967* (pp. 451–473). Vienna, Austria: IAEA.

Deeds, D. A., Vollmer, M. K., Kulongoski, J. T., Miller, B. R., Mühle, J., Harth, C. M., et al. (2008). Evidence for crustal degassing of CF4 and SF6 in Mojave Desert groundwaters. *Geochimica et Cosmochimica Acta*, *72*(4), 999–1013.

Delbart, C., Barbecot, F., Valdes, D., Tognelli, A., Fourre, E., Purtschert, R., et al. (2014). Investigation of young water inflow in karst aquifers using SF6-CFC-3H/He-85Kr-39Ar and stable isotope components. *Applied Geochemistry*, *50*(0), 164–176.

Dotsika, E., Poutoukis, D., Kloppmann, W., Guerrot, C., Voutsa, D., & Kouimtzis, T. H. (2010). The use of O, H, B, Sr and S isotopes for tracing the origin of dissolved boron in groundwater in Central Macedonia, Greece. *Applied Geochemistry*, *25*(11), 1783–1796.

Duvert, C., Stewart, M. K., Cendón, D. I., & Raiber, M. (2015). Time-series of tritium, stable isotopes and chloride reveal short-term variations in groundwater contribution to a stream. *Hydrology and Earth System Sciences Discussion*, *12*(8), 8035–8089.

Emblanch, C., Zuppi, G. M., Mudry, J., Blavoux, B., & Batiot, C. (2003). Carbon 13 of TDIC to quantify the role of the unsaturated zone: The example of the Vaucluse karst systems (southeastern France). *Journal of Hydrology*, *279*(1–4), 262–274.

Eriksson, E. (1958). The possible use of tritium for estimating groundwater storage. *Tellus*, *X(4)*, 472–478.

Fiorillo, F., Leone, G., Pagnozzi, M., & Esposito, L. (2021). Long-term trends in karst spring discharge and relation to climate factors and changes. *Hydrogeology Journal*, *29*(1), 347–377.

Fleury, P., Plagnes, V., & Bakalowicz, M. (2007). Modelling of the functioning of karst aquifers with a reservoir model: Application to Fontaine de Vaucluse (South of France). *Journal of Hydrology*, *345*(1–2), 38–49.

Flook, S., Fawcett, J., Cox, R., Pandey, S., Schöning, G., Khor, J., et al. (2020). A multidisciplinary approach to the hydrological conceptualisation of springs in the Surat Basin of the Great Artesian Basin (Australia). *Hydrogeology Journal*, *28*(1), 219–236.

Flury, M., & Wai, N. N. (2003). Dyes as tracers for vadose zone hydrology. *Reviews of Geophysics*, *41*(1).

Frei, S., & Gilfedder, B. S. (2021). Quantifying residence times of bank filtrate: A novel framework using radon as a natural tracer. *Water Research*, 117376.

Friedrich, R., Vero, G., von Rohden, C., Lessmann, B., Kipfer, R., & Aeschbach-Hertig, W. (2013). Factors controlling terrigenic SF6 in young groundwater of the Odenwald region (Germany). *Applied Geochemistry*, *33*(0), 318–329.

Frind, E. O., Duynisveld, W. H. M, Strebel, O., & Böttcher, J. (1990). Modeling of multicomponent transport with microbial transformation in groundwater: The Fuhrberg case. *Water Resources Research*, *26*(8), 1707–1719.

Froehlich, K. (2013). Dating old groundwater using uranium isotopes: Principles and applications. In A. Suckow et al. (Eds.), *Isotope methods for dating old groundwater*. (pp. 153–177). Vienna: International Atomic Energy Agency.

Gäbler, H.-E., & Bahr, A. (1999). Boron isotope ratio measurements with a double-focusing magnetic sector ICP mass spectrometer for tracing anthropogenic input into surface and ground water. *Chemical Geology*, *156*(1), 323–330.

Gabriel, G., Kirsch, R., Siemon, B., & Wiederhold, H. (2003). Geophysical investigation of buried Pleistocene subglacial valleys in northern Germany. *Journal of Applied Geophysics*, *53*(4), 159–180.

Gat, J. R., Mook, W. G., & Meijer, H. A. J. (2001). Volume II: Atmospheric water. In W. G. Mook (Ed.), *Environmental isotopes in the hydrological cycle: Principles and applications* (p. 113). Paris: UNESCO International Hydrological Programme (IHP-V).

Gerber, C. (2012). *Development of a new degassing system for groundwater dating with radio-Noble gases and application of 85Kr in two case studies*. Master thesis, Universitaet Bern.

Gerber, C., Purtschert, R., Kralik, M., Humer, F., Sültenfuss, J., Darling, G. W., et al. (2012). *Suitability and potential of environmental tracers for base-flow determination in streams: EGU2012-14066*. Paper presented at EGU 12, European Geosciences Union, Vienna

Gerke, K. M., Sidle, R. C., & Mallants, D. (2013). Criteria for selecting fluorescent dye tracers for soil hydrological applications using Uranine as an example. *Journal of Hydrology and Hydromechanics*, *61*(4), 313–325.

Geyh, M. A. (1970). Carbon-14 concentration of lime in soils and aspects of the carbon-14 dating of groundwater. In *Isotope hydrology, 1970* (pp. 215–223). Vienna: IAEA.

Geyh, M. A. (2001). Groundwater, saturated and unsaturated zone. In W. G. Mook (Ed.) *Environmental isotopes in the hydrological cycle: Principles and applications* (p. 196). Paris: UNESCO International Hydrological Programme (IHP-V).

Gillon, M., Barbecot, F., Gibert, E., Corcho Alvarado, J. A., Marlin, C., & Massault, M. (2009). Open to closed system transition traced through the TDIC isotopic signature at the aquifer recharge stage, implications for groundwater 14C dating. *Geochimica et Cosmochimica Acta*, *73*(21), 6447–6696.

Gillon, M., Barbecot, F., Gibert, E., Plain, C., Corcho-Alvarado, J. A., & Massault, M. (2012). Controls on ^{13}C and ^{14}C variability in soil CO_2. *Geoderma*, *189–190*, 431–441.

Godwin, H. (1962). Half-life of radiocarbon. *Nature*, *195*(4845), 984–984.

Gupta, S. K., & Sharma, P. (1984). Soil moisture transport through the unsaturated zone: Tritium tagging studies in Sabarmati basin, western India. *Journal des Sciences Hydrologiques*, *29*, 177–178.

Heaton, T. H. E., & Vogel, J. C. (1981). "Excess air" in groundwater. *Journal of Hydrology*, *50*, 201–216.

Hofmann, T., Darsow, A., Gröning, M., Aggarwal, P., & Suckow, A. (2010). Direct-push profiling of isotopic and hydrochemical vertical gradients. *Journal of Hydrology*, *385*(1–4), 84–94.

Holocher, J., Peeters, F., Aeschbach-Hertig, W., Hofer, M., Brennwald, M., Kinzelbach, W., et al. (2002). Experimental investigations on the formation of excess air in quasi-saturated porous media. *Geochimica et Cosmochimica Acta*, *66*(23), 4103–4117.

Houben, G. J., Koeniger, P., Schloemer, S, Gröger-Trampe, J., & Sültenfuß, J. (2018). Comparison of depth-specific groundwater sampling methods and their influence on hydrochemistry, isotopy and dissolved gases: Experiences from the Fuhrberger Feld, Germany. *Journal of Hydrology*, *557*, 182–196.

IAEA (2006). *Use of chlorofluorocarbons in hydrology. A guidebook*. Vienna: International Atomic Energy Agency.

IAEA (2013). *Isotope methods for dating old groundwater*. Vienna: International Atomic Energy Agency.

IAEA/WMO (2016). *Global network of isotopes in precipitation*. The GNIP Database. https://www.iaea.org/services/networks/gnip

IHP (2000). *Environmental isotopes in the hydrological cycle: Principles and applications*. Paris: UNESCO International Hydrological Programme (IHP-V).

Ingram, R. G. S., Hiscock, K. M., & Dennis, P. F. (2007). Noble gas excess air applied to distinguish groundwater recharge conditions. *Environmental Science & Technology*, *41*(6), 1949–1955.

Jung, M., & Aeschbach, W. (2018). A new software tool for the analysis of noble gas data sets from (ground) water. *Environmental Modelling & Software*, *103*, 120–130.

Jurgens, B. C., Böhlke, J.-K., & Eberts, S. M. (2012). *TracerLPM (version 1): An Excel workbook for interpreting groundwater age distributions from environmental tracer data*. U.S. Geological Survey.

Kaufman, S., & Libby, W. F. (1954). The natural distribution of Tritium. *Physical Review*, *93*.

Kazemi, G. A., Lehr, J. H., & Perrochet, P. (2006). *Groundwater age*. Hoboken, NJ: John Wiley.

Kendall, C., & McDonnell, J. J. (1998). *Isotope tracers in catchment hydrology*. Amsterdam: Elsevier.

Keppel, M., Gotch, T., Inverarity, K., Niejalke, D., & Wohling, D. (2016). *A hydrogeological and ecological characterization of springs near Lake Blanche, Lake Eyre Basin, South Australia*. Adelaide: Government of South Australia, through the Department of Environment, Water and Natural Resources.

Keppel, M., Wohling, D., Love, A., & Gotch, T. (2020). Hydrochemistry highlights potential management issues for aquifers and springs in the Lake Blanche and Lake Callabonna region, South Australia. *Proceedings of The Royal Society of Queensland*, *126*, 65–89.

Kirchner, J. W. (2016). Aggregation in environmental systems, part 1: Seasonal tracer cycles quantify young water fractions, but not mean transit times, in spatially heterogeneous catchments. *Hydrology and Earth System Sciences*, *20*(1), 279–297.

Klump, S., Cirpka, O. A., Surbeck, H., & Kipfer, R. (2008). Experimental and numerical studies on excess-air formation in quasi-saturated porous media. *Water Resources Research*, *44*(W05402).

Konishi, S. (2021). Friday essay: Creation, destruction and appropriation, the powerful symbolism of the Rainbow Serpent. *The Conversation*.

Kupfersberger, H., Rock, G., & Draxler, J. C. (2020). Combining groundwater flow modeling and local estimates of extreme groundwater levels to predict the groundwater surface with a return period of 100 years. *Geosciences, 10*(9), 373.

Lamontagne, S., & Cook, P. G. (2007). Estimation of hyporheic water residence time in situ using 222Rn disequilibrium. *Limnology and Oceanography: Methods, 5*, 407–416.

Lamontagne, S., Taylor, A. R., Batlle-Aguilar, J., Suckow, A., Cook, P. G., Smith, S. D., et al. (2015). River infiltration to a subtropical alluvial aquifer inferred using multiple environmental tracers. *Water Resources Research, 51*(6), 4532–4549.

Leblanc, M., Tweed, S., Lyon, B. J., Bailey, J., Franklin, C. E., Harrington, G., et al. (2015). On the hydrology of the bauxite oases, Cape York Peninsula, Australia. *Journal of Hydrology, 528*, 668–682.

Lee, S.-G., Kim, T.-K., & Lee, T. J. (2011). Strontium isotope geochemistry and its geochemical implication from hot spring waters in South Korea. *Journal of Volcanology and Geothermal Research, 208*(1), 12–22.

Leibundgut, C., Maloszewski, P., & Külls, C. (2009). *Tracers in hydrology*.

Lu, Z.-T. (2016). Radiokrypton dating coming of age. *National Science Review, 3*(2), 172–173.

Lu, Z.-T., & Mueller, P. (2010). Atom trap trace analysis of rare noble gas isotopes. In P. Berman et al. (Eds.), *Advances in atomic, molecular, and optical physics* (pp. 173–205). Amsterdam: Elsevier.

Lu, Z. T., et al. (2014). Tracer applications of noble gas radionuclides in the geosciences. *Earth-Science Reviews, 138*(0), 196–214.

Luhmann, A. J., Covington, M. D., Alexander, S. C., Chai, S. Y., Schwartz, B. F., Groten, J. T., et al. (2012). Comparing conservative and nonconservative tracers in karst and using them to estimate flow path geometry. *Journal of Hydrology, 448–449*(0), 201–211.

Lyons, W. B., Tyler, S. W., Gaudette, H. E., & Long, D. T. (1995). The use of strontium isotopes in determining groundwater mixing and brine fingering in a playa spring zone, Lake Tyrrell, Australia. *Journal of Hydrology, 167*(1), 225–239.

Maiss, M., Ilmberger, J., & Münnich, K. (1994a). Vertical mixing in Überlingersee (Lake Constance) traced by SF6 and heat. *Aquatic Sciences, 56*(4), 329–347.

Maiss, M., Ilmberger, J., Zenger, A., & Münnich, K. (1994b). A SF6 tracer study of horizontal mixing in Lake Constance. *Aquatic Sciences, 56*(4), 307–328.

Maloszewski, P., & Zuber, A. (1982). Determining the turnover time of groundwater systems with the aid of environmental tracers, I. Models and their applicability. *Journal of Hydrology, 57*, 207–231.

Małoszewski, P., Rauert, W., Stichler, W., & Herrmann, A. (1983). Application of flow models in an alpine catchment area using tritium and deuterium data. *Journal of Hydrology, 66*(1), 319–330.

Manning, A. H., & Solomon, D. K. (2003). Using noble gases to investigate mountain-front recharge. *Journal of Hydrology, 275*(3), 194–207.

Matsumoto, T., Han, L.-F., Jaklitsch, M., & Aggarwal, P. K. (2013), A portable membrane contactor sampler for analysis of noble gases in groundwater. *Groundwater, 51*(3), 461–468.

Mazor, E. (1972). Paleotemperatures and other hydrological parameters deduced from noble gases dissolved in groundwaters; Jordan Rift Valley, Israel. *Geochimica et Cosmochimica Acta, 36*(12), 1321–1336.

McCallum, J. L., Cook, P. G., & Simmons, C. T. (2015). Limitations of the use of environmental tracers to infer groundwater age. *Groundwater, 53*(S1), 56–70.

McCallum, J. L., Cook, P. G., Simmons, C. T., & Werner, A. D. (2014a). Bias of apparent tracer ages in heterogeneous environments. *Groundwater, 52*(2), 239–250.

McCallum, J. L., Engdahl, N. B., Ginn, T. R., & Cook, P. G. (2014b). Nonparametric estimation of groundwater residence time distributions: What can environmental tracer data tell us about groundwater residence time? *Water Resources Research, 50*(3), 2022–2038.

Meier, C., Osenbrück, K., Seitz, H.-M., & Weise, S. M. (2017). First lithium isotope data from rivers and subsurface water in the Pamirs. *Procedia Earth and Planetary Science, 17*, 574–577.

Meredith, K. T., Han, L. F., Hollins, S. E., Cendón, D. I., Jacobsen, G. E., & Baker, A. (2016). Evolution of chemical and isotopic composition of inorganic carbon in a complex semi-arid zone environment: Consequences for groundwater dating using radiocarbon. *Geochimica et Cosmochimica Acta, 188*, 352–367.

Meredith, K. T., Moriguti, T., Tomascak, P., Hollins, S., & Nakamura, E. (2013). The lithium, boron and strontium isotopic systematics of groundwaters from an arid aquifer system: Implications for recharge and weathering processes. *Geochimica et Cosmochimica Acta, 112*(0), 20–31.

Millot, R., Guerrot, C., Innocent, C., Négrel, P., & Sanjuan, B. (2011). Chemical, multi-isotopic (Li–B–Sr–U–H–O) and thermal characterization of Triassic formation waters from the Paris Basin. *Chemical Geology, 283*(3), 226–241.

Momoshima, N., et al. (2011). Application of ^{85}Kr dating to groundwater in volcanic aquifer of Kumamoto Area, Japan. *Journal of Radioanalytical and Nuclear Chemistry, 287*(3), 761–767.

Mook, W. G. (2001). Introduction: Theory, methods, review. In W. G. Mook (Ed.), *Environmental isotopes in the hydrological cycle: Principles and applications* (p. 280). Paris: UNESCO International Hydrological Programme (IHP-V).

Mook, W. G., & Plicht, J. (1999). Reporting ^{14}C activities and concentrations. *Radiocarbon, 41*(3), 227–239.

Müller, K., Vanderborght, J., Englert, A., Kemna, A., Huisman, J. A., Rings, J., et al. (2010). Imaging and characterization of solute transport during two tracer tests in a shallow aquifer using electrical resistivity tomography and multilevel groundwater samplers. *Water Resources Research, 46*(3).

Münnich, K.-O. (1957). Messung des C^{14}-gehaltes von hartem grundwasser. *Naturwissenschaften, 34*, 32–33.

Musy, S., Meyzonnat, G, Barbecot, F., Hunkeler, D., Sültenfuss, J., Solomon, D. K., et al. (2021). In-situ sampling for krypton-85 groundwater dating. *Journal of Hydrology X, 11*, 100075.

Ohta, T., Mahara, Y., Momoshima, N., Inoue, F., Shimada, J., Ikawa, R., et al. (2009). Separation of dissolved Kr from a water sample by means of a hollow fiber membrane. *Journal of Hydrology, 376*(1–2), 152–158.

Oliveira, J., Burnett, W. C., Mazzillia, B. P., Braga, E. S., Farias, L. A., Christoff, J., et al. (2003). Reconnaissance of submarine groundwater discharge at Ubatuba coast, Brazil, using ^{222}Rn as a natural tracer. *Journal of Environmental Radioactivity, 69*(1–2), 37–52.

Ollivier, C. (2019). *Caractérisation et spatialisation de la recharge des hydrosystèmes karstiques: Application à l'aquifère de Fontaine de Vaucluse, France: Milieux et changements globaux*. PhD thesis, Avignon Université, 2019.

Ollivier, C., Chalikakis, K., Mazzilli, N., Kazakis, N., Lecomte, Y., Danquigny, C., et al. (2019). Challenges and limitations of karst aquifer vulnerability mapping based on the PaPRIKa method: Application to a large European karst aquifer (Fontaine de Vaucluse, France). *Environments, 6*(3), 39.

Ollivier, C., Mazzilli, N., Olioso, A., Chalikakis, K., Carrière, S. D., Danquigny, C., et al. (2020). Karst recharge-discharge semi distributed model to assess spatial variability of flows. *Science of the Total Environment, 703*, 134368.

Ozyurt, N. N. (2008). Residence time distribution in the Kirkgoz karst springs (Antalya-Turkey) as a tool for contamination vulnerability assessment. *Environmental Geology, 53*(7), 1571–1583.

Ozyurt, N. N., & Bayari, C. S. (2003). LUMPED: A visual basic code of lumped-parameter models for mean residence time analyses of groundwater systems. *Computers & Geosciences, 29*, 79–90.

Phillips, F. M. (2013). Chlorine-36 dating of old groundwater. In A. Suckow et al. (Eds.), *Isotope methods for dating old groundwater* (pp. 125–152). Vienna: International Atomic Energy Agency.

Pinti, D. L., & Marty, B. (2000). Noble gases in oil and gas fields: Origins and processes. In K. Kyser (Ed.), *Fluids and basin evolution* (pp. 160–196). Mineralogical Society of Canada Short Course.

Plummer, L. N., & Glynn, P. D. (2013). Radiocarbon dating in groundwater systems. In A. Suckow et al. (Eds.), *Isotope methods for dating old groundwater* (pp. 33–89). Vienna: International Atomic Energy Agency.

Purtschert, R., Yokochi, R., & Sturchio, N. C. (2013). ^{81}Kr dating of old groundwater. In A. Suckow et al. (Eds.), *Isotope methods for dating old groundwater* (pp. 91–124). Vienna: International Atomic Energy Agency.

Rank, D., & Papesch, W. (2003). Determination of groundwater flow velocity in the southern Vienna Basin from long-term environmental isotope record. In M. Kralik et al. (Eds.), *First conference on applied environmental geology in central and eastern Europe* (pp. 206–207). Vienna: Umweltbundesamt.

Rank, D., Wyhlidal, S., Heiss, G., Papesch, W., & Schott, K. (2016). Arsenal environmental-isotope laboratories 1964–2010: More than 45 years production/application/interpretation of basic isotopehydrological data for central Europe. *Austrian Journal of Earth Sciences, 109*/1, 4–28.

Rivett, M. O., Feenstra, S., & Cherry, J. A. (1994). Transport of a dissolved-phase plume from a residual solvent source in a sand aquifer. *Journal of Hydrology, 159*(1), 27–41.

Robertson, W. D., & Schiff, S. L. (1994). Fractionation of sulphur isotopes during biogenic sulphate reduction below a sandy forested recharge area in south-central Canada. *Journal of Hydrology, 158*(1), 123–134.

Rohden, C., Kreuzer, A., Chen, Z., & Aeschbach-Hertig, W. (2010). Accumulation of natural SF6 in the sedimentary aquifers of the North China Plain as a restriction on groundwater dating. *Isotopes in Environmental and Health Studies, 46*, 279–290.

Rozanski, K., Froehlich, K., & Mook, W. G. (2001). Surface water. In W. G. Mook (Ed.), *Environmental isotopes in the hydrological cycle: Principles and applications* (p. 117). Paris: UNESCO International Hydrological Programme (IHP-V).

Rutherford, J., Ibrahimi, T., Munday, T., Markey, A., Viezzoli, A., Rapiti, A., et al. (2021). An assessment of water sources for heritage listed organic mound springs in NW Australia using airborne geophysical (electromagnetics and magnetics) and satellite remote sensing methods. *Remote Sensing, 13*(7), 1288.

Sahraei, A., Kraft, P., Windhorst, D., & Breuer, L. (2020). High-resolution, in situ monitoring of stable isotopes of water revealed insight into hydrological response behavior. *Water, 12*(2), 565.

Salcher, B. C., & Wagreich, M. (2010). Climate and tectonic controls on Pleistocene sequence development and river evolution in the Southern Vienna Basin (Austria). *Quaternary International, 222*(1), 154–167.

Salcher, B. C., Meurers, B., Decker, K., Hölzel, M., & Wagreich, M. (2012). Strike-slip tectonics and quaternary basin formation along the Vienna Basin fault system inferred from Bouguer gravity derivatives. *Tectonics, 31*(3).

Santos, I. R., Lechuga-Deveze, C., Peterson, R. N., & Burnett, W. C. (2011). Tracing submarine hydrothermal inputs into a coastal bay in Baja California using radon. *Chemical Geology, 282*(1–2), 1–10.

Schäfer, T., Buckau, G., Artinger, R., Kim, J. I., Geyer, S., Wolf, M., et al. (2005). Origin and mobility of fulvic acids in the Gorleben aquifer system: Implications from isotopic data and carbon/sulfur XANES. *Organic Geochemistry, 36*(4), 567–582.

Schlosser, P., Shapiro, S. D., Stute, M., Aeschbach-Hertig, W., Plummer, L. N., & Busenberg, E. (1998). Tritium/³He measurements in young groundwater. Chronologies for environmental records. In IAEA (Ed.), *Isotope techniques in the study of environmental change. Proceedings of a symposium, Vienna, April 1997* (pp.165–189). International Atomic Energy Agency.

Schlosser, P., Stute, M., Dörr, H., Sonntag, C., & Münnich, K.-O. (1988). Tritium/3He-dating of shallow groundwater. *Earth and Planetary Science Letters, 89*, 353–362.

Schlosser, P., Stute, M., Sonntag, C., & Münnich, K.-O. (1989). Tritiogenic ³He in shallow groundwater. *Earth and Planetary Science Letters, 94*, 245–256.

Seiler, K.-P. (2001). Man's impact on groundwater systems. In W. G. Mook (Ed.), *Environmental isotopes in the hydrological cycle. Principles and applications* (p. 105). Paris: UNESCO International Hydrological Programme (IHP-V).

Seltzer, A. M., Krantz, J. A., Ng, J., Danskin, W. R., Bekaert, D. V., Barry, P. H., et al. (2021a). The triple argon isotope composition of groundwater on ten-thousand-year timescales. *Chemical Geology*, 120458.

Seltzer, A. M., Ng, J., Aeschbach, W., Kipfer, R., Kulongoski, J. T., Severinghaus, J. P., et al. (2021b). Widespread six degrees

Celsius cooling on land during the Last Glacial Maximum. *Nature*, *593*(7858), 228–232.

Seltzer, A. M., Ng, J., Danskin, W. R., Kulongoski, J. T., Gannon, R. S., Stute, M., et al. (2019). Deglacial water-table decline in Southern California recorded by noble gas isotopes. *Nature Communications*, *10*(1), 5739.

Sidle, W. C. (2006). Apparent ^{85}Kr ages of groundwater within the Royal watershed, Maine, USA. *Journal of Environmental Radioactivity*, *91*, 113–127.

Sidle, W. C. (2008). Comparison of ^{85}Kr and ^{3}H apparent ground-water ages for source water vulnerability in the Collyer River catchment, Maine. *JAWRA Journal of the American Water Resources Association*, *44*(1), 14–26.

Solomon, D. K., & Sudicky, E. A. (1991). Tritium and helium 3 isotope ratios for direct estimation of spatial variations in groundwater recharge. *Water Resources Research*, *27*, 2309–2319.

Solomon, D. K., Cook, P. G., & Plummer, L. N. (2006). Models of groundwater ages and residence times. In *Use of chlorofluorocarbons in hydrology: A guidebook* (pp. 73–88). Vienna: International Atomic Energy Agency (IAEA).

Solomon, D. K., Gilmore, T. E., Solder, J. E., Kimball, B., & Genereux, D. P. (2015). Evaluating an unconfined aquifer by analysis of age-dating tracers in stream water. *Water Resources Research*, *51*(11), 8883–8899.

Solomon, D. K., Rank, D., Aggarwal, P., Suckow, A., Stolp, B., Gröning, M., et al. (2007). Tritium/helium-3 dating of baseflow: Southern Vienna Basin. In *International symposium on advances in isotope hydrology and its role in sustainable water resources management*. Vienna, Austria.

Stanley, R. H. R., Baschek, B., Lott, D. E., III, & Jenkins, W. J. (2009). A new automated method for measuring noble gases and their isotopic ratios in water samples. *Geochemistry Geophysics Geosystems*, *10*(5).

Stevanović, Z. (2015). *Karst aquifers: Characterization and engineering*. Springer International Publishing.

Stolp, B. J., Solomon, D. K., Suckow, A., Vitvar, T., Rank, D., Aggarwal, P. K., et al. (2010). Age dating base flow at springs and gaining streams using helium-3 and tritium: Fischa-Dagnitz system, southern Vienna Basin, Austria. *Water Resources Research*, *46*.

Strebel, O., Böttcher, J., & Fritz, P. (1990). Use of isotope fractionation of sulfate-sulfur and sulfate-oxygen to assess bacterial desulfurication in a sandy aquifer. *Journal of Hydrology*, *121*(1), 155–172.

Sturchio, N. C., et al. (2004). One million year old groundwater in the Sahara revealed by krypton-81 and chlorine-36. *Geophysical Research Letters*, *31*(5).

Suckow, A. (2012). Lumpy: An interactive lumped parameter modeling code based on MS Access and MS Excel. Paper presented at EGU 12, European Geosciences Union, Vienna.

Suckow, A. (2014a). The age of groundwater: Definitions, models and why we do not need this term. *Applied Geochemistry*, *50*, 222–230.

Suckow, A. (2014b). *Lumpy: Lumped parameter modelling of age distributions using up to two parallel black boxes. Software manual*. Leibniz Institute for Applied Geophysics (LIAG), S3, Geochronology and Isotope Hydrology, Hannover.

Suckow, A., Deslandes, A., Gerber, C., Mallants, D., & Smith, S. D. (2019). *Commissioning of the CSIRO noble gas facilities at Waite*. Adelaide: CSIRO Land and Water.

Suckow, A., Gerber, C., Kralik, M., Sueltenfuss, J., & Purtschert, R. (2013). *The Fischa-Dagnitz Spring, southern Vienna Basin: A multi tracer time series study reassessing earlier conceptual assumptions*. Paper presented at EGU 2013, European Geosciences Union, Vienna.

Tadros, C. V., Hughes, C. E., Crawford, J., Hollins, S. E., & Chisari, R. (2014). Tritium in Australian precipitation: A 50 year record. *Journal of Hydrology*, *513*(0), 262–273.

Taylor, A. R., Smith, S. D., Lamontagne, S., & Suckow, A. (2018). Characterising alluvial aquifers in a remote ephemeral catchment (Flinders River, Queensland) using a direct push tracer approach. *Journal of Hydrology*, *556* (Supplement C), 600–610.

Torgersen, T., & Clarke, W. B. (1985). Helium accumulation in groundwater, I: An evaluation of sources and the continental flux of crustal ^{4}He in the Great Artesian Basin, Australia. *Geochimica et Cosmochimica Acta*, *49*, 1211–1218.

Torgersen, T., & Clarke, W. B. (1987). Helium accumulation in groundwater, III. Limits on helium transfer across the mantle-crust boundary beneath Australia and the magnitude of mantle degassing. *Earth and Planetary Science Letters*, *84*, 345–355.

Torgersen, T., & Ivey, G. N. (1985). Helium accumulation in groundwater, II: A model for the accumulation of the crustal ^{4}He degassing flux. *Geochimica et Cosmochimica Acta*, *49*, 2445–2452.

Torgersen, T., & Stute, M. (2013). Helium (and other noble gases) as a tool for understanding long timescale groundwater transport. In A. Suckow et al. (Eds.), *Isotope methods for dating Old Groundwater* (pp. 179–216). Vienna: International Atomic Energy Agency.

Unterweger, M. P., & Lucas, L. L. (2000). Calibration of the National Institute of Standards and Technology tritiated-water standards. *Applied Radiation and Isotopes*, *52*, 527–531.

Vengosh, A., & Spivak, A. J. (2000). Boron isotopes in groundwater. In P. Cook & A. Herczeg (Eds.), *Environmental tracers in subsurface hydrology* (pp. 479–485). Boston, MA: Kluwer Academic Press.

Vengosh, A., Kloppmann, W., Marei, A., Livshitz, Y., Gutierrez, A., Banna, M. C., et al. (2005). Sources of salinity and boron in the Gaza strip: Natural contaminant flow in the southern Mediterranean coastal aquifer. *Water Resources Research*, *41*(1).

Vereecken, H., Döring, U., Hardelauf, H., Jaekel, U., Hashagen, U., Neuendorf, O., et al. (2000). Analysis of solute transport in a heterogeneous aquifer: The Krauthausen field experiment. *Journal of Contaminant Hydrology*, *45*(3), 329–358.

Vogel, J. C. (1967). Investigation of groundwater flow with radiocarbon. In *Isotopes in hydrology* (pp. 355–368). Vienna.

Wang, X., Li, H., Yang, J., Zheng, C., Zhang, Y., An, A., et al. (2017). Nutrient inputs through submarine groundwater discharge in an embayment: A radon investigation in Daya Bay, China. *Journal of Hydrology*, *551*, 784–792.

Wilske, C., et al. (2020). A multi-environmental tracer study to determine groundwater residence times and recharge in a structurally complex multi-aquifer system. *Hydrology and Earth System Sciences, 24*(1), 249–267.

Wood, C., Cook, P. G., Harrington, G. A., Meredith, K., & Kipfer, R. (2014). Factors affecting carbon-14 activity of unsaturated zone CO_2 and implications for groundwater dating. *Journal of Hydrology, 519, Part A*(0), 465–475.

Yurtsever, Y., Zuber, A., Maloszewski, P., Campana, M. E., Harrington, G. A., Tezcan, L., et al. (2001). Modelling. In W. G. Mook & Y. Yurtsever (Eds.), *Environmental isotopes in the hydrological cycle: Principles and applications* (p. 127). Paris: UNESCO International Hydrological Programme (IHP-V).

Zimmermann, U., Münnich, K. O., & Roether, W. (1967). *Downward movement of soil moisture traced by means of hydrogen isotopes*. Geophysical Monographs, American Geophysical Union 11.

Zoellmann, K., Aeschbach-Hertig, W., & Beyerle, U. (2002). BOXMODEL: An Excel workbook for the interpretation of transient tracer data (^{3}H, ^{3}He, CFCs, ^{85}Kr) with the box-model approach. In W. Kinzelbach, et al. (Eds.), *A survey of methods for groundwater recharge in arid and semi-arid regions*. UNESCO.

8

Assessment of Water Quality and Quantity of Springs at a Pilot-Scale: Applications in Semiarid Mediterranean Areas in Lebanon

Joanna Doummar[1], Marwan Fahs[2], Michel Aoun[1], Reda Elghawi[1], Jihad Othman[1], Mohamad Alali[1], and Assaad H. Kassem[1]

ABSTRACT

This work presents an integrated methodology for the assessment of threats on spring quality and quantity in poorly investigated Mediterranean semiarid karst catchments in Lebanon. Pilot investigations, including (1) high-resolution monitoring of spring water and climate; (2) artificial tracer experiments; and (3) analysis of micropollutants in surface water, groundwater, and wastewater samples, were conducted to assess flow and transport in three karst catchments of El Qachqouch, El Assal, and Laban springs. First, the high-resolution in situ spring data allow the quantification of available water volumes, as well as their seasonal and yearly variability in addition to shortages and floodwaters. Moreover, the statistical analysis of hydrographs and chemographs helps assess the karst typology, spring type, and hydrodynamic behavior (storage versus fast flow). Furthermore, a series of artificial tracer experiments provides information about key-transport parameters related to the intrinsic vulnerability of the pilot springs, while the analysis of micropollutants gives insight into the specific types of point-source pollution as well as contaminant types and loads. On the one hand, the tracer experiments reveal that any potential contamination occurring in snow-governed areas can be observed at the spring for an extensive time due to its intermittent release by gradual snowmelt, even with enough dilution effect. On the other hand, the assessment of persistent wastewater indicators shows that springs in the lower catchment (including El Qachqouch) are highly vulnerable to a wide range of pollutants from point source (dolines and river) and diffuse percolation. Such contaminant breakthrough is challenging to predict because of the heterogenous duality of infiltration and flow, typical of karst systems. Finally, this set of investigations is essential for the proper characterization of poorly studied systems in developing areas, whereby results can be integrated into conceptual and numerical models to be used by decision makers as support tools in science-evidenced management plans.

8.1. INTRODUCTION

Freshwater, notably groundwater, is presently under tremendous stresses due to climate change and variability in addition to the increase of urbanization, contamination,

[1]Department of Geology, American University of Beirut, Beirut, Lebanon
[2]Institut Terre et Environnement de Strasbourg, University of Strasbourg, Strasbourg, France

water needs, and demands (Hou et al., 2013; Jongman et al., 2014; Kløve et al., 2014; Van Loon et al., 2017; Luo et al., 2020). The Mediterranean region has been identified as one of the most vulnerable areas in terms of increase in forecasted air temperatures and precipitation (Diffenbaugh & Giorgi, 2012; Goderniaux et al., 2015; Nerantzaki & Nikolaidis, 2020), expected to affect drastically water resources in semiarid regions (Iglesias et al., 2007; Doummar & Aoun, 2018a; Dubois et al., 2020; Marin et al., 2021; Sivelle et al., 2021). Karst aquifers

Threats to Springs in a Changing World: Science and Policies for Protection, Geophysical Monograph 275, First Edition.
Edited by Matthew J. Currell and Brian G. Katz.
© 2023 American Geophysical Union. Published 2023 by John Wiley & Sons, Inc.
DOI:10.1002/9781119818625.ch08

predominant in the Mediterranean (Chen et al., 2017) provide about 25% of the water supply worldwide (Ford & Williams, 2007; Stevanović, 2019b). Mediterranean karst catchment areas characterized by limited surface runoff and high infiltration rates reaching 70% are drained by one or multiple springs (Doummar et al., 2012a, 2018a,b; Hartmann et al., 2014a,b, 2015). Due to the duality of flow and heterogeneities in karst systems, flow and transport occur in highly permeable conduits draining a low permeability matrix (Geyer et al., 2008; Mudarra et al., 2010). Generally, the breakthrough of contaminants in karst springs varies according to the dynamic conditions in the aquifer (Doummar et al., 2018b) and the type of pollutant (Hillebrand et al., 2012; Doummar et al., 2014; Doummar & Aoun, 2018b). While transport may occur rapidly because of fast flow velocities (Pronk et al., 2006; Bailly-Comte et al., 2010), the flow rates increasing exponentially during a short period as a response to high rain events will induce a high dilution, thus reducing the concentration of contaminants in the spring (Chang et al., 2021). Given this complexity, the spring responsive behavior, often highly variable to climatic conditions and hydraulic properties of the aquifer, is very difficult to predict in the long run and short run (Sivelle et al., 2021). Chen et al. (2018) show that the total flow rates of a karst spring in an Alpine setting will substantially decrease under the different climatic scenarios mostly because of the shift in snowmelt patterns. Nerantzaki and Nikolaidis (2020) found that multiyear droughts are expected after 2059 for three investigated springs in Greece under all varying climatic scenarios. Moreover, Hartmann et al. (2012) show that a 10% to 30% decrease in flow rate is expected in a large karst spring in the West Bank after 2068. Furthermore, in Lebanon, it is expected under forecasted climatic scenarios (2020–2100; e.g., IPSL_CM5; GCM; RCP, 6.0) to witness a high variability of spring flow rates with more pronounced extremes and periods of droughts (Doummar et al., 2018b) in snow-governed mountainous areas. Additionally, sensitivity studies show the high influence of varying climatic factors (notably precipitation) on spring hydrographs (Dubois et al., 2020; Sivelle et al., 2021) and water availability.

Karst springs are an important component of the groundwater systems and play a significant role in the development of civilizations (Luo, et al., 2020; Stevanović, 2019a). In particular, springs have been investigated as vital resources for social and economic development (Andreo et al., 2006; He et al., 2019) in many areas around the world, such as Jinan Springs (Gao et al., 2020), Gallusquelle Spring in the Swabian Alps (Sauter, 1992; Heinz et al., 2009; Doummar et al., 2012b), and notably around the Mediterranean area (Nerantzaki & Nikolaidis, 2020). In these semiarid environments where water scarcity is rapidly increasing (Hartmann et al., 2014b), springs have been regarded as important resources for large- to small-scale local water supply, such as El Gran Sasso Springs in Italy (Barbieri et al., 2005; Pettita et al., 2020), the Lez Spring in southern France for Montpellier (Fleury et al., 2009;, Marechal et al., 2013; Sivelle et al., 2021), and the Eastern Ronda Springs and Ubrique Spring for Malaga Province in southern Spain (Barberá and Andreo, 2012; Hartmann et al., 2013a; Marin et al., 2021). In these areas, the continuous monitoring of spring quality and quantity is performed to ensure sustainability in the supply and the preservation of the water quality at the source. The increasing urbanization, especially in areas that lack wastewater treatment plants (mostly in rural developing countries), has resulted in a growing level of unpredictable contamination (Gao et al., 2020). Furthermore, a sturdy understanding of the hydrological processes and the factors influencing groundwater dynamics including climatic ones are needed to develop well-informed water management tools and policies (Luo et al., 2020, Stevanović & Stevanović, 2021). Numerical models have been proposed as successful decision support tools for water management in karst (Sivelle et al., 2021). Flow in these systems has been simulated using lumped and distributed approaches depending on the level of surface and subsurface characterization (Worthington, 1999; Doummar et al., 2012b; Hartmann et al., 2013c; Duran & Gill, 2021). However, the suitability of the model highly relies on the amount and quality of the available data and their temporal extent (Hartmann et al., 2017). In some instances, robust sensitivity analysis allows decreasing the uncertainty in the model output (Hartmann et al., 2013b; Chen & Goldscheider, 2014; Mazilli et al., 2017; Dubois et al., 2020). To overcome the challenges of transport assessment in karst, even with a calibrated and validated flow model, the analysis of spring responses provides insights into the dynamics of a karst system. First, the breakthrough of conservative (or reactive) contaminants and/or spring signatures (stable isotopes) has been used in spring high-resolution time series to understand the dynamic response of springs to a variation in input and extent of dilution (Frank et al., 2018;, Hillebrand et al., 2015; Wang et al., 2020; Ahmed et al., 2021). Moreover, tracer experiments have been implemented to assess the connection between karst springs and a contamination point source and estimate transport velocities and dispersivity (Goeppert & Goldscheider, 2008; Marin et al., 2015; Doummar et al., 2018b; Benischke, 2021) in the aquifer and its intrinsic vulnerability to contamination (Epting et al., 2018). Additionally, emerging micropollutants (MPs) such as pharmaceuticals and personal care products were revealed to be suitable transport indicators for persistent and degradable contamination of different origins (Hillebrand et al., 2012; Doummar et al., 2014; Zirlewagen et al., 2016; Stange & Tiehm, 2020). They can be detected to various extents in raw wastewater and treated wastewater, if persistent, or in groundwater and surface water, notably in areas lacking proper wastewater treatment systems (Gasser et al., 2010; Schmidt et al., 2013; Doummar & Aoun, 2018b; Clemens et al., 2020).

Lebanon counts more than 409 springs with discharges ranging between 0.001 m³/s to more than 10 m³/s on average (ElGhawi et al., 2021). Some are used for local supply, while others may serve as an alternative decentralized water source for selected villages to overcome the forecasted water scarcity. The high urbanization and lack of an effective wastewater system in Lebanon (Massoud et al., 2010), pose a significant contamination risk on springs located below 1,600 m above sea level even in rural settings. Therefore, there is a need for thorough investigations on selected potential springs, to understand their vulnerability against contamination events or climatic parameters (Epting et al., 2018). The latter can be achieved only with a robust monitoring network, the collection of high-resolution data, the analysis of representative temporal and spatial water quality samples, and the assessment of contamination indicators (Torresan et al., 2020). The objective of this work is to highlight some of the important methods used for the conceptualization of flow and transport in pilot karst springs and the identification of their inherent resilience to contamination hazards and potential future threats in a rural groundwater catchment in Lebanon. The proposed methodology can be applied to other case studies in poorly investigated spring catchment areas. The results are further discussed in terms of policy enforcement and drafting of guidelines and laws to ensure sustainable protection of spring water resources (Fleury, 2009).

8.2. FIELD SITE

The investigated springs are located in middle east Lebanon, north of the capital Beirut in a Mediterranean semiarid snow-governed climatic region (Fig. 8.1; Table 8.1). They belong to the Nahr El Kalb River rural catchment (Fig. 8.1). The catchment extends from sea level to an elevation of 2,600 m above sea level and is characterized by karstified Jurassic and Cretaceous rock sequences disturbed by complex structural deformations (Bakalowicz, 2005). The catchment is drained by important springs heavily relied upon for water supply; for instance, the Jeita Spring with mean flowrates of 8 m³/s is used as a water supply source for the capital Beirut (1.5 million inhabitants; Doummar et al., 2014; Koeniger et al., 2017). Based on two recording stations, the total precipitation recorded on the catchment area varies between 800 mm (closer to the coast) and 1,800 mm as snow in the mountains (Fayad et al., 2017; Koeniger et al., 2017; Doummar et al., 2018a; Dubois et al., 2020). The Qachqouch Spring located at 64 m above sea level, emerges from Jurassic age rocks composed of fractured limestone and basalts. During low flow periods, the spring is used to complement the water deficit in the capital city Beirut and surrounding areas. Its total yearly discharge reaches 35–55 mm³ based on high-resolution monitoring of the spring (2014–ongoing; Dubois et al., 2020). Flow maxima reach a value of 10 m³/s for short periods following flood events; it is about 2 m³/s during high flow periods and 0.2 m³/s during recession periods. On the other hand, Laban and Assal springs are located in the highlands of Nahr El Kalb catchment, at respective altitudes of 1,552 and 1,600 m above sea level. The main source of recharge is the snowmelt over a total groundwater basin of about 25 km². They partially drain the Albian-Cenomanian rock formations composed of limestone and dolostones. The upper catchment area is a plateau characterized by a high-density doline distribution (exceeding 19 dolines/km²), which enables relatively fast infiltration of snowmelt and rain. The dolines mapped during various campaigns (2012, by the BGR: BundesAnstalt Fuer Geowissenschaften and Rohstoffe) by AUB 2015 and AUB 2020–21 are buried ones with nondiscernable 20–50 cm swallets buried by rock debris. Fast infiltration of rain or snowmelt can occur in the buried holes within the doline while diffuse infiltration happens within the soil depending on the rock facies in the Cenomanian rock sequence. The thickness of the soil in the dolines may exceed 5 m as portrayed by representative auger excavations. The Assal Spring has an annual volume of 22–30 mm³ (Doummar et al., 2018a) and is used locally for water supply of 24,000 m³/day, while water from the Laban Spring, with a total annual volume of 20–25 mm³, is conveyed to the Chabrouh Dam in Faraya (Fig. 8.1). The overflow from both springs feeds the two tributaries of the Dog River (Nahr El Kalb) in the highlands (Doummar and Aoun 2018b/a), while that of Qachqouch Spring is discharged into the river closer to the coast. The highest observed volumes in the three springs recorded during the high-flow periods (December to April) exceed the water demand and supply, while available water volumes drop substantially during low-flow periods extending from May to November where water supply is mostly needed. Therefore, the three karst springs are currently not exploited to their full potential, due to the presence of alternative resources and the natural deficit in water availability during low-flow owing to their karstic nature. The Qachqouch Spring is a karst spring characterized by a duality of flow in a low permeability matrix and high permeability phreatic conduit system (Dubois et al., 2020). As such, it is highly reactive

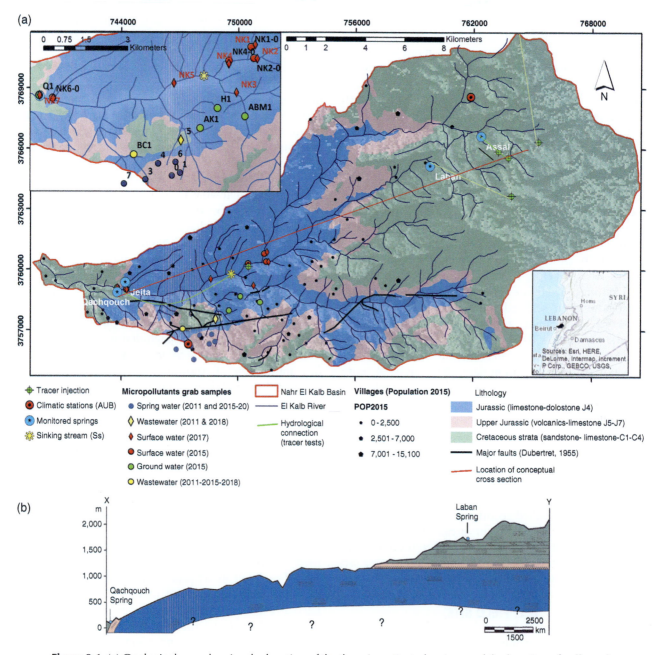

Figure 8.1 (a) Geological map showing the location of the three investigated springs and the location of collected samples on the Nahr El Kalb catchment. (b) A conceptual cross section shows the different lithologies and thicknesses of the main aquifers.

to rain events with a recession coefficient ranging between 0.005 and 0.1 depending on the event responses (Dubois et al., 2020). Assal Spring is less responsive to snowmelt and is characterized by a storage capacity that allows it to sustain a flow rate of 240 l/s during the dry season (August to October). The Laban Spring is highly reactive to snowmelt due to its higher level of karstification and has a limited storage capacity, thus it runs almost dry during the summer period (August to October). Wastewater treatment plants are absent in most areas of Lebanon, especially in rural areas because of the difficulty of continuous operation and maintenance (Karnib, 2014). Additionally, wastewater effluents and solid waste may be disposed of directly on the river flanks, or in bottomless septic tanks in areas that are not connected to the public wastewater network (Massoud et al., 2010). Furthermore, outdated generic groundwater protection guidelines or laws (for wells, springs, and river) are not reinforced or rigorous enough to ensure the protection of spring catchment areas. As a consequence,

Table 8.1 Geological and hydrogeological characteristics of the three investigated springs

Spring	XYZ (m above sea level) (WGS, 1984)	Formation	Major threats	Major opportunities
Qachqouch	33.943985°N 35.637690°E 64 m	Jurassic limestone; aquifer highly karstified and responsive to rain events (the flow rates increase shortly following precipitation events; low to moderate storage	Water contamination from wastewater effluents surface-water interaction (Doummar & Aoun, 2018 a, b)	Floodwaters to be used for managed aquifer recharge or to compensate for the water shortage
Assal	34.009710° N 35.838760°E 1,552 m	Cenomanian dolostones and limestone, medium karstified high storage capacity, reactive to snowmelt events	Increasing waste water contamination, climate change, snow cover decrease, longer recession period because of earlier snowmelt and increasing temperatures (Doummar et al., 2018 b)	Limited land use/ landcover leads to limited contamination. Snowmelt ensures a longer-lasting recession. Springs can be used to compensate low-flow shortages at lower altitudes
Laban	33.994961°N 35.828203°E 1,662 m	Cenomanian limestone, highly karstified, low storage capacity, reactive to snowmelt		

land use and cover expand in highly vulnerable areas of spring recharge without proper mitigation measures (Korfali & Jurdi, 2009). The major threats faced by the investigated springs are related to intrinsic or anthropogenic factors influencing both the pristine spring water quality and quantity yielding an increase in the karst disturbance factor (van Beynen & Townsend, 2005; North et al., 2009). Such threats can decrease the available water of good quality for supply over the short- and long-term scale (especially in countries that have adhered to the United Nations Sustainable Developmental Goals (SDG VI). On the other hand, the opportunities are infrastructural, technological, or policy-making changes that can enhance the sustainable supply of water quantity over time, mitigate the contamination risk, and achieve a potential recovery of the pristine status of the spring.

8.3. INVESTIGATION METHODS

Extensive hydrogeological studies are performed on spring water, including (1) high-resolution monitoring to analyze flow dynamics and hydrochemical variations (Gao et al., 2020), (2) assessment of intrinsic transport from tracer experiments, and (3) evaluation of specific transport of selected micropollutants. These methods aim at collecting sufficient data about water quality and quantity, and at constructing conceptual models that feed into process-based numerical models to be calibrated based on continuous data acquisition and system characterizations. Such models if validated can be used for prediction purposes (Hartmann et al., 2014a, 2020), ultimately to anticipate the forthcoming threats to existing springs.

8.3.1. High-Resolution Data Collection

A rigorous monitoring network in groundwater basins is lacking in Lebanon due to the high cost of maintenance and operation. Therefore, since 2014, a monitoring network was set up to collect high-resolution data on a pilot catchment area (in the framework of international research projects). This constant monitoring of spring flow and quality allows quantifying the water volumes and the flow rates, and their variation during the hydrological regimes (Mudarra et al., 2012) and consequently ensure the evidence-supported protection of the investigated springs in terms of quality and quantity. The collected monitoring data in the three springs and catchment consist of automatic data, grab samples, and automatic sampling entailing climate and flow data, as well as physicochemical data. The rate and frequency of sampling along with the type of data are directly linked to the information required for spring characterization (Table 8.2).

Information comes from the following sources.

1. Climatic stations (HOBO and alpine Campbell brands) for the measurement of precipitation (including snow), humidity, wind direction, and magnitude, temperature, and radiation at two different altitudes (950 and 1,700 m above sea level). The data are used for the assessment of potential evapotranspiration and input precipitation (frequency of 15–60 min) and snowmelt based on temperature variation.

2. Flow monitoring in the river and springs is done using pressure transducers for water level measurement. Discharge is estimated using rating curves based on the monthly measurements of discharge. Where and when accessibility to both spring and river is constrained,

Table 8.2 Information provided by the measured parameters/experiments on springs used for spring characterization and construction of conceptual models

Spring parameters and frequency of measurements	Indicator for spring response used as input for conceptual and numerical models	Selected literature
Flow rates/ Water level 2014–ongoing 30-min interval 2014–ongoing	Time series correlative analysis and statistical correlation between input and output Evaluation of aquifer type, storage, and recession Insights into aquifer geometry Analysis of snowmelt on volume and spring dependence of climate variability and change Applications: Water availability and alternatives	Cinkus et al., 2021 Olarinoye et al., 2020 Dubois et al., 2020
Bacteriological analysis, turbidity and particle distribution Periodically 2019–ongoing Occasionally or event based prior to 2019	Indicative of potential bacteriological contamination Application in early warning systems and contamination indicator	Pronk et al., 2006 Stedmon et al., 2011 Frank et al., 2018
Electrical conductivity Temperature 2014–ongoing	Indicative of fast infiltrated water from point source origin and snow component Degree of karstification	Cinkus et al., 2021 Wang et al., 2020 Torresan et al., 2020 Ahmed et al., 2021
Chemical analysis Periodically 2019–ongoing Occasionally or event based, prior to 2019	Insights into water-rock interaction, identification and estimation of anthropogenic contamination indicator	Gasser et al., 2010 Schmidt et al., 2013
Micropollutants (pharmaceuticals and personal care products; PPCPs) Occasionally or event based	Origin of contaminants Transport and persistence of compounds in the matrix Type of infiltration (diffuse versus point source) Specific vulnerability	Einsiedl et al., 2010 Doummar et al., 2014 Doummar & Aoun 2018a, 2018b Warner et al., 2019
Artificial tracer experiments Occasionally or event based. Constant monitoring with field fluorometers	Intrinsic vulnerability Spring protection Contaminant outbreak management	Geyer et al., 2007 Goeppert & Goldscheider, 2008 Doummar et al., 2018b Beniscke, 2021
Stable isotopes (oxygen and hydrogen) Since 2019 Occasionally or event based prior to 2019	Recharge assessment (quantification of fast infiltration, elevation, etc.)	Perrin et al., 2003 Barbieri et al., 2005 Koeniger et al., 2017 Rusjan et al., 2019

uncertainties in high-flow periods may lead to an overestimation of the annual budget, and consequently the quantities of water available for supply in addition to the calculation of contaminant/tracer masses. The error in flood flow rates can be reduced or quantified based on the analysis of a longer time-series data set due to the high variability and seasonality of flow in addition to a statistical analysis of numerical modeling output. However, these errors are not detrimental to the conceptualization of flow and early numerical modeling calibration.

3. In situ physicochemical parameters are measured with a periodically calibrated (when needed) multiparameter probe (Aquatroll 600-In situ) for electrical conductivity (EC), temperature (T), turbidity (TU), and pH, as well as dissolved oxygen (DO) installed on the spring.

4. Two automatic samplers are scheduled every 3 days to collect samples for the analysis of indicator parameters in spring water (El Qachqouch) for the following analysis:

a. Stable isotopes (oxygen and deuterium) collected in glass bottles analyzed using a PICARRO isotopic analyzer L2130-i cavity ring-down spectrometer (CRDS) with a VAP A0211 vaporizer and automatic sampler. The standards for oxygen and hydrogen used for calibration and routine checking of the measurements are well preserved standards of known isotopic composition (values versus VSMOW for $\delta^{18}O$: 0.3, -20.6, -29.6 ‰, and for $\delta^{2}H$: 1.8, -159, and -235.0 ‰).

b. Major ions for water types and pollution assessment collected occasionally to weekly analyzed using ion chromatography (IC) with the appropriate concentration standards for the respective measured ions.

c. Particle size distribution for the assessment of suspended particles leading to turbidity using a Coulter counter (Multisizer 4e, Brand Beckman) with different apertures used according to the grain size measurements (ranging from 2μm to 2,000 μm)

5. Field fluorometers (GGUN-FL30-Albilia) are installed at each spring for the measurements of natural fluorescence in pristine waters and tracer experiments conducted on injection points on the catchment area. The fluorometers were calibrated in the field following each experiment with the injected tracer and the spring water.

6. Bacteriological analysis (fecal and total coliform, *Pseudomonas aeruginosa,* and enterococci analyzed occasionally) at a local certified bacteriological laboratory to detect the variation of fecal indicators in the spring and associate them with continuous point-source contamination on the catchment.

8.3.2. Tracer Experiments and Identification of Transport Parameters

Tracer Experiments

Tracer experiments were undertaken on the catchment area, whereby artificial dyes (uranine, sodium fluorescein, and occasionally amidorhodamine-G) considered conservative and nontoxic to humans and the environment (Kass, 1998) were injected into the river (Qachqouch catchment) and specific dolines on the catchment of the Laban and Assal springs. Tracer concentrations were simultaneously monitored and recorded every 15 min in the spring using a field fluorometer (GGUN-FL30; Schnegg, 2002) that was calibrated for the applied tracers. Discharge measurements were estimated based on the recorded water level and rating curve under different flow conditions (high flow, low flow, snowmelt, and medium flow). Restitution of the tracer allows identifying a connection between the injection point, considered as a vulnerable contribution point of the recharge area, and the spring (Benischke, 2021). Furthermore, the transport parameters such as mean velocity and dispersivity can be estimated for conservative pollutants based on the analysis of the breakthrough curve using a 1-D transport model (Toride et al., 1999; Geyer et al., 2007; Goeppert et al., 2020; Sivelle et al., 2020). Tracer experiments reflect the intrinsic vulnerability of a spring system as well as the transport mechanisms of a conservative pollutant from different origins (fast preferential flow; doline, or a sinking allochthonous stream; river). While the duration of tracer recovery provides insights into the duration of the breakthrough of a nonreactive pollutant, the observed tracer concentration is indicative of the intensity of the contamination above admissible limits for a certain contaminant load (Doummar et al., 2018a).

Micropollutant Analysis

Micropollutants (MPs) were analyzed in 75 grab samples collected from surface water, wastewater, wells, and springs in 2015–2019 to characterize the point source contaminants existing on the catchment area. Additional limited samples collected and analyzed in 2011 (Doummar et al., 2012a) serve for comparison purposes of the variation of MPs over the last decade. The selected micropollutants span from pharmaceuticals to personal care products present in domestic, industrial, and hospital wastewater effluents such as nonsteroidal anti-inflammatory drug (ibuprofen), lipid regulators (gemfibrozil), artificial sweeteners (sucralose and acesulfame-K), epileptic drugs (carbamazepine), bronchodilators (albuterol), hospital contrast media (iohexol), detergents (nonylphenol), dyes manufacturing (quinoline), and others (metformin and caffeine). Furthermore, these pharmaceuticals were also monitored in one of the springs to detect the variations in concentrations and loads in spring water (Zuccato et al., 2005; Hillebrand et al., 2012). Moreover, the breakthrough of MP loads and concentrations in the spring was related to the origin of effluents (surface water or point source infiltration) according to the flow dynamics based on binary mixing models (Buerge et al, 2009; Gasser et al., 2010; Doummar et al., 2014; Mawhinney et al., 2011; Oppenheimer et al., 2011; Wolf et al, 2012; van Stempvoort et al., 2013, Liu et al., 2014; Nödler et al., 2016). The MPs that are revealed to be persistent and least degradable in the system independent of recharge are the ones that can be used as viable wastewater indicators (Oppenheimer et al., 2011) and can be of threat to the spring water quality if they exceed the maximum admissible limits (MAL). Additionally, the distribution and evolution of MPs concentrations can provide insights into the type of wastewater and water usage on the rural catchment (Warner et al., 2019).

A correlation between easily monitored parameters at the spring (i.e., turbidity and electrical conductivity) with the breakthrough of these emerging micropollutants (i.e., ibuprofen, sucralose, acesulfame-K, carbamazepine, gemfibrozil) was used to assess the contamination arrival at the spring with easily measured indicator parameters (Doummar et al., 2018a). Binary and ternary mixing models (based on mass-fluxes of chloride, micropollutants, and stable isotopes) are used to identify the percentage of wastewater inflow into pristine groundwater (from wastewater effluents discarded on the catchment and in the river; Gasser et al., 2010; Schmidt et al., 2013; Doummar & Aoun 2018b). The analysis of the breakthrough curve of micropollutants with other measured parameters (electrical conductivity, chloride, calcium, etc.) allows identifying mass fluxes of these persistent pharmaceuticals (Doummar et al., 2014; Doummar & Aoun, 2018a).

8.4. DISCUSSION AND RESULTS

8.4.1. High-Resolution Data As Insight to Systems Hydrodynamics

The high-resolution data (until 2018) from the three existing springs were partially included in the global data set on karst springs (Olarinoye et al., 2020). The three springs show an uneven distribution of flow rates during the year indicating a shortage in the summertime following the hydrograph recession. The total water volumes throughout the 6 years show a high variability from dry to intermediate to wet years, typical of semiarid regions, which adds to the lack of predictability in water availability for supply (Fig. 8.2, Table 8.3). The Qachqouch Spring is characterized by flood waters that can reach up to 50 mm^3 during high-flow periods in wet years (the discharge rates exceeding its average discharge during low-flow periods as recorded in July–October).

Detailed correlative analysis of time series reveals information about the geometry and the parameterization of a subsurface karst system (Mangin, 1975; Dubois et al., 2020). For instance, the number of groundwater reservoirs and the spring recession coefficients are inferred from statistical analysis of time series (discharge, electrical conductivity, precipitation). Furthermore, the analysis of the flow time series can yield a classification of the spring typology (Mangin, 1975; El Hakim & Bakalowicz, 2007; Stevanović, 2015), which unravels the type of flow, indicative of the behavioral response of the spring, its reactivity versus storage capacity, and its porosity type (equivalent porous, versus dual and triple porosity). Additionally, the springs response (discharge and other monitored parameters) to rain or snowmelt events provides valuable information about the total volume of fast-point source infiltration (such as dolines). In the Qachqouch Spring, the correlation of electrical conductivity, stable isotopes, and chloride along with flow show that the newly recharged water ranges between 10% and 70% of the total volume per event, the latter being highly dependent on the saturation of the system. On the other hand, the Assal Spring is classified as type 3, characterized by a slow infiltration and a higher storage. The discharge curve displays a fluctuation indicative of diurnal snowmelt and nocturnal freezing. The volume of freshly infiltrated snowmelt directly related to the number of dolines on the catchment is also estimated daily based on the high-resolution data series. The relevance of newly infiltrated water or snow has an important implication if it is closely related to a point source contamination (Jodar et al., 2020). On the other hand, the Laban Spring located 2 km away from El Assal Spring within the same aquifer, displays a different hydrodynamic behavior, as shown from a k value closer to 1, and an I value closer to 0. The latter can be attributed to the variation of facies within the Cenomanian aquifer (lower dolomitic member versus the upper more karstified limestone member).

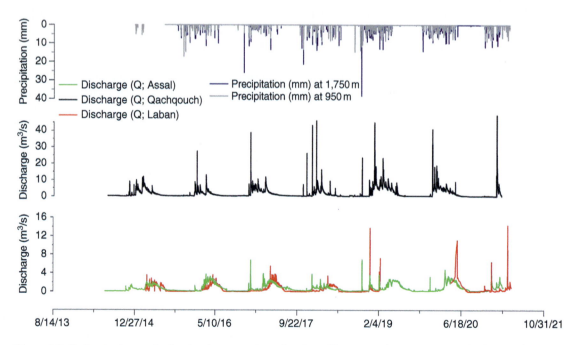

Figure 8.2 Spring hydrographs for the three monitored springs illustrating the variation in discharge along with total amount of precipitation (snow and rain).

Table 8.3 Spring volumes illustrating the variation from wet (2018–2020), intermediate (2016–2017), and dry years (2014–2017 and 2017–2018)

	2014–15	2015–16	2016–17	2017–18	2018–19	2019–20	Spring type (k, i)[a]
Precipitation (950 m)	NA	921.6	1,034.0	1,090.0	1,764.0	1,319.0	
Precipitation (1,700 m)	NA	1,110.0	1.005.0	1,084.0	1,838.0	1,405.0	
V Qachqouch (mm^3)	49.80	35.8	61.5	43.40	100.1	69.2	Type 1 (0.11, 0.77)[b]
V Laban (mm^3)	NA	16.1	18.9	12.50	NA	20.0	Type 2 (0.09, 0.24)[c]
V Assal (mm^3)	NA	24.2	24.6	16.90	30.9	27.1	Type 3 (0.4, 0.44)[c]

Note: Spring type and coefficient estimated from spring hydrographs.
[a] After Mangin (1975); k: characterizes the extent of the phreatic zone and its regulating capacity; its storage and discharge of fast infiltrated water (k>0.5 is characteristic of porous aquifer); i close to 0 implies a fast infiltration, compared with a value closer to 1.
[b] From Dubois et al. (2020).
[c] Calculated based on time series following the method by Mangin (1975) in Dubois et al. (2020).

8.4.2. Assessment of Spring Intrinsic Vulnerability

Qachqouch Spring

The tracer breakthrough curve (BTC) recorded at the Qachqouch Spring reveals a connection between the river and the spring (Fig. 8.3, Table 8.4). the breakthrough curve was characterized by one main peak and other minor peaks or discontinuous recovery of the tracer for a longer period. The first peak in May and November dyes appear as a composite peak, which may be due to a considerable longitudinal dispersion in the river before infiltration occurs during snowmelt and low-flow periods. The injected uranine was first detected at the spring between 5 to 20 h after injection, which corresponds to maximum transport velocities of 0.07 m/s, 0.17 m/s, 0.25 m/s, and 0.77 m/s in, May, November, June, and February, respectively (Fig. 8.3). The mean velocities and dispersivities are estimated for the first main peak of the BTC based on 1-D analytical solutions for transport Two-Region Nonequilibrium model (2NREM) depending on the extent of tailing and shape of the BTC (Fig. 8.3). The dispersivities ranging between 39 m and 780 m are indicative of transport in karst (Doummar et al., 2018b) and imply the duration of recovery of a tracer and a potential conservative contaminant from the river.

The tracer experiment indicates a relationship between the heavily polluted river and the spring under different flow conditions. During high-flow conditions, as the base level in the river is higher, infiltration through fast flow pathways along the river (sinking streams and karst flooding) is more prominent (Gutierrez et al., 2014), as indicated by the higher mass recovery during February (8% compared with 1% during the lowest flow). Despite a highest infiltration of river water in the spring during February, the dilution attenuates the maximum concentration of the BTC. On the other hand, in November, the observed concentrations of conservative contaminants are expected to be highest despite the limited mass loads from the river. The first peak in the BTC recovered in the Qachqouch lasts between 2.5 and 6.5 days, while the tracer continues to appear in the spring for more than 40 days. The duration and number of peaks depend on the tracer injection and mass, as well as on the flow regimes in both the river and the spring. On the one hand, the relatively long duration of breakthrough indicates the lingering effect of a contaminant transport between the river and the spring (Fig. 8.3). The multipeak BTC shows that the river and the spring are linked by continuous infiltration or through multiple conduits or multiple sinking streams. On the other hand, phreatic diameters calculated based on the total volume of water during the mean transit time of the first BTC peak provide information about the subsurface conduit dimensions, and revealed to range between 1.4 m and 4.8 m under varying flow periods (Table 8.4). Moreover, the subsurface complexity and heterogeneity can be further assessed with a more advanced modeling of the multiple-peak BTC using a convolution of multiple-step input breakthrough curves (Siirila-Woodburn et al., 2015).

Upper Catchment Springs

Springs located above 1,500 m, namely Laban and Assal springs, are mostly governed by snowmelt and show a high vulnerability both in terms of quantity and quality. The mean transport velocities in Assal range between 0.001 and 0.003 m/s during low-flow periods (Fig. 8.4). The BTCs recorded at the Assal and Laban springs during a snowmelt event (May 2015 and July 2020, respectively) were modeled using a simple advection dispersion model (ADM), since the tailing effect is due to a superposition of various melting signals. A more advanced model is being implemented to illustrate the convolved tracer arrival with subsequent melting events. Since the tailing was not prominent in the BTCs recorded in low-flow periods (June 2014 and 2016), they are also

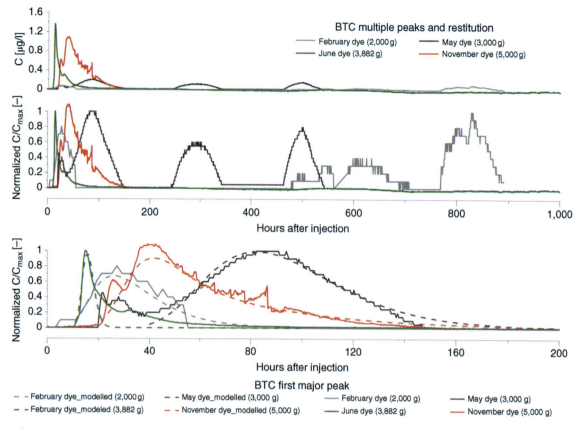

Figure 8.3 Tracer breakthrough curves (BTC) recorded at Qachqouch Springs from four injections in the Nahr El Kalb sinking stream (2016–2020). First peaks in the BTCs are modeled using 2NRE model to account for the tailing effect.

modeled using the ADM for comparison purposes among tracer results. Tracer experiments undertaken under snowmelt conditions in both spring catchments reveal velocities of 0.01–0.013 m/s, indicating a fast response during snowmelt. The tracer experiment breakthrough curve is restituted for a duration exceeding 500 h (or 21 days) because of the reactivation of the tracer in the saturated zone during daily snowmelt events, therefore any conservative contamination occurring at the catchment areas in the investigated dolines will result in a relatively long breakthrough at the spring, which will require subsequent treatment if the catchment area is not well protected. Moreover, the melting of the snow highly affects the discharge at the spring, therefore these springs can be considered highly vulnerable to climate variability, as they are highly responsive to daily snowmelt events. Longitudinal dispersivities for both Laban and Assal systems, calculated based on longitudinal dispersion and mean velocities, are 6.5 m and 13.3 m, respectively (Table 8.4). Phreatic diameters between injection points and springs range between 0.55 m and 3.91 m for Assal Spring and 1.71 m for Laban Spring, depending on the flow period.

8.4.3. Assessment of Spring Specific Vulnerability

The concentrations of pharmaceuticals in the springs were below toxic limits in the range of nanograms per liters (Chiffre et al., 2016; Doummar et al., 2018a,b). Even if not considered of a great threat to human consumption, the identified micropollutants can serve as an indicator for the different types of pollution on the catchment area and of the vulnerability of the springs, depending on their source of recharge and their hydrodynamic characteristics (Warner et al., 2019). On the one hand, three types of pollution sources were identified on the catchment of each spring, the predominant one being related to domestic wastewater effluents (caffeine, diclofenac, gemfibrozil, artificial sweeteners, carbamazepine, cotinine). Minor ones related to industrial (quinoline), hospital (iohexol), or agricultural practices were also detected, where lipid regulators are used in poultry farms (Doummar & Aoun, 2018b) on the Qachqouch catchment. On the other hand, the raw wastewater collected on the catchment shows a similar composition, to the exception of quinoline used in rural industrial zones (BIK sample). Table 8.5 displays the range of concentrations for different MPs.

Table 8.4 Characteristics of the tracer experiments, results of the graphical interpretation and estimation of transport parameters based on BTC analysis and inverse modeling

Name		May dye			Feb dye			Nov dye	June dye	Asl-1	Asl-2	Asl-3	Lab-1
Injection points		\multicolumn{8}{l}{Sinking stream (Nahr El Kalb)}				Doline	
Observation point		\multicolumn{8}{l}{Qachqouch spring}				Assal		Laban
Date		05/08/16			02/19/17			11/24/17	06/04/20	07/01/14	06/01/16	05/01/15	07/01/20
Type of BTC		Multipeak (3 peaks)			Multipeak (3 peaks)			One major peak	One major peak	One major peak	One major peak	One major peak	One major peak
Tracer type		SF			AR			SF	SF	SF	SF	AR	SF
M	g	3,000			2,000			5,000	3,883	400	2,000	1,500	2,408
x	m	8,920			8,700			8,920	8,920	1,330	2,330	1,330	3,140
M_R	g	71			150			58	213	112	360	135	241
M_R	%	2%			8%			1%	5%	28%	18%	9%	10%
Breakthrough peaks (P)		P1	P2	P3	P1	P2	P3	P1	P1	P1	P1	P1	P1
Qmean	m³/s	0.78	0.65	0.62	4.86	3.16	3.98	0.37	0.72	0.25	0.31	1.05	0.27
Cp	µg/l	0.2	0.12	0.16	0.09	0.03	0.04	1.07	1.36	1.5	2	0.9	5
Tf	h	35	242	461	3	480	768	15	10	288.0	175.2	21.0	65.4
tcp	h	87	293	500	22	604	829	42	15	336.0	197.0	29.0	87.0
vf	m/h	255	37	19	2788	18	11	595	893	4.62	13.30	63.30	48.00
vf	m/s	0.07	0.01	0.01	0.77	0.01	0.00	0.17	0.25	0.0013	0.0037	0.0176	0.0133
Duration BTC	hours	112	233	81	55	226	121	156	110	182	71	>220	535
	days	4.68	9.73	3.36	2.29	9.42	5.04	6.50	4.58	8	2.95	>10	22.3
Model		2NRE		NA	2RNE		NA	2RNE	2RNE	AD	AD	AD	AD
v_m	m/h	66			241			140	421	3.8	11.4	45.7	34.4
v_m	m/s	0.018			0.067			0.039	0.117	0.001	0.003	0.013	0.010
D_m	m²/h	27,000			188,000			50,000	16,600	24	79	300	458
D_m	m²/s	8			52			14	5	0.007	0.022	0.083	0.127
tm	h	135			36			64	21	1.40	7.38	16.88	30.00
α	m	408			780			357	39	6.32	6.89	6.56	13.31
φ	m	3.66			4.80			1.74	1.40	0.55	1.10	3.91	1.71

Note: M = Total injected mass; x = Distance of transport; Qmean = Mean flow during test; MR = Recovered mass; φ = Phreatic diameter; SF = Sodium Fluorescein; AR = Amidorhodamine-G; Cp = Peak concentration; Tf = First time of arri val; NA = Not modeled with conventional methods; Tcp = Time of peak concentration; vf = Maximum velocity; 2NRE = Two region nonequilibrium model; AD = Advection dispersion model; vm = Mean velocity; D = Longitudinal dispersion; tm = mean transit time; α = Longitudinal dispersivity.

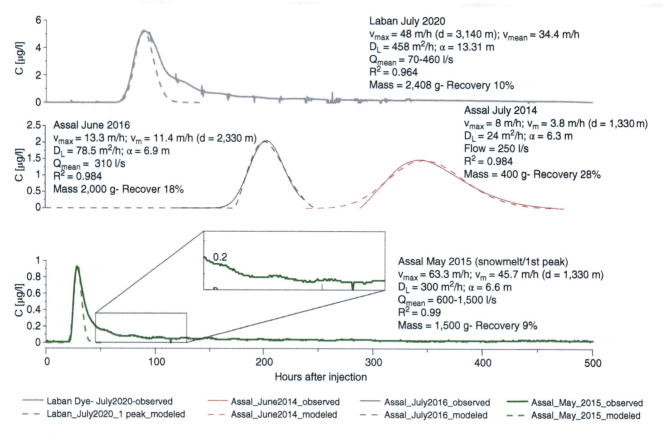

Figure 8.4 Tracer breakthrough curves (BTC) recorded at Assal and Laban springs from four injections in dolines on their respective catchment areas (2014–2020).

The MPs found in the Qachqouch Spring water samples are infiltrating via various pathways: (1) surface water infiltration, (2) fast infiltration point source such as doline, as well as (3) diffuse infiltration (Doummar & Aoun, 2018a). The most persistent are the ones that are found in the spring during periods where the river is dry and no recharge occurs, such as carbamazepine and gemfibrozil. Therefore, these two micropollutants can be used as wastewater indicators (Doummar et al., 2014). Mass loads of MPs estimated from flow rates and concentrations allow the backtrack calculations of used drugs and number of users and loads of wastewater on the catchment (Zuccato et al., 2005, for cocaine). For instance, daily mass loads of carbamazepine (CBZ) used on a smaller subcatchment (small-scale springs including three villages) vary between 0.26 and 81 mg/l, while this load increases to 63–5,075 mg/l per day in the Qachqouch Spring because of the inflow of additional point source contamination infiltrating to the spring from the entire catchment.

In the upper springs, only caffeine was detected, implying a rather limited amount of contamination due to the lack of urbanization in the highlands of the area (two mountain huts, restricted settlements mostly above Laban Spring, and the ski resort). Additionally, the fecal coliform and enterococci are also limited (Assal: 012 CFU/100 ml and 0–130 CFU/100 ml), in samples collected during baseflow (July 2020–February 2021) depending on the sampling time, and show a lower level of fecal contamination in the upper catchment springs. However, it is expected that breakthrough of contaminants occurs from wastewater stored in bottomless pits from mountain huts. The latter require further high-resolution monitoring and tracer experiments in specific locations to test the contribution of such point-sources of pollution.

8.4.4. From Conceptual to Numerical Models

Conceptual and process-based numerical models are developed based on the synthesis of field experiments and statistical correlative analysis of the time series collected on the spring catchment areas (Mudarra et al., 2019; Dubois et al., 2020). Additionally, insights into the intrinsic and specific vulnerability allow for understanding the recharge mechanisms and point-source potential pollution. First, a semidistributed linear reservoir numerical model has been parameterized and constructed for the complex catchment of Qachqouch Spring and further calibrated based on the continuous analysis of field and time-series data (Dubois et al., 2020). A lumped model was selected for the

Table 8.5 Concentrations of micropollutants and bacteriological analysis in surface water, spring, well, and wastewater samples collected on the Nahr El Kalb catchment

Type of sample	Surface water			Groundwater (springs and wells)											Upper catchment springs					Wastewater		
Name	Nahr El Kalb			Qachqouch								Other springs		Wells	Laban			Assal		BC1		BlK
Approximate discharge range (l/s)	NA	NA	250–1,810	1,097–4,800	2,000	NA	NA	700	300–7,500	290			0.14–1.06	NA	NA			NA		.1–2			
Bacteriological analysis																							
Date (bacteriological)	November 2019–Feb 2021			November 2019–Feb 2021								Apr-17		NA	July 2020–February 2021					NA		
Number of samples	12			13								7		NA	5			6				
Fecal coliform (CFU/100 ml)	4,303–100,000			102–3,783								3–80		NA	2–500			0–40				
Total Coliform (CFU/100 ml)	1,686–9,799			0–968								NA			0–130			0–12				
Enterococci (CFU/100 ml)	450–2,500			6–255											0–33			0–3				
Micropollutant analysis (ng/l)																							
Date (Mps)	3/1	9/1	5/1	4/1	6/1 3/1	9/1	5/1	1/1	6/20	4/1		5/1			3/1	9/1		3/1	9/1	3/1	5/1	4/1
Number of samples (81)	5	1	4	7	3	1	1	20	19	7		2			2	1		2	2	1	1	1
Gemfibrozil (lipid regulators)	21	–	17	47	NA	–	–	5.2–38	NA	NA		0			–	–		–	–	1,424	3,500	780
Iohexal (contrast media)	–	–	10	33	NA	–	12	19–47	NA	NA		0–13			–	–		–	–	–	290	0
Acesulfame-K (artificial sweetener)	NA	NA	85	80	NA	NA	170	s	NA	NA		180–640			–	–		–	–	–	210,000	10,000
Ibuprofen (nonsteroidal anti-inflammatory drug)	24	–	13	64	NA	–	–	–	NA	NA		0			–	–		–	–	>7,500	2,100	3,000
Diclofenac (analgesic)	–	–	7	5	NA	–	–	–	NA	–		0			–	–		–	–	45	–	310
Caffeine (stimulants)	297	190	304	330	22–120	17	9.9	–	15–54	14–380		–			6	–		0	246	>17,000	–	81,000
Cotinine (nicotine metabolite)	16	–	23	47	20–23	13	–	NA	16–38	11–890		–			–	–		–	–	1,255	7,100	1,900
Carbamazepine (anticonvulsants)	5	25	NA	8	8.1–15	43	16.0	NA	–	–		–			–	–		–	–	227	370	710
Metformin (blood sugar control)	NA	NA	NA	NA	NA	NA	NA	NA	–	6–11		NA			–	–		–	–	–	NA	360
Quinoline (manufacturing dyes)	NA	NA	14	–	NA	NA	–	NA	–	7.6–12		–			–	–		–	–	–	–	330
Atenolol (antihypertensive agent)	10	–	6	13	–	–	–	0.0	–	19		–			–	–		–	–	0	0	500

Note: A value of implies BMRL = Below mean reporting limit; NA = not analyzed.

Qachqouch Spring catchment, following the detailed statistical and correlative time-series analysis and spring classification as a complex Jurassic highly karstified aquifer. Such high variability of flow and response to rain events and karst heterogeneity, complex dynamics could not be simulated using an equivalent porous medium continuum model. On the other hand, an integrated distributed process-based flow model was constructed with MIKE SHE (Danish Hydrological Institute in 2017) for the less karstified Cenomanian aquifer of El Assal Spring (using an equivalent porous approach with a bypass function along dolines for fast infiltration; Doummar et al., 2018a). Similarly, a 2-D variable saturated flow (VSF) model was developed by Koohbor et al. (2020) to simulate flow in El Assal Spring while accounting for discrete fractures (DFM). Both models were used to simulate future flow rates to forecasted climate change scenarios and show a drastic change in water availability after 2070 and a high variability between wet, dry, and intermediate years after 2030 with a steady increase of recession duration of 5.5 days per decade. The VSF model shows that neglecting the fractures leads to an overestimation of the flow. However, the integrated distributed model can still serve in the case of El Assal Spring to simulate peak and recession flow and help in the understanding of spring response to snowmelt for future management purposes (Kassem et al., 2022). Furthermore, these models allow qualifying the water availability for supply with some degree of uncertainties based on predicted future flow under various climate conditions, and identify water resources alternatives (Hartmann et al., 2012; Doummar et al., 2018a; Sivelle et al., 2021). The results have shown highest sensitivity of the model output (mean and minimum discharge) to temperature in the snow-governed springs, while a decreasing precipitation will mostly impact the availability of flood waters in the Qachqouch Spring at lower elevation. While these numerical models can be further validated and potentially upscaled to aquifers of similar reservoir characteristics, with a collection of additional data, they can act as a decision support tool to test the response and sensitivity of the springs to a variation in input (temperature, rain, snowmelt) or contamination (Kresic & Stevanović, 2009; Parise et al., 2015). Nonetheless, the selection of the most suitable model depends not only on the type and available data, but also on the degree of heterogeneity of the system and on the model's efficiency of portraying complex karst processes (Scanlon et al., 2003). To overcome this challenge and select the proper modeling approach (lumped, distributed, semidistributed models), it is primordial to achieve a quantitative conceptualization of the karst system in space and time based on a proper long-term hydrogeological assessment and monitoring.

8.5. CONCLUSIONS

This work illustrates a selection of methods used in karst catchments for the proper characterization of spring systems in poorly studied semiarid regions facing a high risk of water scarcity and pollution. On the one hand, the characterization of flow and transport in these pilot springs allows gathering scientific evidence to support changes in policy and for the establishment of guidelines for the protection of the sustainable quantity and quality of these valuable resources (Fleury, 2013; Parise et al., 2018). For instance, the relationship established from tracer experiments between the river and the spring and the occurrence of indicator micropollutants implies the need for the protection of the El Kalb River against pollution and potential contamination from point source domestic, industrial, and hospital wastewater effluents. Additionally, the groundwater catchment is characterized by point source contamination as portrayed by the persistent pharmaceuticals detected in the sampled upstream springs and wells. On the other hand, the response of the spring to contamination is not homogenous throughout the year, as it highly depends on the saturation of the system, the intensity of the precipitation event, the discharge of the river, and the type and frequency of effluents. Therefore, guidelines for spring protection should account for the variability of the vulnerability and resilience of the karst spring.

The investigation at the pilot scale is essential for upscaling purposes, where the integration of field data into similar and/or regional scale models can allow for a better management of spring water resources at a larger scale. In the future, the water deficit typical of these semiarid regions will be exacerbated as forecasted by forward-flow simulations because of climate change constraints. Therefore, the high-resolution monitoring yields a quantification of yearly volumes of floodwaters that can be used for potential managed recharge (storage and recovery; ASR) during the high flow to decrease shortage during low-flow periods. Additionally, the identification of indicator parameters and spring response to different types of contamination hazards can be used in the development of early warning systems to be implemented for water treatment and supply. The investigation of the pilot karst springs since 2014 allowed the classification of the springs and the identification of potential threats, while providing models to act as support tools for decision makers to alleviate the risks and ensure the sustainable exploitation of spring water or the identification of alternative water resources. This study highlights the importance of setting up monitoring networks and collecting high-resolution data in poorly studied spring catchments to understand karst systems, calibrate and validate models, and predict potential spring responses to variable input (recharge or contamination).

ACKNOWLEDGMENTS

This work was funded by multiple grants and awards since 2014: USAID and National Academy of Science (Peer Science; project award: 102881; Cycle 3); UNICEF (Award #103924; Project #25778); AUB University Research Board (Award #103951; Project #25512); and KARMA project (L-CNRS in the framework of the PRIMA program; Award #103895; Project #25713). Moreover, we thank the Beirut and Mount Lebanon Water Establishment and The Litani Water Authority for facilitating the installation of instruments and access to field sites. We thank Jason J. Gurdak for his contribution to the work. Our thanks also to Michele Citton, Fouad Andari, and Emmanuel Dubois for field data collection. Additionally, the committee members of the municipalities of Sakiet El Misk-Bhersaf, Bikfaya-Mhaidseh, Ballouneh, and Kfardebbiane, among others, are acknowledged for their kind collaboration.

REFERENCES

Ahmed, N., Ye, M., Wang, Y., Greenhalgh, T., & Fowler, K. (2021). Using δ18O and δ2H to detect hydraulic connection between a sinkhole lake and a first-magnitude spring. *Groundwater*. https://doi.org/10.1111/gwat.13105

Andreo, B., Goldscheider, N., Vadillo, I., Vías, J. M., Neukum, C., Sinreich, M., et al. (2006). Karst groundwater protection: First application of a Pan-European approach to vulnerability, hazard and risk mapping in the Sierra de Líbar (southern Spain). *Science of the Total Environment*, 357 (1–3), 54–73. https://doi.org/10.1016/j.scitotenv.2005.05.019

Bailly-Comte, V., Martin, J. B., Jourde, H., Screaton, E. J., Pistre, S., & Langston, A. (2010). Water exchange and pressure transfer between conduits and matrix and their influence on hydrodynamics of two karst aquifers with sinking streams, *Journal of Hydrology*, 386 (1–4), 55–66. https://doi.org/:10.1016/j.jhydrol.2010.03.005.

Bakalowicz, M. (2005). Karst groundwater: A challenge for new resources. *Hydrogeology Journal*, 13(1), 148–160. https://doi.org/10.1007/s10040-004-0402-9

Barberá, J. A., & Andreo, B. (2012). Functioning of a karst aquifer from S Spain under highly variable climate conditions, deduced from hydrochemical records. *Environmental Earth Sciences*, 65, 2337–2349.

Barbieri, M., Boschetti, T., Petitta, M., & Tallini, M. (2005). Stable isotopes (^{2}H, ^{18}O and ^{87}Sr/^{86}Sr) and hydrochemistry monitoring for groundwater hydrodynamics analysis in a karst aquifer (Gran Sasso, Central Italy). *Applied Geochemistry*, 20, 2063–2081.

Benischke, R. (2021). Review: Advances in the methodology and application of tracing in karst aquifers. *Hydrogeology Journal*, 29, 67–88. https://doi.org/10.1007/s10040-020-02278-9

Buerge, I. J., Keller, M., Buser, H. R., Müller, M. D., & Poiger, T. (2011), Saccharin and other artificial sweeteners in soils: Estimated inputs from agriculture and households, degradation, and leaching to groundwater. *Environmental Science & Technology*, 45(2), 615–621.

Chang, Y., Hartmann, A., Liu, L., Jiang, G., & Wu, J. (2021). Identifying more realistic model structures by electrical conductivity observations of the karst spring, *Water Resources Research*, 57(4). https://doi.org/10.1029/2020WR028587

Chen, Z., & Goldscheider, N. (2014). Modeling spatially and temporally varied hydraulic behavior of a folded karst system with dominant conduit drainage at catchment scale, Hochifen-Gottesacker Alps, *Journal of. Hydrology*, 514, 41–52. https://doi.org/10.1016/j.jhydrol.2014.04.005

Chen, Z., Auler, A. S., Bakalowicz, M., Drew, D., Griger, F., Hartmann, J., et al. (2017). The world karst aquifer mapping project: Concept, mapping procedure and map of Europe. *Hydrogeology Journal*, 25(3), 771–785. https://doi.org/10.1007/s10040-016-1519-3

Chen, Z., Hartmann, A., Wagener, T., & N. Goldscheider (2018). Dynamics of water fluxes and storages in an Alpine karst catchment under current and potential future climate conditions. *Hydrology and Earth System Sciences*, 22(7), 3807–3823. https://doi.org/10.5194/hess-22-3807-2018

Chiffre, A., Degiorgi, F., Buleté, A., et al. (2016). Occurrence of pharmaceuticals in WWTP effluents and their impact in a karstic rural catchment of Eastern France. *Environmental Science and Pollution Research*, 23, 25427–25441. https://doi.org/10.1007/s11356-016-7751-5

Cinkus, G., Mazzilli, N., & Jourde, H. (2021). Identification of relevant indicators for the assessment of karst systems hydrological functioning: proposal of a new classification. *Journal of Hydrology*, 603 (Part C), 127006.

Clemens, M., Khurelbaatar, G., Merz, R., Siebert C., van Afferden, M., & Rödiger, T. (2020). Groundwater protection under water scarcity; from regional risk assessment to local wastewater treatment solutions in Jordan. *Science of the Total Environment*, 706, 136066. https://doi.org/10.1016/j.scitotenv.2019.136066

Diffenbaugh, N. S., & Giorgi, F. (2012). Climate change hotspots in the CMIP5 global climate model ensemble. *Climate Change*, 114, 813–822. https://doi.org/10.1007/s10584-012-0570-x

Doummar, J., & Aoun, M. (2018a). Assessment of the origin and transport of four selected emerging micropollutants sucralose, acesulfame-K, gemfibrozil, and iohexol in a karst spring during a multi-event spring response, *Journal of Contaminant Hydrology.*, 215, 11–20. https://doi.org/10.1016/j.jconhyd.2018.06.003

Doummar, J., & Aoun, M. (2018b). Occurrence of selected domestic and hospital emerging micropollutants on a rural surface water basin linked to a groundwater karst catchment, *Environmental Earth Sciences*, 77(9), 351. https://doi.org/10.1007/s12665-018-7536-x

Doummar, J., Geyer, T., Baierl, M., Nödler, K., Licha, T., & M. Sauter (2014). Carbamazepine breakthrough as indicator for specific vulnerability of karst springs: Application on the Jeita spring, *Lebanon, Applied Geochemistry*, 47, 150–156.

Doummar, J., Hassan Kassem, A., & Gurdak, J. J. (2018a). Impact of historic and future climate on spring recharge and discharge based on an integrated numerical modeling approach: Application on a snow-governed semiarid karst catchment area *Journal of Hydrology*, 565, 636–649. https://doi.org/10.1016/j.jhydrol.2018.08.062

Doummar, J., Margane, A., Geyer, T., & Sauter, M. (2018b). Assessment of key transport parameters in a karst system under different dynamic conditions based on tracer experiments: The Jeita karst system, Lebanon. *Hydrogeology Journal*. https://doi.org/10.1007/s10040-018-1754-x

Doummar, J., Noedler, K., Geyer, T. & Sauter, M. (2012a). *Assessment and analysis of micropollutants (2010–2011)*. Special Report No. 13 of Technical Cooperation Project "Protection of Jeita Spring," Prepared by Department of Applied Geology, University of Göttingen, Germany.

Doummar, J., Sauter, M., & T. Geyer (2012b). Simulation of flow processes in a large scale karst system with an integrated catchment model (Mike She): Identification of relevant parameters influencing spring discharge. *Journal of Hydrology*, 426–427, 112–123, https://doi.org/10.1016/j.jhydrol.2012.01.021

Dubois, E., Doummar, J., Pistre, S., & Larocque, M. (2020). Calibration of a semidistributed lumped model of a karst system using time series data analysis: The example of the Qachqouch karst spring. *Hydrology and Earth System Sciences* 24, 4275–4290. https://doi.org/10.5194/hess-24-4275-2020

Duran, L., & Gill, L. (2021). Modeling spring flow of an Irish karst catchment using Modflow-USG with CLN. *Journal of Hydrology*, 597, 125971. https://doi.org/10.1016/j.jhydrol.2021.125971

Einsiedl, F., Radke, M., & Maloszewski, P. (2010). Occurrence and transport of pharmaceuticals in a karst groundwater system affected by domestic wastewater treatment plants. *Journal of Contaminant Hydrology*, 117, 26–36.

ElGhawi, R., Pekhazis, K., & Doummar, J. (2021). Multi-regression analysis between stable isotope composition and hydrochemical parameters in karst springs to provide insights into groundwater origin and subsurface processes: Regional application to Lebanon. *Environmental Earth Sciences*, 80, 230. https://doi.org/10.1007/s12665-021-09519-4

El-Hakim, M., & Bakalowicz, M. (2007). Significance and origin of very large regulating power of some karst aquifers in the Middle East. Implication on karst aquifer classification. *Journal of Hydrology*, 333(2–4), 329–339. https://doi.org/10.1016/j.jhydrol.2006.09.003

Epting, J. M., Page, R., Auckenthaler, A., & Huggenberger, P. (2018). Process-based monitoring and modeling of Karst springs: Linking intrinsic to specific vulnerability. *Science of the Total Environment*, 625, 403–415. https://doi.org/10.1016/j.scitotenv.2017.12.272

Fayad, A., Gascoin, S., Faour, G., López-Moreno, J. I., Drapeau, L., Le Page, M., & R. Escadafal (2017). Snow hydrology in Mediterranean mountain regions: A review. *Journal of Hydrology*, 551, 374–396.

Fleury, S. (2009). *Land use policy and practice in karst terrains*. Springer.

Fleury, P., Ladouche, B., Conroux, Y. Jourde, H., & Dörfliger, N. (2009). Modeling the hydrologic functions of a karst aquifer under active water management: The Lez spring. *Journal of Hydrology*, 365 (3–4), 235–243. https://doi.org/10.1016/j.jhydrol.2008.11.037

Fleury, P., Maréchal, J. C., & Ladouche, B. (2013). Karst flash-flood forecasting in the city of Nîmes (southern France). *Engineering Geology*, 164, 26–35.

Ford, D., & Williams, P. (2007). *Karst hydrogeology and geomorphology*. West Sussex, England: John Wiley & Sons Ltd. https://doi.org/10.1002/9781118684986

Frank, S., Goeppert, N., & Goldscheider, N. (2018). Fluorescence-based multi-parameter approach to characterize dynamics of organic carbon, faecal bacteria and particles at alpine karst springs. *Science of the Total Environment*, 615, 1446–1459.

Gao, Z., Liu, J., Xu, X., Wang, Q., Wang, M., Feng, J., et al. (2020). Temporal variations of spring water in karst areas: A case study of Jinan spring area, northern China. *Water*, 12 (4).

Gasser, G., Rona, M., Voloshenko, A., Shelkov, R., Tal, N., Pankratov, I., Elhanany, S., & Lev, O. (2010). Quantitative evaluation of tracers for quantification of wastewater contamination of potable water sources. *Environmental Science & Technology*, 44 (10), 3919–3925. https://doi.org/10.1021/es100604c

Geyer, T., Birk, S., Licha, T., Liedl, R., & Sauter, M. (2007). Multitracer test approach to characterize reactive transport in karst aquifers. *Ground Water*, 45(1), 36–45.

Geyer, T., Birk, S., Licha, T., Liedl, R., & Sauter, M. (2008). Quantification of temporal distribution of recharge in karst systems from spring hydrographs. *Journal of Hydrology*, 348, 452–463.

Goderniaux, P., Brouyère, S., Wildemeersch, S., Therrien, R., & Dassargues, A. (2015). Uncertainty of climate change impact on groundwater reserves: Application to a chalk aquifer. *Journal of Hydrology*, 528, 108–121.

Goeppert, N., & Goldscheider, N. (2007). Solute and colloid transport in karst conduits under low- and high-flow conditions. *Ground Water*, 46(1), 61–68. https://doi.org/10.1111/j.1745-6584.2007.00373.x

Goeppert, N., Goldscheider, N., & Berkowitz, B. (2020). Experimental and modeling evidence of kilometer-scale anomalous tracer transport in an alpine karst aquifer. *Water Research*, 178, 115755. https://doi.org/10.1016/j.watres.2020.115755

Gutierrez, F., Parise, M., De Waele, J., & Jourde, H. (2014). A review on natural and human-induced geohazards and impacts in karst. *Earth Science Reviews*, 138, 61–88. https://doi.org/10.1016/j.earscirev.2014.08.002

Hartmann, A., Barberá, J. A., & Andreo, B. (2017). On the value of water quality data and informative flow states in karst modeling. *Hydrology and Earth System Sciences*, 21, 5971–598.

Hartmann, A., Barberá, J. A., Lange, J., Andreo, B., & M. Weiler (2013a). Progress in the hydrologic simulation of time variant recharge areas of karst systems: Exemplified at a karst spring in Southern Spain. *Advances in Water Resources*, 54, 149–160. https://doi.org/10.1016/j.advwatres.2013.01.010

Hartmann, A., Gleeson, T., Rosolem, R., Pianosi, F., Wada, Y., & T. Wagener (2015). A large-scale simulation model to assess karstic groundwater recharge over Europe and the Mediterranean. *Geoscientific Model Development*, 8(6), 1729–1746. https://doi.org/10.5194/gmd-8-1729-2015

Hartmann, A., Goldscheider, N., Wagener, T., Lange, J., & Weiler, M. (2014a). Karst water resources in a changing world: review of hydrological modeling approaches. *Reviews of Geophysics*, 52, 218–242. https://doi.org/10.1002/2013RG000443

Hartmann, A., Lange, J., Vivó, À., Mizyed, N., Smiatek, G., Mizyed, N., et al. (2012). A multi-model approach for improved simulations of future water availability at a large Eastern Mediterranean karst spring. *Journal of Hydrology. 468–469*, 130–138. https://doi.org/10.1016/j.jhydrol.2012.08.024

Hartmann, A., Liu, Y., Olarinoye, T., Berthelin, R., & Marx, V. (2020). Integrating field work and large-scale modeling to improve assessment of karst water resources. *Hydrogeology Journal*. https://doi.org/10.1007/s10040-020-02258-z

Hartmann, A., Mudarra, M., Andreo, B., Marín, A., Wagener, T., & Lange, J. (2014b). Modeling spatiotemporal impacts of hydroclimatic extremes on groundwater recharge at a Mediterranean karst aquifer. *Water Resources Research, 50*(8), 6507–6521.

Hartmann, A., Wagener, T., Rimmer, A., Lange, J., Brielmann, H., & Weiler, M. (2013c). Testing the realism of model structures to identify karst system processes using water quality and quantity signatures. *Water Resources Research, 49*(6), 3345–3358. https://doi.org/10.1002/wrcr.20229

Hartmann, A., Weiler, M., Wagener, T., Lange, J., Kralik, M., Humer, F., et al. (2013b). Process-based karst modeling to relate hydrodynamic and hydrochemical characteristics to system properties. *Hydrology and Earth System Sciences, 17*(8), 3505–3521. https://doi.org/10.5194/hess-17-3305-2013

He, X., Wu, J., & Guo, W. (2019). Karst spring protection for the sustainable and healthy living: The examples of Niangziguan Spring and Shuishentang Spring in Shanxi, China. *Exposure and Health, 11*, 153–165. https://doi.org/10.1007/s12403-018-00295-4

Heinz, B., Birk, S., Liedl, R., Geyer, T., Straub, K. L., Andresen, J., et al. (2009). Water quality deterioration at a karst spring (Gallusquelle, Germany) due to combined sewer overflow: Evidence of bacterial and micro-pollutant contamination. *Environmental Geology, 57* (4), 797–808.

Hillebrand, O., Nödler, K., Licha, T., Sauter, M., & Geyer, T. (2012). Caffeine as an indicator for the quantification of untreated wastewater in karst systems. *Water Research, 46*(2), 395–402.

Hillebrand, O., Nödler, K., Sauter, M., & Licha, T. (2015). Multitracer experiment to evaluate the attenuation of selected organic micropollutants in a karst aquifer. *Science of the Total Environment, 506–507*, 338–343. https://doi.org/10.1016/j.scitotenv.2014.10.102

Hou, D., Song, X., Zhang, G., et al. (2013). An early warning and control system for urban, drinking water quality protection: China's experience. *Environmental Science and Pollution Research, 20*, 4496–4508. https://doi.org/10.1007/s11356-012-1406-y

Iglesias, A., Garrote, L., Flores, F., & Moneo, M. (2007). Challenges to manage the risk of water scarcity and climate change in the Mediterranean. *Water Resource Management, 21*, 775.

Jodar, J., González-Ramón, A., Martos-Rosillo, S., Heredia, J., Herrera, C., Urrutia, J., et al. (2020). Snowmelt as a determinant factor in the hydrogeological behaviour of high mountain karst aquifers: The Garcés karst system, Central Pyrenees (Spain). *Science of the Total Environment, 748*, 141363. https://doi.org/10.1016/j.scitotenv.2020.141363

Jongman, B., Hochrainer-Stigler, S., Feyen, L., Aerts, J., Mechler, R., Wouter Botzen, W., et al. (2014). Increasing stress on disaster-risk finance due to large floods. *Nature Climate Change, 4(4)*, 264–268.

Karnib, A. (2014). A methodological approach for quantitative assessment of the effective wastewater management: Lebanon as a case study. *Environmental Processes, 1*(4), 483–495. https://doi.org/10.1007/s40710-014-0032-8

Kass, W. (1998). *Tracing technique in geohydrology*. Balkema, Rotterdam, The Netherlands.

Kassem, H. A., Doummar, J., Gurdak, J., J. (2022). Sensitivity of an integrated groundwater flow model to model parameters—application to vulnerability assessment of karst aquifers. *Groundwater for Sustainable Development*, 17. 100737. https://doi.org/10.1016/j.gsd.2022.100737

Kløve, B., Pertti Ala-Aho, P., Bertrand, G., Gurdak, J. J., Kupfersberger, H., Kværner, J., et al. (2014). Climate change impacts on groundwater and dependent ecosystems. *Journal of Hydrology, 518*(B), 250–266. https://doi.org/10.1016/j.jhydrol.2013.06.037

Koeniger, P., Margane, A., Abi-Rizk, J., & Himmelsbach, T. (2017). Stable isotope-based mean catchment altitudes of springs in the Lebanon Mountains. *Hydrological Processes, 31*(21), 3708–3718. https://doi.org/10.1002/hyp.11291

Koohbor, B., Fahs, M., Hoteit, H., Doummar, J., Younes, A., & Belfort, B. (2020). An advanced discrete fracture model for variably saturated flow in fractured porous media. *Advances in Water Resources, 140*, 103602. https://doi.org/10.1016/j.advwatres.2020.103602

Korfali, S. I., & Jurdi, M. (2009). Provision of safe domestic water for the promotion and protection of public health: A case study of the city of Beirut, Lebanon. *Environmental Geochemistry and Health, 31*, 283–295.

Kresic, N., & Stevanović, Z. (2009). *Groundwater hydrology of springs: Engineering, theory, management and sustainability*. Oxford, U.K.: Butterworth-Heinemann.

Liu, Y., Blowes, D., Groza, L., Sabourin, M. J., & Ptacek, C. J. (2014). Acesulfame-K and pharmaceuticals as co-tracers of municipal wastewater in a receiving river. *Environmental Science: Processes and Impacts, 16*, 2789.

Luo, Q., Yang, Y., Qian, J., Wang, X., Chang, X., Ma, L., et al. (2020). Spring protection and sustainable management of groundwater resources in a spring field. *Journal of Hydrology, 582*, 124498. https://doi.org/10.1016/j.jhydrol.2019.124498

Mangin, A. (1975). Contribution à l'étude hydrodynamique des aquifères karstiques. Ph.D. dissertation, Université de Dijon, Dijon, France.

Marechal, J. -C., Vestier, A., Jourde, H. & Dorfliger, N. (2013). L'hydrosysteme du Lez: Une gestion active pour un karst à enjeux. *Karstologia, 62*, 1–6.

Marin, A. I., Andreo, B., & Mudarra, M. (2015). Vulnerability mapping and protection zoning of karst springs. Validation by multitracer tests. *Science of the Total Environment, 435–446*, https://doi.org/10.1016/j.scitotenv.2015.05.029

Marin, A. I., Martín Rodríguez, J. F., Barberá, J. A., et al. (2021). Groundwater vulnerability to pollution in karst aquifers, considering key challenges and considerations: Application to the Ubrique springs in southern Spain. *Hydrogeology Journal, 29*, 379–396. https://doi.org/10.1007/s10040-020-02279-8

Massoud, M. A., Tareen, J., Tarhini, A., et al. (2010). Effectiveness of wastewater management in rural areas of developing countries: a case of Al-Chouf Caza in Lebanon. *Environmental Monitoring and Assessment, 161,* 61–69. https://doi.org/10.1007/s10661-008-0727-2

Mawhinney, D. B., Young, R. B., Vanderford, B. J., Borch, T., & Snyder, S. A (2011). Artificial sweetener sucralose in U.S. drinking water systems. *Environnebtal Science and Technology, 45*(20), 8716–8722.

Mudarra, M., Andreo, B., & Mudry, J. (2010). Hydrochemical heterogeneity in the discharge zone of a karstic aquifer. In B. Andreo et al. (Eds.), *Advances in research in Karst media* (pp. 163–168). Berlin: Springer.

Mudarra, M., Andreo, B. & Mudry, J. (2012). Monitoring groundwater in the discharge area of a complex karst aquifer to assess the role of the saturated and unsaturated zones. *Environmental Earth Sciences, 65*(8), 2321. https://doi.org/10.1007/s12665-011-1032-x

Mudarra, M., Hartmann, A., & Andreo, B. (2019). Combining experimental methods and modeling to quantify the complex recharge behavior of karst aquifers. *Water Resources Research, 55*(2), 1384–1404.

Nerantzaki, S. D., & Nikolaidis, N. P. (2020). The response of three Mediterranean karst springs to drought and the impact of climate change. *Journal of Hydrology, 591,* 125296. https://doi.org/10.1016/j.jhydrol.2020.125296

Nödler, K., Tsakiri, M., Aloupi, M., Gatidou, G., Stasinakis, A. S., & Licha, T (2016). Evaluation of polar organic micropollutants as indicators for wastewater-related coastal water quality impairment. *Environmental Pollution, 211,* 282–290.

North, L. A., van Beynen, P. E., & Parise, M. (2009). Interregional comparison of karst disturbance: west-central Florida and southeast Italy. *The Journal of Environmental Management, 9* (5), 1770–1781.

Olarinoye, T., Gleeson, T., Marx, V., et al. (2020). Global karst springs hydrograph dataset for research. *Scientific Data, 7,* 59. https://doi.org/10.1038/s41597-019-0346-5

Oppenheimer, J., Eaton, A., Badruzzaman, M., Haghani, A. W., & Jacangelo, J. G (2011). Occurrence and suitability of sucralose as an indicator compound of wastewater loading to surface waters in urbanized regions. *Water Research, 45*(13), 4019–4027.

Parise, M., Closson, D., Gutiérrez, F. & Stevanovic, Z. (2015). Anticipating and managing engineering problems in the complex karst environment. *Environmental Earth Sciences, 74,* 7823–7835. https://doi.org/10.1007/s12665-015-4647-5

Parise, M., Gabrovsek, F., Kaufmann, G., & N. Ravbar (2018). Recent advances in karst research: From theory to fieldwork and applications. In M. Parise et al. (Eds.), *Advances in Karst research: Theory, fieldwork and applications* (pp. 1–24). Geological Society, London, Special Publications, 466. https://doi.org/10.1144/SP466.26

Petitta, M., Barberio, D. M., Barbieri, M., Billi, A., Doglioni, C., Passaretti, S., et al. (2020). Groundwater monitoring in regional discharge areas selected as "Hydrosensitive" to seismic activity in Central Italy. In *Advances in natural hazards and hydrological risks: Meeting the challenge* (pp. 21–25). New York: Springer.

Pronk, M., Goldscheider, N., & Zopfi, J. (2006). Dynamics and interaction of organic carbon, turbidity and bacteria in a karst aquifer system. *Hydrogeology Journal, 14,* 473–484. https://doi.org/10.1007/s10040-005-0454-5

Rusjan, S., Sapač, K., Petrič, M., Lojen, S., & Bezak, N. (2019). Identifying the hydrological behavior of a complex karst system using stable isotopes. *Journal of Hydrology, 577,* 123956.

Sauter, M. (1992). *Quantification and forecasting of regional groundwater flow and transport in a karst aquifer (Gallusquelle, Malm, SW Germany).* Tubinger Geowissenschaftliche Arbeiten, Part C (13) (1992), 151.

Scanlon, B. R., Mace, R. E., Barrett, M. E., & Smith, B. (2003). Can we simulate regional groundwater flow in a karst system using equivalent porous media models? Case study, Barton Springs Edwards aquifer USA, *Journal of Hydrology., 276,* 137–158. https://doi.org/10.1016/S0022-1694(03)00064-7

Schmidt, S., Geyer, T., Marei, A., Guttman, J., & Sauter, M. (2013). Quantification of long-term wastewater impacts on karst groundwater resources in a semi-arid environment by chloride mass balance methods, *Journal of Hydrology, 502,* 177–190. https://doi.org/10.1016/j.jhydrol.2013.08.009

Schnegg, P. A. (2002). An inexpensive field fluorometer for hydrogeological tracer tests with three tracers and turbidity measurement. In E. Bocanegra et al. (Eds.), *Groundwater and Human Development, Mar Del Plata, Argentina, October 2002* (pp. 1484–1488).

Siirila-Woodburn, E. R., Fernandez Garcia, D., & Sanchez-Vila, X. (2015). Improving the accuracy of risk prediction from particle-based breakthrough curves reconstructed with kernel density estimators. *Water Resources Research, 51,* 4574–4591.

Sivelle, V., Jourde, H., Bittner, D., Mazzilli, N., & Tramblay, Y. (2021). Assessment of the relative impacts of climate changes and anthropogenic forcing on spring discharge of a Mediterranean karst system. *Journal of Hydrology, 598,* 126396. https://doi.org/10.1016/j.jhydrol.2021.126396

Sivelle, V., Renard, P., & Labat, D. (2020). Coupling SKS and SWMM to solve the inverse problem based on artificial tracer tests in karstic aquifers. *Water, 12,* 1139. https://doi.org/10.3390/w12041139

Stange, C. & Tiehm, A. (2020). Occurrence of antibiotic resistance genes and microbial source tracking markers in the water of a karst spring in Germany. *Science of the Total Environment, 742,* 140529. https://doi.org/10.1016/j.scitotenv.2020.140529

Stedmon, C. A., Seredyńska-Sobecka, B., Boe-Hansen, R, Le Tallec, N., Christopher, K., Waul, C. K., et al. (2011). A potential approach for monitoring drinking water quality from groundwater systems using organic matter fluorescence as an early warning for contamination events. *Water Resources, 45,* 6030–6038. https://doi.org/10.1016/j.watres.2011.08.066

Stevanović, Z. (2015). *Karst aquifers: Characterization and engineering.* Cham, Switzerland: Springer.

Stevanović, Z. (2019a). Karst aquifers in the arid world of Africa and the Middle East: Sustainability or humanity? In T. Younos et al. (Eds.), *Karst water environment: The handbook of environmental chemistry, vol. 68.* Cham: Springer. https://doi.org/10.1007/978-3-319-77368-1_1

Stevanović, Z. (2019b). Karst waters in potable water supply: A global scale overview. *Environmental Earth Sciences, 78,* 1–12. https://doi.org/10.1007/s12665-019-8670-9

Stevanović, Z., & Stevanović, A. M. (2021). Monitoring as the key factor for sustainable use and protection of groundwater in karst environments: An overview. *Sustainability, 13*(10), 5468. https://doi.org/10.3390/su13105468

Toride, N., Leij, F., & van Genuchten, M. (1999). *The CXTFit code for estimating transport parameters from laboratory or field tracer experiments. Version 2.1*. Research Report 137, USDA, Riverside, California.

Torresan, F., Fabbri, P., Piccinini, L., Dalla Libera, N., Pola, M., & Zampieri, D. (2020). Defining the hydrogeological behavior of karst springs through an integrated analysis: A case study in the Berici Mountains area (Vicenza, NE Italy). *Hydrogeology Journal,* 1–19.

Van Beynen, P. E., & Townsend, K. M. (2005). A disturbance index for karst environments. *Environmental Management, 36* (1), 101–116.

Van Loon, A. F., Kumar, R., & Mishra, V. (2017). Testing the use of standardised indices and GRACE satellite data to estimate the European 2015 groundwater drought in near-real time. *Hydrology Earth System Sciences, 21*(4), 1947–1971.

Van Stempvoort, D. R., Roy, J. W., Grabuski, J., Brown, S. J., Bickerton, G. & Sverko, E. (2013). An artificial sweetener and pharmaceutical compounds as co-tracers of urban wastewater in groundwater. *Science of the Total Environment, 46–462*, 348–359.

Wang, F., Chen, H., Lian, J., Fu, Z., & Nie, Y. (2020). Seasonal recharge of spring and stream waters in a karst catchment revealed by isotopic and hydrochemical analyses. *Journal of Hydrology, 591*, 125595. https://doi.org/10.1016/j.jhydrol.2020.125595

Warner, W., Licha, T., & Nödler, K. (2019). Qualitative and quantitative use of micropollutants as source and process indicators. A review. *Science of the Total Environment, 686,* 75–89. https://doi.org/10.1016/j.scitotenv.2019.05.385

Wolf, L., Zwiener, C., & Zemann, M. (2012). Tracking artificial sweeteners and pharmaceuticals introduced into urban groundwater by leaking sewer networks. *Science of the Total Environment., 430*, 8–19

Worthington, S. R. H. (1999). A comprehensive strategy for understanding flow in carbonate aquifers. In A. N. Palmer et al. (Eds.), *Karst modeling* (pp. 30–37). Karst Water Institute, Special Publication 5.

Zirlewagen, J., Licha, T., Schiperski, F., Nödler, K., & Schyett, T. (2016). Use of two artificial sweeteners, cyclamate and acesulfame, to identify and quantify wastewater contributions in a karst spring, *Science of the Total Environment, 547*. 356–365.

Zuccato, E., Chiabrando, C., Castiglioni, S., et al. (2005). Cocaine in surface waters: A new evidence-based tool to monitor community drug abuse. *Environmental Health, 4*, 14. https://doi.org/10.1186/1476-069X-4-14

9
Uncertainties in Understanding Groundwater Flow and Spring Functioning in Karst

Francesco Fiorillo[1], Mauro Pagnozzi[1], Rosangela Addesso[2], Simona Cafaro[2], Ilenia M. D'Angeli[3], Libera Esposito[1], Guido Leone[1], Isabella S. Liso[3], and Mario Parise[3]

ABSTRACT

In karst environments, typically characterized by peculiar hydrogeological features and high heterogeneity and anisotropy, the connection between the recharge areas and the springs is often not straightforward. Rapid infiltration underground and the resulting network of karst conduits are frequently at the origin of a lack of correspondence among topographic divides and underground watersheds. As a consequence, in many karst areas there is still much work to do to fully understand the groundwater flow, with the only "underground truth" often being provided by cave data. In this chapter, we start from general considerations about the difficulty in comprehending hydrogeology in karst and use them to analyze one of the most important karst areas of southern Italy, the Alburni Massif in Campania. In detail, we present data about the main karst features at the surface (dolines, endorheic basins, etc.), the most important cave systems (reaching maximum depth of about 450 m below the surface), and the main basal springs coming out at the massif borders. Integration of the different sources of data allows us to hypothesize the main directions of groundwater flows, and to perform the first attempts in correlating recharge and discharge data, but such hypotheses then often prove to be wrong by data from cave and diving explorations.

9.1. PECULIARITIES OF KARST HYDROGEOLOGY

Karst is an extremely peculiar setting, with unique landscapes characterized by a variety of landforms (such as dolines, swallets, shafts, karrenfields, poljes, etc.), which act as sites of concentrated recharge for the aquifers and, together with the main geological and hydrogeological features of soluble materials, are at the origin of the turbulent flow of water within the karst rock masses (Worthington et al., 2001; Brinkmann & Parise, 2012).

[1] Department of Science and Technology, University of Sannio, Benevento, Italy
[2] MIDA Foundation, Pertosa, Italy
[3] Department of Earth and Environmental Sciences, University Aldo Moro, Bari, Italy

Such peculiarities cause the need to approach hydrogeological studies in karst with dedicated methods and techniques, since implementation of the classical hydrogeological laws and procedures is not significant (Goldscheider & Drew, 2007; Jourde et al., 2007). Starting from the noncorrespondence among hydrographic boundaries at the surface and hydrogeological boundaries underground (Gunn, 2007; Parise, 2016a), the whole issue of infiltration, transfer, and discharge of water in karst is extremely complex (Stevanovic, 2015, and references therein). In such a context, mapping some of the most typical karst landforms such as dolines/sinkholes, endorheic basins, and poljes (Angel et al., 2004; Dorsaz et al., 2013; Miao et al., 2013; Fragoso-Servon et al., 2014; Wu et al., 2016; Pagnozzi et al., 2019; Zumpano et al., 2019), understanding their mechanisms of formation

(Waltham et al., 2005; Del Prete et al., 2010; Gutierrez et al., 2014; Parise, 2019, 2022) and their hydraulic role as well (Bonacci 1995, 2001; Fiorillo et al., 2015; Parise et al., 2015), is of crucial importance to gain insights into the actual hydrogeological regime in karst areas.

In this chapter, through illustration of the Alburni case study (S Italy), one of the most significant karst areas in the country, we intend to point out the difficulties inherent in understanding karst hydrogeology, the crucial importance of cooperation with direct explorations by cavers, and the need to approach the issue with specifically designated approaches. With this aim, we analyze the karst depressions at the summit plateau, estimate the related recharge, and compare it to the total amount coming from the main springs surrounding the massif. Then, through information derived from cave surveys, including diving explorations through some of the sumps within the cave systems, we point out the still-open problems regarding hydrogeology in the Alburni Massif.

9.2. MATERIALS AND METHODS

Mapping of dolines and endorheic areas on the Alburni Massif plateau was carried out through an integrated methodology, consisting of bounding their limits on 1:5000 scale topographic maps, supported by field survey, and uploading in GIS environment the geomorphological data together with those regarding strata attitude and presence of tectonic faults, as mapped from the official geological maps (Cestari, 1971; Scandone, 1971; De Riso & Santo, 1997).

The regional inventory of karst caves in Campania (managed by the Campanian Speleological Federation, available at http://www.fscampania.it/catasto-2/catasto/) was the starting point for the analysis of the main characters of the caves in the area: namely, through scrutiny of the individual cave surveys, in the form of plan map and profiles, the presence of water within each inventoried cave was checked. Typically, this corresponds to a stop in exploration of the cave, except those few cases where it is possible to continue through diving explorations. When the condition above (presence of water) was satisfied, its altitude within the cave system (corresponding to the maximum depth of the cave) was extracted as water level reference at the site. Collecting all these data, a preliminary attempt in reconstructing the Alburni water table was carried out. In addition, the outcomes of several tracing experiments, particularly cave-to-spring multi-tracer tests, carried out during the last 10 years in the area, were considered to prove some connections among caves and springs.

Data about the main springs in the area derive from detailed analysis of the existing scientific literature, but without any doubt they represent still a pitfall in the overall analysis, due to lack of continuity in recording the spring discharges. Rainfall and temperature data were taken from the official reports by the Italian Hydrography Service during the last decades. Eventually, the groundwater recharge in the long-term scale was estimated by applying the annual model proposed by Fiorillo et al. (2015), which can be implemented especially for wide areas with strong morphological irregularities not entirely covered by hydrological monitoring. Based on long-term mean annual data, the total amount of meteoric precipitation, runoff, and recharge are computed in GIS environment in the model, estimating the recharge and the runoff coefficient for both open and endorheic areas. The annual model provides a mean long-term estimation of the recharge.

Based on a 20 × 20 m Digital Elevation Model, the spatial annual mean rainfall and annual mean temperature have been estimated by GIS tools; temperature and rainfall data were collected for the time period 1971–1999, then a reliable correlation was found using annual mean rainfall and annual mean temperature regression lines (Pagnozzi et al., 2019). The equations provided were implemented using raster data, and raster calculator tools in the GIS environment. Then, using the Turc (1954) formula, the long-term annual mean of the actual evapotranspiration was estimated; this grid has been subtracted from the annual mean rainfall distribution grid, providing the long-term annual mean effective rainfall distribution grid.

In the endorheic area, A_E, as the runoff cannot escape, the recharge amount, R, can be considered equal to the effective afflux, F_{eff}:

$$(R)_{A_E} = (F_{eff})_{A_E}. \tag{9.1}$$

In the open areas, A_O, the recharge amount R can be estimated. This assumes that all the groundwater flow feeding the spring discharges, Q_S, and no-flow boundaries occurs toward the argillaceous, terrigenous, and flysch sequences (impervious terrains). Following this assumption, the total discharge, Q_s, from springs is

$$Q_s = (R)_{A_E} + (R)_{A_O}, \tag{9.2}$$

which allows us to obtain the recharge in the open areas in the case of null groundwater abstraction:

$$(R)_{A_O} = Q_s - (F_{eff})_{A_E}, \tag{9.3}$$

and the total recharge on the catchment area, A_C, is

$$(R)_{A_C} = (R)_{A_O} + (R)_{A_E} = Q_s, \tag{9.4}$$

valid if no groundwater occurs in the spring catchment, as for the Alburni karst massif.

The model assumes that all the amount of recharge reaches the basal water table, even though the vadose zone may present local saturated zones (i.e., sumps within karst systems, perched water tables, etc.).

The most common hydrologic parameter used to estimate aquifer recharge is the ratio between the volume of spring discharge and the rainfall. This is computed annually, assuming that cross boundary flow does not occur (Drogue, 1971; Bonacci & Magdalenic, 1993; Bonacci, 2001). Such a rough estimation can be improved considering the evapotranspiration processes and distinguishing the areas characterized by different recharge conditions. Among these latter, there are endorheic basins that are closed depressions where the runoff is completely adsorbed (internal runoff; White, 2002; Sauro, 2005) and are generally hydraulically connected to one or more springs.

The recharge coefficient used is expressed in terms of the fraction of the effective afflux, F_{eff}, providing the effective recharge coefficient, C_R; if water pumping does not occur, the following equations can be deducted (Fiorillo et al., 2015):

$$(C_R)_{A_E} = 1; \quad (C_R)_{A_o} = \frac{(R)_{A_o}}{(F_{eff})_{A_o}}; \quad (C_R)_{A_c} = \frac{(R)_{A_c}}{(F_{eff})_{A_c}}. \quad (9.5)$$

The same coefficients can be expressed in function of total afflux, F; in a generic area, A, the recharge coefficient is

$$(C'_R)_A = \frac{(R)_A}{(F)_A}. \quad (9.6)$$

Finally, another evaluation is the contribution of endorheic areas to spring discharge. In this case, as all the recharge amounts inside endorheic areas (minus the pumping amount, Q_P) are assumed to reach basal springs, the effective contribution to spring discharge, C_S, can be expressed by

$$(C_S)_{A_E} = \frac{(F_{eff} - Q_P)_{A_E}}{Q_s}. \quad (9.7)$$

As a consequence, the effective contribution to spring discharge of open areas, A_O, is

$$(C_S)_{A_o} = 1 - (C_S)_{A_E}. \quad (9.8)$$

In terms of total afflux, F, the total contribution to spring discharge in a generic area, A, could be estimated by the following equation:

$$(C'_S)_A = \frac{(F - Q_P)_A}{Q_s}. \quad (9.9)$$

Further details of the method are described in Fiorillo et al. (2015).

9.3. THE ALBURNI MASSIF

The Alburni Massif (Campania region of S Italy) extends over 270 km² reaching a maximum altitude of 1,742 m a.s.l. It is characterized by steep slopes bounding a mostly flat and undulating summit plateau. Two rivers bound the massif: namely, the Calore Lucano to the southwest, and the Tanagro River to the northeast, their valleys being filled by heterogeneous alluvial deposits, slope breccias, sands, and conglomeratic deposits (Fig. 9.1).

The massif can be described as a monoclinal southwest-dipping ridge marked by faults and composed of a Mesozoic carbonate sequence of Jurassic-Cretaceous age (Sartoni & Crescenti, 1962); these soluble rocks are covered by a Miocene flysch sequence consisting of clays and sandstones (Scandone, 1972; Ippolito et al., 1973; Patacca & Scandone, 2007). During the Pliocene and Pleistocene, several faults caused the uplift of the massif (Gioia et al., 2011; Cafaro et al., 2016) and the development of deep karst processes (Santangelo & Santo, 1997). The summit plateau shows a variety of sites of concentrated water infiltration, typical of karst settings, such as dolines and shafts (Klimchouk, 2000; Ford & Williams, 2007; Palmer, 2007; Williams, 2008), which rapidly transfer the runoff into a complex network of caves and conduits (Del Vecchio et al., 2013; Cafaro et al., 2016), and then to the saturated zone of the aquifer. This concentrated recharge occurs mainly after intense rainstorms and snowmelt, while during normal rainfall events the recharge shows a diffuse modality, in function of the epikarst characters at the summit plateau.

The main springs (Basso Tanagro and Pertosa on the north side, Castelcivita and Auso to the south) drain the saturated zone of the aquifer, and are distributed in the areas surrounding the massif; a systematic record of their discharge is missing, with only sporadic measurements available (Brancaccio et al., 1973; Celico et al., 1994; Ducci, 2007). Overall, the total discharge can be estimated being in the order of 7–8 mc/s (Table 9.1).

In karst settings, due to scarcity or limited length of the surface runoff, endorheic areas play a prominent role in the recharge processes (Denizman, 2003; Palmer, 2010; Heidari et al., 2011; Parise et al., 2015; Zumpano et al., 2019). Their size and spatial distribution are typically linked to the structural control by faults and the main discontinuity systems in the rock mass (Palmer, 1991, 2007; Hauselmann et al., 1999; Parise, 2011).

Mapping of dolines and endorheic areas on the Alburni Massif was carried out through an integrated approach (Fig. 9.2), consisting of bounding their limits on 1:5000 scale topographic maps, supported by field survey, and

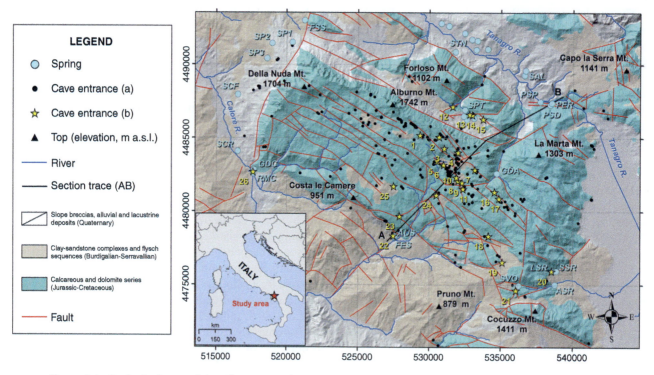

Figure 9.1 Geological map of the Alburni Massif.

Table 9.1 Springs surrounding the Alburni Massif, and related discharge values (If available)

Label	Spring name	Elevation (m a.s.l.)	Mean annual discharge (m³/s)
RMC	Risorgenza del Mulino di Castelcivita	65	nd
GDC	Grotta di Castelcivita	94	1.50
SCR	Controne	100	0.10
SP1	Postiglione1	570	0.10
SP2	Postiglione2	570	0.10
SP3	Postiglione 3	570	0.10
SCF	Sorgenti Cafaro	180	nd
FSS	Fontana Scorzo Sicignano	363	0.01
STN	Sorgenti del Tanagro	204	3.5
SAL	Sorgenti Auletta	235	nd
PSP	Polle sorgive Pertosa	195	nd
SPT	Sorgenti Petina	647	0.10
PSD	Polle Santa Domenica	243	nd
PER	Grotta di Pertosa	263	1.10
LSR	Lavatoio San Rufo	669	0.01
SSR	Sorgente San Rufo	636	nd
ASR	Abbotituro San Rufo	672	0.01
SVO	Sorgente Valetorno	848	nd
AUS	Risorgenza dell'Auso	280	1.00
FES	Sorgente Festola	280	nd
GDA	Grotta dell'acqua	875	nd

uploading in GIS environment the geomorphological data together with those regarding strata attitude and presence of tectonic faults, as mapped from the official geological map.

The morphometric analysis proved that closed depressions (extending up to a few square kilometers) developed on strata mostly characterized by horizontal or near-to-horizontal attitude; differently from other karst areas in Campania (Matese and Picentini mountains), the high density of sinkholes on the Alburni karst plateau has to be related, therefore, to the mostly horizontal bedding.

Recharge can be defined as the downward flow of water reaching the water table (De Vries & Simmers, 2002). In order to assess the recharge on the karst system at the Alburni, the hydrological analysis was preceded by a detailed geomorphological investigation of the karst landforms (dolines and depressions) on the summit plateau; both hydrological and morphometric analyses allowed depiction of a specific overview of recharge processes in which such karst landforms play a predominant role because the effective meteoric water falling on it contributes to feed the springs.

About 400 caves, with several reaching depth of around 450 m and with development of some kilometers, characterize the Alburni Massif (Bellucci

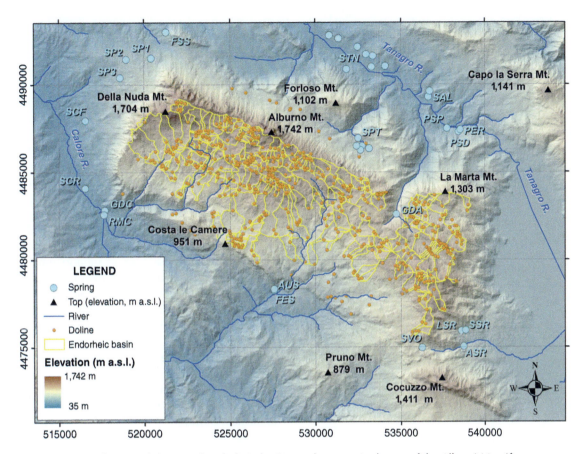

Figure 9.2 Map showing dolines and endorheic basins on the summit plateau of the Alburni Massif.

et al., 1991, 1995). This remarkable karst is essentially related to the presence of the wide, high plateau, bounded by fault systems, and with a variety of infiltration sites, mainly corresponding to blind valleys and small catchments on the flysch deposits, which surface hydrology feeds the many swallets at the contact with the limestone (Santangelo & Santo, 1997; Del Vecchio et al., 2013; Cafaro et al., 2016). Through scrutiny of the data about the Alburni caves, all those where water was found were selected (Fig. 9.1 and Table 9.2). It must be pointed out that in these caves generally the presence of water corresponds to the end of explorations, given the impossibility (in some cases) and the difficulty (in others) to pass the flooded passages. Further, presence of water does not necessarily mean that the saturated zone has been reached; actually, some of the water could be related to perched groundwater, due to less-permeable intercalations within the stratigraphy, or to local clogging by debris and breakdown deposits. Nevertheless, we used the elevations at which water was documented into caves to build the hydrogeological profile shown in Figure 9.3 by assuming water as representative of the base water table.

9.4. RESULTS

The Alburni karst massif can be considered a wide karst system, where surficial and groundwater hydrology are strictly linked but still unclear. Surficial hydrology appears controlled by the wide summit plateau, which has been assumed as a wide closed area, where the runoff infiltrates in sinking points, providing a concentrated recharge. Outside of it, along the steep slopes bounding the plateau, the runoff can escape from the catchment and feed the rivers directly.

All karst landforms mapped and digitalized in a GIS environment provided a total number of 539 dolines, with average density of 5.97 depressions per km^2 (Fig. 9.2); 62% of these closed depressions have area less than 0.1 km^2 (Pagnozzi et al., 2019). Their pattern distribution highlights that the central plateau is mostly affected by dolines of small size, while only along the north-western, eastern, and southern borders endorheic areas are generally ≥1 km^2. The statistical approach adopted in the study area allowed assessment of the pitting index (total karst area/plateau area), which represents a measure of superficial karst development,

Table 9.2 Caves where water has been found within the karst systems

ID	Label	Cave name	Cave entrance elevation (m a.s.l.)	Water elevation (m a.s.l.)
1	MAR	Grava di Maria	1,300	1,097
2	VEN	Grava del Vento	1,270	1,231
3	ISC	Inghiottitoio sotto Serra Carpineto	1,230	1,076
4	INV	Grava d'Inverno	1,150	949
5	VIT	Grotta dei Vitelli	1,120	735
6	FUM	Grotta del Fumo	1,058	615
7	PAR	Grava II del Parchitiello	1,112	907
8	SM2	Inghiottitoio Piani di Santa Maria II	1,096	1,094
9	SM3	Inghiottitoio Piani di Santa Maria III	1,076	656
10	SM1	Inghiottitoio Piani di Santa Maria I	1,086	807
11	OSS	Grava delle Ossa	1,060	769
12	LAU	Grotta del Lauro	550	532
13	POE	Grava del Poeta	635	590
14	MIL	Grotta Milano	640	600
15	IMP	Inghiottitoio di Mastro Peppe	680	595
16	FAL	Grotta del Falco	1,105	944
17	CAM	Grotta II di Campitelli	1,099	993
18	MIN	Grava del Minollo	888	577
19	SER	Grava del Serrone	970	754
20	GSR	Grotta di san Rufo	698	672
21	GPA	Grotte del Piano di Allaga	912	870
22	GAO	Grotta dell'Auso di Ottati	280	260
23	MEL	Grava di Melicupo	674	415
25	GEN	Grava dei Gentili	841	404
24	GAT	Grava dei Gatti	943	541
26	GAU	Grotta dell'Ausino	69	49

Note: Caves are designated by yellow stars in Figure 9.1. Labels are as in Figure 9.3.

providing information about the extent of karstification (Denizman, 2003; Haryono et al., 2017). At Alburni, the ratio between karst area and plateau is 2.96.

To estimate the recharge, a preliminary delimitation of the catchment spring area, A_c, has to be provided. Definition of the spring catchment area is a challenge in karst settings (Gunn, 2007; Parise, 2016b), especially if a wide karst system is drained by several springs, as at the Alburni Massif. A useful approach is to associate the whole mountain or karst system to a lumped system, and to consider the overall output from spring outlets, without focusing the analysis on a single spring and its relative catchment. In the Alburni case, the karst terrains are bounded by impervious terrains, which make the delimitation of the lumped spring catchment easier; only along the southeast sector, the spring catchment cannot be accurately defined.

Figure 9.3 provides the hydrogeological cross section along the Alburni Massif considering some of the main springs (Auso, 277 m a.s.l., to the south; and Pertosa, 250 m a.s.l., to the northeast); the different elevation between these springs is coherent with fault systems affecting the carbonate hydrostructure. Dolines and endorheic basins drain the meteoric water on the summit plateau through the network of shafts and conduits below. Looking at Figure 9.3, the cave profiles, redrawn from the Regional Inventory of Caves of Campania, and adding bedding information, highlight that development of the karst systems is highly controlled by the prevailing discontinuity systems in the rock mass, both as subhorizontal passages (bedding) and as vertical pits (fractures or faults).

However, it is very arduous to assess the groundwater flowpath in the shafts (Jouves et al., 2017), so that in many cases scholars refer to indirect methods in order to gain insights about the karst flow system (geophysics, geodesy, etc.; Martel et al., 2018). In our case, detailed studies were carried out on the Alburni catchment area, based on a methodical collection of available data about hydrology, water geochemistry, and piezometric data of the aquifer with its main outflows. Being that the karst environment is intersected by a complex system of conduits, passages, and shafts (only partly known), the most reliable approach to propose a valid hydrological model is represented by tracing experiments, particularly the cave-to-spring multitracer tests (Goldscheider & Drew, 2007; Filippini et al., 2018).

At Alburni, looking at the karstified limestone outcrops and the morphological features of the calcareous area with an elevation higher than that of the springs, the estimated recharge area is 267 km². This wide area includes the karst plateau, considered as a unique closed area, A_E, extended 90.09 km². The catchment zones outside the internal runoff area constitute the open areas ($A_O = A_C - A_E$).

The main results are shown in Table 9.3; taking into account the effective rainfall distribution and the temperature values, the mean actual evapotranspiration at Alburni Massif can be estimated (545 mm/year). This value is comparable to evapotranspiration rates for nearby karst massifs of southern Italy (Fiorillo et al., 2015; Fiorillo & Pagnozzi, 2015), while the amount of recharge is higher in Alburni, due to concentrated recharge at the summit plateau and to runoff being limited along the steep slopes bounding the massif.

Looking at the numbers listed in Table 9.3, the annual effective afflux (P_{eff}) of the whole catchment area is 246×10^6 m³, the annual spring discharge (Q) is 230.6×10^6 m³, and the ratio Q/P_{eff} provides the effective recharge coefficient of 0.94. The difference between the effective recharge from precipitation (7.8 m³/s) and the spring discharge (7.4 m³/s), estimated in 0.4 m³/s, could be

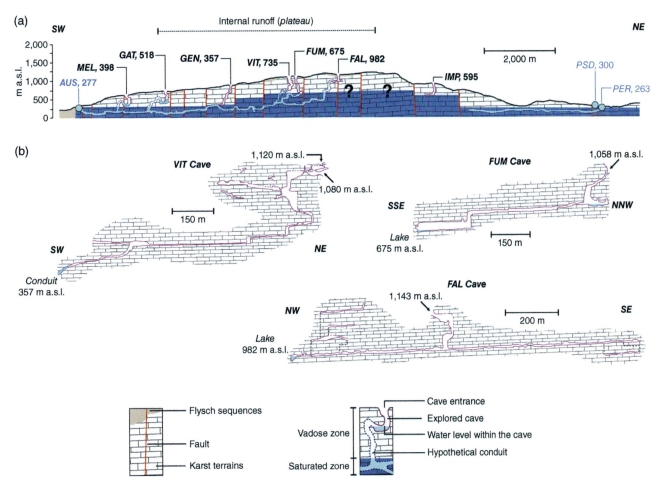

Figure 9.3 Hydrogeological schematic cross section across the Alburni Massif, based upon speleological data from the Regional Inventory of Caves of Campania, managed by the Campanian Speleological Federation; trace of section in Figure 9.1. Some profiles of selected caves are also shown, after the surveys from Campanian Speleological Federation (modified from http://www.fscampania.it/catasto-2/catasto/), with addition of the strata attitude.

Table 9.3 Hydrological parameters obtained from the recharge analysis for the Alburni Massif

Category	Mean elevation m a.s.l.	Area km²	F m³×10⁶/y	F mm/y	T °C	AET mm/y	F_{eff} m³×10⁶/y	F_{eff} mm/y	RO m³×10⁶	Q_p m³×10⁶/y	R m³×10⁶/y	C_R	C'_R	C_s	C'_s
Plateau area, A_E	1,175	90	149	1,658	7.9	500	104	1,157	0.0	0.0	104	1.00	0.69	0.450	0.646
Open area, A_O	828	177	243	1,375	10.5	569	142	805	13.4	0.0	128.6	0.90	0.53	0.550	0.354
Alburni, A_C	945	267	392	1,470	9.6	545	246	923	13.4	0.0	232.6	0.94	0.59	1.000	1.000

Source: Adapted from Fiorillo et al. 2019.
Note: F, afflux (mean precipitation on the catchment); T, temperature; AET, actual evapotranspiration; F_{eff}, effective afflux (mean effective precipitation on the catchment); RO, runoff; Q_p, groundwater abstracted; R, recharge; C_R, effective recharge coefficient; C'_R, total recharge coefficient; C_s, effective contribution to spring discharge; C'_s, total contribution to spring discharge.

associated with runoff losses and/or to minor springs, for which discharge data are unavailable.

An high effective recharge coefficient ($C_R = 0.90$) has been found for the open area (zone outside the summit plateau) where the runoff amount is only 13.4×10^6 m³. Even if the runoff amount is believed to be a very limited component in the hydrological balance in karst areas, this value could be considered as underestimated if compared with other areas of the southern Apennines (cf. Fiorillo et al., 2015), due to poor knowledge of the total discharge amount and spring catchment area boundaries of the Alburni Massif.

Considering only the summit plateau (90 km²), this area totally contributes to spring discharge, as all the recharge amounts inside the endorheic areas are assumed to reach the basal springs; in particular it represents 34% of the total Alburni catchment, but provides about half of the effective contribution to spring discharge ($C_s = 0.45$), and is even higher in terms of total rainfall ($C'_s = 0.65$).

The above estimations refer to a long-term scale (annual mean rainfall over a time span of several decades), though annual recharge changes yearly, typically concentrating in specific seasons. Kessler (1967) highlighted the role of the first 4 months of the year in controlling the recharge in a karst environment of Hungary, and its dependence on the amount of rainfall recorded in the previous year (during the last 4 months). These characteristics are even exacerbated in Mediterranean climate areas, especially within the framework of the climate changes we are experiencing. At the Alburni Massif, recharge occurs mainly during the winter and spring seasons, and depends on the previous autumn rainfall and the snowmelt as well, which are needed to satisfy the retention water of the soil cover.

9.5. DISCUSSION AND CONCLUSIONS

As repeatedly demonstrated worldwide, anthropogenic activities may produce significant changes in the hydraulic and hydrogeological regimes of karst areas (Bakalowicz, 1995, 2005; Ozanić & Rubinić, 2003; Ravbar & Sebela, 2015; Chen et al., 2017; Parise et al., 2018). This occurs through a variety of human actions, ranging from land use changes (Foley et al., 2005; Quine et al., 2017; Peng et al., 2020), to quarrying and mining (Gunn, 1993, 2003; Hobbs & Gunn, 1998; Formicola et al., 2010; Parise, 2010, 2016b), variations in the amount and distribution of the natural vegetative cover (Ravbar et al., 2011; Huebsch et al., 2014), and overexploitation of groundwater resources (Hartmann et al., 2012; Finger et al., 2013; Musgrove et al., 2016; Jia et al., 2017). All these actions often lead to severe disturbance to the natural karst environment (Calò & Parise, 2009) as proved through the application of the Karst Disturbance Index (Van Beynen & Townsend, 2005; North et al., 2009) to many different karst settings in the globe (Calò & Parise, 2006; Day, 2011). In the Alburni case study, the rural character of the area, which is a mountain setting mostly dedicated to pasture and with a limited human presence essentially distributed at its borders, is not considered to have in the near future a possible role in changing the hydrological regime. Nevertheless, protection and safeguard of karst groundwater and, more in general, of karst ecosystems (Bonacci et al., 2009; Fleury, 2009; Gabrovsek et al., 2018) needs to be continuously pursued. This is one of the main goals of this contribution, hopefully helping to emphasize this remarkable karst area, aimed at improving and spreading its knowledge among the local inhabitants and the scientific community in the effort to increase the awareness of the natural resources it hosts. It is also worth mentioning the fact that the area is included in a national park (see *Parco Nazionale del Cilento, Alburni e Vallo di Diano*; http://www.cilentoediano.it), which was also declared a Geopark by UNESCO in 2010, thus becoming member of the UNESCO network of Global Geoparks (Aloia et al., 2012; Santangelo et al. 2015).

The analysis presented in this chapter, based upon computation of the recharge at the summit plateau of Alburni Massif and its comparison with the total spring discharge, in spite of many assumptions, shows a general agreement of the outcomes. Nevertheless, this cannot be considered as a definitive result, since many issues still remain to be fully examined and understood. Tracer tests in Alburni have shown in the past how the expected outcomes, in terms of sites of emergence, flow directions and velocity, and discharge values, have often been quite different from those forecasted on the basis of previous knowledge.

In the history of Alburni cave explorations, many tracer tests were addressed to prove the links among the karst systems and the basal springs (Del Vecchio et al., 2013; Parise & Santo, 2017). Among the first outcomes, it has been demonstrated since the 1950s the link between the Castelcivita Caves and the Auso Spring, for a total development of more than 6 km (Santo, 1994). These researches were also useful to develop a first conceptual hydrogeological model along the Calore River. During the 1990s, an automatic datalogger installed at Risorgenza del Mulino provided data that indicated a deep circuit for the water at this spring (T 16.5 °C), as also proved by later cave diving explorations. Further, the delay (24 to 48 hr) in temperature changes after intense rainstorms on the Alburni high plain demonstrated the connection between the vertical systems and the basal water table (Santangelo & Santo, 1997). More recently, other tracing tests demonstrated the hydrogeological connection among the active swallow holes in Piana dei Campitelli and at Grotta del Falco with the nearby spring at Grotta dell'Acqua (Bocchino et al., 2014; Cozzolino et al., 2015). At the same time, the fluorescein was detected also at the waterfall within the Pertosa Cave and at some springs in the Tanagro River, outlining a quite complex scenario, which still needs further data to be entirely understood (Pedrali et al., 2015; Pastore, 2016). In particular, cave diving explorations at Grotta del Falco proved the development of the cave system through one of the main tectonic lines of the massif, the Vallone Lontrano-Petina (Gueguen et al., 2012; Cafaro et al., 2016), which seems to transfer the water from this system to the central part of the Alburni Massif, toward Grava del Fumo and the S. Maria karst system, and, in turn, to the Auso Spring on the southwest foothills of the massif (Fig. 9.4). This tectonic line acts certainly as an

Figure 9.4 Karst features of the Alburni Massif: (a) the sump at Grotta del Falco (photo: GSAVD); (b) view of the shafts in the Parchitiello system (photo: GSAVD); (c) downhill sump in the Grave del Minollo (photo: GSAVD); (d) Auso Spring, at the south foothills of the massif (photo: F. Fiorillo).

important draining structure, as actually previously hypothesized by Bellucci and coworkers (1991).

The so far available tracer test data still hold some doubts regarding the central sector of the summit plateau: whether this is in communication with the southwest or the northeast side of the massif, and if there actually is the possibility of some dispersion within the groundwater network, with different functioning during the dry seasons (when the karst conduits may act independently) and during floods.

In conclusion, notwithstanding the efforts and the many continuing explorations, hydrogeology of the Alburni Massif still has several dark points, which need further work. This was also favored by high dispersion of data in the past, due to lack of communication among cave grottos, and to unpublished materials. The few available data, especially those concerning the spring discharges around the Alburni Massif, make any conclusion quite uncertain, since more detailed surveys and monitoring actions are needed.

Nevertheless, through the example of the Alburni Massif, we have pointed out some of the difficulties inherent in carrying out karst hydrogeology research, and the need for a continuous and updated exchange of information with the cavers exploring the cave systems, since they represent the main source of new data ("the underground truth") in such settings.

ACKNOWLEDGMENTS

The authors are grateful to editors and reviewers for their helpful advice.

REFERENCES

Aloia, A., Guida, D., & Valente, A. (2012). Geodiversity in the geopark of Cilento and Vallo di Diano as heritage and resource development. *Rendiconti Online della Società Geologica Italiana, 21*, 688–690.

Angel, J. C., Nelson, D. O., & Panno, S. V. (2004). Comparison of a new GIS-based technique and a manual method for determining sinkhole density: An example from Illinois' sinkhole plain. *Journal of Cave and Karst Studies, 66*, 9–17.

Bakalowicz, M. (1995). La zone d'infiltration des aquifers karstiques. Methodes d'etude. Structure et fonctionnement. *Hydrogeologie, 4*, 3–21.

Bakalowicz, M. (2005). Karst groundwater: A challenge for new resources. *Hydrogeology Journal, 13*, 148–160.

Bellucci, F., Giulivo, I., Pelella, L., & Santo, A. (1991). Carsismo ed idrogeologia del settore centrale dei Monti Alburni (Campania). *Geologia Tecnica, 3*, 5–12.

Bellucci, F., Giulivo, I., Pelella, L., & Santo, A. (1995). *Monti Alburni. Ricerche speleologiche.* De Angelis, Avellino.

Bocchino B., Del Vecchio, U., De Nitto, L., Lo Mastro, F., Marraffa, M., Maurano, F., et al. (2014). Increasing people's awareness about the importance of karst landscapes and aquifers: An experience from southern Italy. In N. Kukuric, Z. Stevanović & N. Kresic (Eds.), *Proceedings International Conference and Field Seminar "Karst without boundaries"* (pp. 398–405).

Bonacci, O. (1995). Groundwater behavior in karst: Example of the Ombla Spring (Croatia). *Journal of Hydrology, 165*, 113–134. https://doi.org/10.1016/0022-1694(94)02577-X

Bonacci, O. (2001). Monthly and annual effective infiltration coefficient in Dinaric karst: Example of the Gradole karst spring catchment. *Hydrological Sciences Journal, 46*(2), 287–300.

Bonacci, O., & Magdalenić, A. (1993). The catchment area of the Sv. Ivan Karst spring in Istria, Croatia. *Ground Water, 31*(5), 767–773.

Bonacci, O., Pipan, T., & Culver, D. C. (2009). A framework for karst ecohydrology. *Environmental Geology, 56*, 891–900.

Brancaccio, L., Civita, M., & Vallario, A. (1973). Prime osservazioni sui problemi idrogeologici dell'Alburno, Campania. *Bollettino della Società dei Naturalisti in Napoli, 82*.

Brinkmann, R., & Parise, M. (2012). Karst environments: Problems, management, human impacts, and sustainability, An introduction to the special issue. *Journal of Cave and Karst Studies, 74*(2), 135–136.

Cafaro, S., Gueguen, E., Parise, M., & Schiattarella, M. (2016). Morphometric analysis of karst features of the Alburni mts, southern Apennines, Italy. *Geografia Fisica e Dinamica Quaternaria, 39*, 121–128.

Calò, F., & Parise, M. (2006). Evaluating the human disturbance to karst environments in southern Italy. *Acta Carsologica, 35*, 47–56.

Calò, F., & Parise, M. (2009). Waste management and problems of groundwater pollution in karst environments in the context of a post-conflict scenario: the case of Mostar (Bosnia Herzegovina). *Habitat International, 33*, 63–72.

Celico, P., Pelella, L., Stanzione, D., & Aquino, S. (1994). Sull'idrogeologia e l'idrogeochimica dei Monti Alburni. *Geologica Romana, 30*, 687–698.

Cestari, G. (1971). *Note illustrative della Carta Geologica d'Italia alla scala 1:100.000. Foglio 198 Eboli.* Nuova Tecnica Grafica, Roma.

Chen, Z., Auler, A. S., Bakalowicz, M., Drew, D., Griger, F., Hartmann, J., et al. (2017). The World Karst aquifer mapping project: Concept, mapping procedure and map of Europe. *Hydrogeology Journal, 25*, 771–785. https://doi.org/10.1007/s10040-016-1519-3

Cozzolino, L., Damiano, N., Del Vecchio, U., Minieri, G., Testa, L., & Trifone, P. (2015). Prove di colorazione e recenti esplorazioni nell'area della Grotta del Falco, Monti Alburni. In *Proceedings of the Twenty-Second International Congress of Speleology*, 323–328.

Day, M., Halfen, A., & Chenoweth, S. (2011). The cockpit country, Jamaica: Boundary issues in assessing disturbance and using a karst disturbance index in protected areas planning. In P.E. Van Beynen (Ed.), *Karst management* (pp. 399–414). Dordrecht: Springer.

Del Prete, S., Iovine, G., Parise, M., & Santo, A. (2010). Origin and distribution of different types of sinkholes in the plain areas of Southern Italy. *Geodinamica Acta, 23*(1/3), 113–127. https://doi.org/10.3166/ga.23.113-127

Del Vecchio, U., Lo Mastro, F., Maurano, F., Parise, M. & Santo, A. (2013). The Alburni Massif, the most important karst area of southern Italy: History of cave explorations and recent developments. In M. Filippi & P. Bosak (Eds.), *Proceedings 16th International Congress of Speleology, Brno, 21–28 July 2013, 1* (pp. 41–46).

Denizman, C. (2003). Morphometric and spatial distribution parameters of karstic depressions, lower Suwannee river basin, Florida. *Journal of Cave and Karst Studies, 65*, 29–35.

De Riso, R., & Santo, A. (1997). Geologia, evoluzione geomorfologica e frane del Bacino del T. Pietra (Campania). *Quaderni di Geologia Applicata, 4*(2), 19–33.

De Vries, J. J., & Simmers, I. (2002). Groundwater recharge: An overview of processes and challenges. *Hydrogeology Journal, 10*, 5–17.

Dorsaz, J. M., Gironás, J., Escauriaza, C., & Rinaldo, A. (2013). The geomorphometry of endorheic drainage basins: Implications for interpreting and modeling their evolution. *Earth Surface Processes and Landforms, 38*, 1881–1896.

Drogue, C. (1971). Coefficient d'infiltration ou infiltration eficace, sur le roches calcaires. *Actes du Colloque d'Hydrologie en Pays Calcaire, Besancon 15*, 121–130.

Ducci, D. (2007). Intrinsic vulnerability of the Alburni karst system southern Italy. In M. Parise & J. Gunn (Eds.), *Natural and anthropogenic hazards in Karst areas: Recognition, analysis, and mitigation* (pp. 137–152). Geological Society, London, Special Publication 279.

Filippini, M., Squarzoni, G., De Waele, J., Fiorucci, A., Vigna, B., Grillo, B., et al. (2018). Differentiated spring behavior under changing hydrological conditions in an alpine karst aquifer. *Journal of Hydrology, 556*, 572–584.

Finger, D., Hugentobler, A., Huss, M., Voinesco, A., Wernli, H., Fischer, D., et al. (2013). Identification of glacial meltwater runoff in a karstic environment and its implication for present and future water availability. *Hydrology and Earth System Sciences, 17*, 3261–3277.

Fiorillo, F., & Pagnozzi, M. (2015). Recharge process of Matese karst massif southern Italy. *Environmental Earth Sciences, 74*, 7557–7570.

Fiorillo, F., Pagnozzi, M., & Ventafridda, G. (2015). A model to simulate recharge processes of karst massifs. *Hydrological Processes, 29*, 2301–2314.

Fiorillo, F., Pagnozzi, M., & Ventafridda, G. (2019). Analysis of annual mean recharge in main karst systems of southern Italy. *Rendiconti Online Societa Geologica Italiana, 47*, 36–40. https://doi.org/10.3301/ROL.2019.07

Fleury, S. (2009). *Land use policy and practice on karst terrain, living on limestone*. Dordrecht: Springer.

Foley, J. A., DeFries, R., Asner, G. P., Barford, C., Bonan, G., Carpenter, S. R., et al. (2005). Global consequences of land use. *Science, 309* (5734), 570–574. https://doi.org/10.1126/science.1111772

Ford, D. C., & Williams, P. W. (2007). *Karst geomorphology and hydrology*, 2nd ed. Chichester, U.K.: John Wiley & Sons. https://doi.org/10.1002/9781118684986

Formicola, W., Gueguen, E., Martimucci, V., Parise, M., & Ragone, G. (2010). Caves below quarries and quarries above caves: Problems, hazard and research. A case study from southern Italy. *Geological Society of America Abstracts with Programs, 42* (5).

Fragoso-Servon, P., Bautista, F., Frausto, O., & Pereira, A. (2014). Caracterización de las depresiones kársticas (forma, tamaño y densidad) a escala 1:50,000 y sus tipos de inundación en el Estado de Quintana Roo, México. *Revista Mexicana de Ciencias Geológicas, 31*(1), 127–137.

Gabrovsek, F., Peric, B. & Kaufmann, G. (2018). Hydraulics of epiphreatic flow of a karst aquifer. *Journal of Hydrology, 560*, 56–74.

Gioia, D., Schiattarella, M., Mattei, M., & Nico, G. (2011). Quantitative morphotectonics of the Pliocene to Quaternary Auletta basin, southern Italy. *Geomorphology, 134*, 326–343.

Goldscheider, N., & Drew, D. (2007). *Methods in karst hydrogeology. International Association of Hydrogeologists 26*. London: Taylor & Francis.

Gueguen, E., Cafaro, S., Schiattarella, M., & Parise, M. (2012). A new methodology for the analysis of morpho-structural data of karstic caves in the Alburni Mountains of southern Italy. *Rendiconti Online Società Geologica Italiana, 21* (1), 614–616.

Gunn, J. (1993). The geomorphological impact of limestone quarrying. *Catena, 25*, 187–198.

Gunn, J. (2003). Quarrying of limestones. In J. Gunn (Ed.), *Encyclopedia of cave and karst science* (pp. 608–611). London: Routledge.

Gunn, J. (2007). Contributory area definition for groundwater source protection and hazard mitigation in carbonate aquifers. In M. Parise & J. Gunn (Eds.), *Natural and anthropogenic hazards in karst areas: Recognition, analysis, and mitigation*. Geological Society, London, Special Publication 279 (pp. 97–109). https://doi.org/10.1144/SP279.9

Gutierrez, F., Parise, M., De Waele, J., & Jourde, H. (2014). A review on natural and human-induced geohazards and impacts in karst. *Earth-Science Reviews, 138*, 61–88. https://doi.org/10.1016/j.earscirev.2014.08.002

Hartmann, A., Lange, J., Vivo Aguado, A., Mizyed, N., Smiatek, G. & Kunstmann, H. (2012). A multi-model approach for improved simulations of future water availability at a large Eastern Mediterranean karst spring. *Journal of Hydrology, 468–469*, 130–138.

Haryono, E., Trijuni Putro, S., & Suratman, S. (2017). Polygonal karst morphology of Karangbolong area, Java-Indonesia. *Acta Carsologica, 46*(1), 63–72.

Hauselmann, P., Jeannin, P. Y., & Bitterli, T. (1999). Relationships between karst and tectonics: Case-study of the cave system North of Lake Thun Bern, Switzerland. *Geodinamica Acta, 12*(6), 377–387.

Heidari, M., Khanlari, G. R., Taleb Beydokhti, A. R., & Momeni, A. A. (2011). The formation of cover collapse sinkholes in North of Hamedan, Iran. *Geomorphology, 132*, 76–86. https://doi.org/10.1016/j.geomorph.2011.04.025

Hobbs, S. L., & Gunn, J. (1998). The hydrogeological effect of quarrying karstified limestone: Options for protection and mitigation. *The Quarterly Journal of Engineering Geology, 31*, 147–157.

Huebsch, M., Horan, B., Blum, P., Richards, K. G., Grant, J., & Fenton, O. (2014). Statistical analysis correlating changing agronomic practices with nitrate concentrations in a karst aquifer in Ireland. *WIT Transactions on Ecology and the Environment, 182*, 99–109.

Ippolito, F., Ortolani, F., & Russo, M. (1973). Struttura marginale tirrenica dell'Appennino campano: Reinterpretazione di dati di antiche ricerche di idrocarburi. *Memorie della Società Geologica Italiana, 12*, 227–250.

Jia, Z., Zang, H., Zheng, X., & Xu, Y. (2017). Climate change and its influence on the karst groundwater recharge in the Jinci Spring Region, Northern China. *Water, 9*, 267.

Jourde, H., Roesch, A., Guinot, V., & Bailly-Comte, V. (2007). Dynamics and contribution of karst groundwater to surface flow during Mediterranean flood. *Environmental Geology, 51*(5)., 725–730. https://doi.org/10.1007/s00254-006-0386-y

Jouves, J., Viseur, S., Arfib, B., Baudement, C., Camus, H., Collon, P., & Guglielmi, Y. (2017). Speleogenesis, geometry, and topology of caves: A quantitative study of 3D karst conduits. *Geomorphology, 298*, 86–106.

Kessler, H. (1967). Water balance investigations in the karstic regions of Hungary. In *Proceedings of the Symposium "Hydrology of fractured rocks"*. Dubrovnik, October 1965, 1 (pp. 91–105). International Association of Scientific Hydrology, UNESCO.

Klimchouk, A. B. (2000). The formation of epikarst and its role in vadose speleogenesis. In A. B. Klimchouk, D. C. Ford, A. N. Palmer, & W. Dreybrodt (Eds.), *Speleogenesis: Evolution of karst aquifers* (pp. 91–99). National Speleological Society, Huntsville, Alabama, USA.

Martel, R., Castellazzi, P., Gloaguen, E., Trépanier, L., & Garfias, J. (2018). ERT, GPR, InSAR, and tracer tests to characterize karst aquifer systems under urban areas: The case of Quebec City. *Geomorphology, 310*, 45–56.

Miao, X., Qiu, X., Wu, S.-S., Luo, J., Gouzie, D. R., & Xie, H. (2013). Developing efficient procedures for automated sinkhole extraction from Lidar DEMs. *Photogrammetric Engineering Remote Sensing, 79*(6), 545–554. https://doi.org/10.14358/PERS.79.6.545

Musgrove, M., Opsahl, S. P., Mahler, B. J., Herrington, C., Sample, T. L. & Banta, J. R. (2016). Source, variability, and transformation of nitrate in a regional karst aquifer: Edwards aquifer, central Texas. *Science of the Total Environment, 568*, 457–469.

North, L. A., Van Beynen, P. E., & Parise, M. (2009). Interregional comparison of karst disturbance: west central Florida and southeast Italy. *Journal of Environmental Management, 90*, 1770–1781.

Ozanić, N., & Rubinić, J. (2003). The regime of inflow and runoff from Vrana Lake and the risk of permanent water pollution. *RMZ: Materials and Geoenvironment, 50*, 281–284.

Pagnozzi, M., Fiorillo, F., Esposito, L., & Leone, G. (2019). Hydrological features of endorheic areas in Southern Italy. *Italian Journal of Engineering Geology and Environment*, Special Issue 1, 85–9. https://doi.org/10.4408/IJEGE.2019-01.S-14

Palmer, A. N. (1991). Origin and morphology of limestone caves. *Geological Society of America Bulletin, 103*, 1–25.

Palmer, A. N. (2007). *Cave geology*. Cave Books.

Palmer, A. N. (2010). Understanding the hydrology of karst. *Geologia Croatica, 63*, 143–148. https://doi.org/10.4154/gc.2010.11

Parise, M. (2010). The impacts of quarrying in the Apulian karst. In F. Carrasco, J. W. La Moreaux, J. J. Duran Valsero, & B. Andreo (Eds.), *Advances in research in karst media* (pp. 441–447). Springer.

Parise, M. (2011). Surface and subsurface karst geomorphology in the Murge (Apulia, southern Italy). *Acta Carsologica, 40*(1), 79–93.

Parise, M. (2016a). How confident are we about the definition of boundaries in karst? Difficulties in managing and planning in a typical transboundary environment. In Z. Stevanovic, N. Kresic, & N. Kukuric (Eds.), *Karst without boundaries* (pp. 27–38). IAH-Selected Papers on Hydrogeology, 23. CRC Press. https://doi.org/10.1201/b21380-4

Parise, M. (2016b). Modern resource use and its impact in karst areas: Mining and quarrying. *Zeitschrift fur Geomorphologie, 60* (X), 199–216.

Parise, M. (2019). Sinkholes. In W. B. White, D. C. Culver, & T. Pipan (Eds.), *Encyclopedia of caves*, 3d ed.(pp. 934–942). Academic Press, Elsevier.

Parise, M. (2022). Sinkholes, Subsidence and Related Mass Movements. In J. J. F. Shroder (Ed.), *Treatise on Geomorphology, vol. 5* (pp. 200–220). Academic Press, Elsevier. https://dx.doi.org/10.1016/B978-0-12-818234-5.00029-8

Parise, M., & Santo, A. (2017). Tracer tests history in the Alburni Massif (southern Italy). *IOP Conference Series: Earth and Environmental Science, 95*, 062006. https://doi.org/10.1088/1755-1315/95/6/062006

Parise, M., Gabrovsek, F., Kaufmann, G., & Ravbar, N. (2018). Recent advances in karst research: from theory to fieldwork and applications. In M. Parise, F. Gabrovsek, G. Kaufmann, & N. Ravbar (Eds.), *Advances in karst research: Theory, Fieldwork and applications* (pp. 1–24). Geological Society, London, Special Publication 466. https://doi.org/10.1144/SP466.26

Parise, M., Ravbar, N., Živanović, V., Mikszewski, A., Krešić, N., Mádl-Szönyi, J., et al. (2015). Hazards in karst and managing water resources quality. In Z. Stevanović (Ed.), *Karst aquifers: Characterization and engineering* (pp. 601–687). Professional Practice in Earth Sciences, Springer. https://doi.org/10.1007/978-3-319-12850-4_17

Pastore, C. (2016). *Analisi idrogeologica dell'area carsica dei Monti Alburni Salerno, Campania*. Master Thesis, University Alma Mater, Bologna, Italy.

Patacca, E., & Scandone, P. (2007). Geology of the southern Apennines. *Italian Journal of Geosciences*, Special Issue 7, 75–119.

Pedrali, L., Buongiorno, V., Antonini, G., Cafaro, S., & De Nitto, L. (2015). Convergenza di dati per l'esplorazione della Grotta del Falco sul Massiccio degli Alburni (Campania). In *Proceedings of the Twenty-Second International Congress of Speleology*, 537–542.

Peng, J., Tian, L., Zhang, Z., Zhao, Y., Green, S.M., Quine, T.A., et al. (2020). Distinguishing the impacts of land use and climate change on ecosystem services in a karst landscape in China. *Ecosystem Services, 46*, 101199. https://doi.org/10.1016/j.ecoser.2020.101199

Quine, T., Guo, D., Green, S. M., Tu, C., Hartley, I., Zhang, X., et al. (2017). Ecosystem service delivery in karst landscapes: Anthropogenic perturbation and recovery. *Acta Geochimica, 36* (3), 416–420. https://doi.org/10.1007/s11631-017-0180-4

Ravbar, N., & Sebela, S. (2015). The effectiveness of protection policies and legislative framework with special regard to karst landscapes: insights from Slovenia. *Environmental Science Policy, 51*, 106–116.

Ravbar, N., Engelhardt, I., & Goldscheider, N. (2011). Anomalous behavior of specific electrical conductivity at a karst spring induced by variable catchment boundaries: The case of the Podstenjšek spring, Slovenia. *Hydrological Processes, 25*, 2130–2140.

Santangelo, N., & Santo, A. (1997). Endokarst processes in the Alburni massif Campania, Southern Italy: Evolution of ponors and hydrogeological implications. *Zeitschrift fur Geomorphologie, 41*(2), 229–246.

Santangelo, N., Romano, P., & Santo, A. (2015). The geo-itineraries in the Cilento Vallo di Diano Geopark: A tool for tourism development in southern Italy. *Geoheritage, 7*, 319–335.

Santo, A. (1994). Idrogeologia dell'area carsica di Castelcivita (M. Alburni – SA). *Geologia Applicata e Idrogeologia, 28*, 663–673.

Sartoni, S., & Crescenti, U. (1962). Ricerche biostratigrafiche nel Mesozoico dell'Appennino meridionale. *Giornale di Geologia, 2*, Bologna.

Sauro, U. (2005). Closed depressions. In D. C. Culver & W. B. White (Eds.), *Encyclopedia of caves* (pp. 108–127). Amsterdam: Elsevier.

Scandone, P. (1971). *Note illustrative della Carta Geologica d'Italia alla scala 1:100.000. Fogli 199 e 210 Potenza e Lauria*. Nuova Tecnica Grafica, Roma.

Scandone, P. (1972). Studi di geologia lucana: Carta dei terreni della serie calcareo-silico-marnosa e note illustrative. *Bollettino della Società dei naturalisti in Napoli, 81*, 225–300.

Stevanovic, Z. (Ed.) (2015). *Karst aquifers: Characterization and engineering*. Professional Practice in Earth Sciences. Springer.

Turc, L. (1954). Le bilan d'eau des sols: Relations entre les precipitations, l'evaporation, et l'ecoulement. *Annales Agronomiques, 5*, 491–595.

Van Beynen, P. E., & Townsend, K. (2005). A disturbance index for karst environments. *Environmental Management, 36,* 101–116.

Waltham, T., Bell, F., & Culshaw, M. (2005). *Sinkholes and subsidence: Karst and cavernous rocks in engineering and construction.* Springer Praxis.

White, W. B. (2002). Karst hydrology: Recent developments and open questions. *Engineering Geology, 65,* 85–105. https://doi.org/10.1016/S0013-7952(01)00116-8

Williams, P. W. (2008). The role of the epikarst in karst and cave hydrogeology: A review. *International Journal of Speleology, 37,* 1–10. https://doi.org/10.5038/1827-806X.37.1.1

Worthington, S., Ford, D., & Beddows, P. (2001). Characteristics of porosity and permeability enhancement in unconfined carbonate aquifers due to the development of dissolutional channel systems. In G. Gunay, D. Ford, P. Williams, & K. Johnson (Eds.), *Present state and future trends of karst studies* (pp. 13–29). Technical Documents in Hydrology. Paris: UNESCO.

Wu, Q., Deng, C., & Chen, Z. (2016). Automated delineation of karst sinkholes from LIDAR-derived digital elevation models. *Geomorphology, 266,* 1–10. https://doi.org/10.1016/j.geomorph.2016.05.006

Zumpano, V., Pisano, L., & Parise, M. (2019). An integrated framework to identify and analyze karst sinkholes. *Geomorphology, 332,* 213–225.

10

The Great Subterranean Spring of Minneapolis, Minnesota, USA, and the Potential Impact of Subsurface Urban Heat Islands

Greg Brick

ABSTRACT

Anthropogenic subsurface urban heat islands (SUHIs) in groundwater under cities are known worldwide. SUHIs are potentially threats to springs because much spring biota, like trout, amphipods, and rare plants, is cold stenothermal. The city of Minneapolis, Minnesota, USA, has a SUHI documented by the temperature of an underground spring, dubbed Little Minnehaha Falls, inside Schieks Cave, which is located 23 m below the central core of the city. In 2000, the temperature of that spring was elevated 11°C above regional background groundwater temperatures (8°C) at this latitude (45°N). A thermometric survey of the cave and nearby tunnel seepages in 2007 found that an abandoned drill hole through the bedrock ceiling of the cave was discharging groundwater with a temperature of 17.9°C. By comparison, groundwater in the deep water table below the cave was closer to natural background temperatures for the region. The unusually warm groundwater was thereby localized to the strata above the cave. This is the strongest signal of anthropogenic groundwater warming in the state of Minnesota and is attributed to vertical heat conduction from basements and pavements. Minneapolis is unique among SUHIs in that a cave forms a natural collection gallery deep below the city surface, whereas the literature is almost exclusively based on data from observation wells.

10.1. INTRODUCTION

Elevated groundwater temperatures are a potentially important threat to springs because much spring biota is cold stenothermal, examples being trout, amphipods, and rare plants (Brick, 2017b). According to Taniguchi et al. (2007, p. 596), "The heat island effect due to urbanization on subsurface temperature is an important global groundwater quality issue because it may alter groundwater systems geochemically and microbiologically." In the Twin Cities, Minneapolis and St. Paul, Minnesota, USA (Fig. 10.1), there are designated trout streams within the metropolitan area, whose springsheds are

Lands and Minerals Division, Minnesota Department of Natural Resources, St. Paul, Minnesota, USA

heavily built over and will be impacted by rising groundwater temperatures (Meersman, 2012).

According to Taylor and Stefan (2008):

> Urban development can influence groundwater temperatures in a number of ways: Paved surfaces become much warmer than sod surfaces on clear, sunny days. Heat is conducted from these surfaces into the soil, and can reach shallow groundwater. Taniguchi and Uemura (2005) provide evidence of conduction-based warming of groundwater due to urbanization. Surface water runoff from warm ground surfaces can infiltrate from ponds, channels, or rain gardens. Percolating warm water may carry its heat, not lost in the soil, into an aquifer. (p. 6)

Elevated groundwater temperatures below cities are well known worldwide. In Asia, Taniguchi et al. (2009) have documented anthropogenic thermal effects on groundwater in Osaka, Japan, and Bangkok, Thailand. In Europe, Epting et al. (2017) confirm the same trends at

---

Threats to Springs in a Changing World: Science and Policies for Protection, Geophysical Monograph 275, First Edition.
Edited by Matthew J. Currell and Brian G. Katz.
© 2023 American Geophysical Union. Published 2023 by John Wiley & Sons, Inc.
DOI:10.1002/9781119818625.ch10

Figure 10.1 Location of the Minneapolis-St. Paul metropolitan area (star) (adapted from Wikimedia/U.S. Geological Survey).

Basel, Switzerland. According to Hemmerle et al. (2019, p. 1), "This is demonstrated for the city of Paris, where measurements from as early as 1977 reveal the existence of a substantial subsurface urban heat island (SUHI) with a maximum groundwater temperature anomaly of around 7 K." In North America, Ferguson and Woodbury (2004, p. 1; 2007) describe the subsurface heat island effect below Winnipeg, Manitoba: "Downward heat flow to depths as great as 130 m has been noted in some areas beneath the city and groundwater temperatures in a regional aquifer have risen by as much as 5°C in some areas." Yalcin and Yetemen (2009), Zhu et al. (2010), Menberg et al. (2013a), and Bayer et al. (2019), among others, go so far as to consider tapping into SUHIs as a source of geothermal energy. In keeping with conventions in the SUHI literature, from this point on C (Celsius) will be used to indicate a measured temperature, whereas K (Kelvin) indicates a temperature difference.

The earliest known record of groundwater temperature in the Platteville Limestone in what became Minneapolis is from Nicollet (1845). At Coldwater Spring (a Platteville spring about 10 km from Schieks Cave and its underground spring), Nicollet reported an average of 7.8°C in July 1836 and 7.5°C in January 1837 (Nicollet, 1845, p. 69).

Taylor and Stefan (2008, 2009) projected a rise in groundwater temperature for the Twin Cities metropolitan area of 3K (and even higher in global-warming scenarios) in research funded by the Minnesota Pollution Control Agency. However, they were most likely unaware that the first measurement indicating elevated groundwater temperature in Minneapolis had already been reported by an undergraduate geology student years earlier from Chalybeate Springs, a mineral water resort that was popular before the American Civil War. These springs emanate from the Platteville Limestone with a temperature of 14°C, 6K above the expected 8°C at latitude of 45°N (Brick, 1993).

A time series of temperature measurements at Coldwater Spring, on the outskirts of Minneapolis, lasting nearly 2 years (2013–2015), recorded fluctuations from 10.7°C to 13.1°C, well above those of Nicollet (1845) already mentioned. Kasahara (2016, p. iii) attributed this to "an anthropogenic source of heat within the spring-shed or spring discharge area." (See also Alexander & Brick, 2021, p. 98.)

However, while these elevated temperatures are notable, they are half the temperature changes in a spring and a free-flowing well located inside a cave under downtown Minneapolis at a depth of 23 m below street level, which is the focus of the remainder of this chapter.

10.2. BACKGROUND

10.2.1. Cave and Spring

Schieks Cave is the largest cave under downtown Minneapolis, underlying half a city block (Fig. 10.2). This maze cave in the St. Peter Sandstone has a ceiling of Platteville Limestone. The cave is most likely anthropogenic in origin (Brick, 2017a).

The earliest document regarding Schieks Cave (or its spring) is the 1904 Nic. Lund map. Although rather crude and incomplete, the map is rich in hydrologic details such as "creeks" and "lakes." At the location of the ceiling spring, the map notes "WIDE CRACK IN LEDGE, LARGE BODY OF WATER COMING THROUGH." Upon first entering the cave, Carl J. Illstrup, city sewer engineer, reported:

> Dripping from the ceiling at one place there was a regular curtain of water 30 feet (10 m) in width. The water in the middle (of a pool at the bottom of the curtain) was 20 feet (6 m) deep at one point and tapered down to inches at the shore line. It was a beautiful sight but we had to drain it to remedy the troubles in the Fourth Street tunnel. (Fitzsimmons, 1931, p. 1)

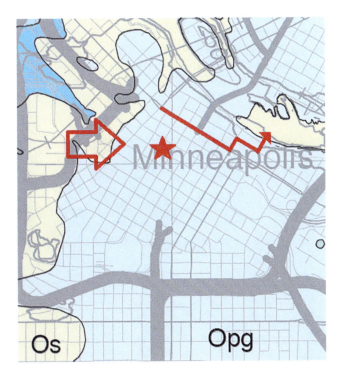

Figure 10.2 Location of Schieks Cave (star) on Hennepin County bedrock map, north at top (Retzler, 2018). Opg = Ordovician Platteville-Glenwood Formation; Os = Ordovician St. Peter Sandstone. Big arrow indicates general direction of groundwater flow in the shallow Quaternary aquifer, adapted from Kanivetsky (1989). Thin arrow represents the Washington Avenue tunnel, where temperature measurements were taken in 2007, and its flow direction to Mississippi River (see Table 10.1).

A second map, by J. E. Lawton in 1929, based on a detailed survey, depicts Schieks Cave extensively modified by the construction of piers, walls, and artificial drainage systems, the latter to prevent further erosion of the soft sandstone. The ceiling spring was shown as "WATER FALLS," but this time enclosed in a separate concrete chamber provided with floor drains. Both cave maps are reproduced as Plate 3 in Kress and Alexander (1980).

Longnecker (1907) described the spring and well in print for the first time and Brick (2021) gives an extended analysis of his narrative from a geological perspective. Longnecker included a photo captioned, "The Subterranean Falls," showing how "the crystal-pure spring water gushes forth between the ledges of limestone and falls into a concrete basin built by the city engineer's staff." Dornberg (1939, p. 8) was the first to refer to the ceiling spring as "Little Minnehaha Falls" (LMF) and label it as such on his map (Fig. 10.3). This was a jocular reference to Minnehaha Falls, a well-known Minneapolis landmark. Zalusky (1953b, p. 6) described LMF as "a falls which I estimated in the darkness to be about 10 feet (3 m) wide and a drop of 5 feet (1.5 m)." So Illstrup's 10 m curtain of water had apparently dwindled to a third of its former length. Kress and Alexander (1980, pp. 65–66) state that "in view of the almost complete cover of the surface by buildings or pavement and the inevitable disruption of the near-surface groundwater flow by the excavation of building foundations, it is not surprising that 'Little Minnehaha Falls' is drying up."

10.2.2. Discovery of Thermal Anomaly

No substantive new information about the spring was forthcoming until my own visit of 28 May 2000. An extended account of the trip is found in Brick (2009, pp. 191–203). Schieks Cave is normally accessed by a 23 m shaft from Fourth Street, and traffic must be diverted to open the manhole, which is why the Minneapolis Sewer Department rarely visits the cave; so it's difficult to acquire additional data.

Spring water pours from a bedding plane in the Platteville Limestone, depositing vertically striped flowstone on the walls of the concrete chamber built to contain the spring (Fig. 10.4). Judging from Dornberg's 1939 photos, the discharge did not appear to have diminished much since then, contrary to other reports. The water fell as an extended sheet about 3 m long (matching Zalusky's dimensions of half a century earlier) and the flow rate was visually estimated at 5 gallons per minute (GPM) (=19 liters per minute). Floor drains convey the water to the North Minneapolis Tunnel, a deep-level sanitary sewer.

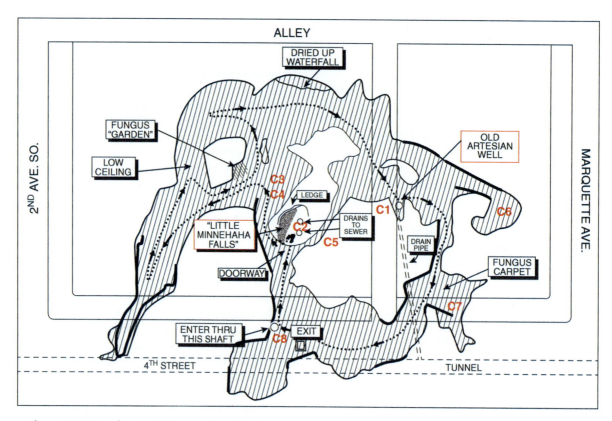

Figure 10.3 Dornberg's 1939 map of Schieks Cave, a maze cave in the St. Peter Sandstone, showing the ceiling spring, dubbed Little Minnehaha Falls, and the Old Artesian Well, a drill hole through the Platteville Limestone forming the ceiling of the cave. Dotted line indicates Dornberg's "route of camera exploration." C1 to C8 are temperature measurement points (see Table 10.1).

Figure 10.4 A concrete chamber inside Schieks Cave hosts the anthropogenically warmed ceiling spring Little Minnehaha Falls, which issues from a bedding plane in the Platteville Limestone, depositing zebra flowstone on the walls (photo by author, 2000).

Figure 10.5 Spring issuing from fissure in the Washington Avenue tunnel, location T-1 (see Table 10.1). Seepage entrains whitish St. Peter sand grains from outside the tunnel lining (photo by author).

Figure 10.6 The Old Artesian Well flowing in the "wrong" direction, showing bottom of drill hole in the Platteville ceiling of Schieks Cave (photo by author).

Upon equilibrating a calibrated SPER Scientific® mercury thermometer (±1%) in the spring orifice for several minutes, I was surprised to note that the groundwater temperature was 19°C, higher than the expected 8°C at this latitude (Brick, 2014).

I also examined a drill hole in the limestone ceiling of the cave, which discharged an estimated 50 GPM (=190 liters per minute), falling into a shallow concrete basin (Fig. 10.6). However, I mistakenly thought this was storm water at the time, only later recognizing it as a water well, so I did not measure its temperature during the first trip. There's no manifestation of this abandoned well at street level, the site being entirely covered with buildings and a car park. Dornberg (1939, p. 1) refers to it as the "Old Artesian Well" (OAW).

10.2.3. Geology

The bedrock layers are Ordovician units making up the Twin Cities basin, overlain by unconsolidated Quaternary glacial sediments (Retzler, 2018). A diagrammatic cross-section of the geology at Schieks Cave is shown as Fig. 10.7. Relative thickness of the layers above the cave are depicted in accordance with Zalusky (1953a), who apparently had access to the driller's field notes from the construction of the Fourth Street entrance shaft.

The Platteville Limestone is karstified, acting as aquifer or aquitard depending on the setting (Steenberg et al., 2011). The largest and most abundant springs in Minneapolis emanate from this layer (Brick, 1997). While a meter thick layer of Glenwood Shale, intercalated between the Platteville and St. Peter, has been omitted for clarity, this layer acts as an aquitard within the Twin Cities basin. Inside Schieks Cave, however, LMF issues from a bedding plane separating the two lowermost members of the Platteville Limestone: the thin Pecatonica (which usually falls away in open voids) and the thicker Mifflin (which remains to form flat ceilings) members. The shale does not appear to play a role here.

Two water tables are depicted in Fig. 10.7: a perched water table in the shallow Quaternary aquifer and Ordovician limestone above the cave, and a deep water table in the St. Peter Sandstone below the cave. The uppermost sandstone has a separate unsaturated zone. Groundwater infiltrates the tunnels through breaks in the concrete lining.

10.3. METHODS

A search for existing well water temperature data was conducted. Two drilled wells penetrate Schieks Cave, one of which is the Old Artesian Well described above. The second is the M. L. & T. Co well, with an intact steel casing that passes entirely through the cave to an unknown depth below, about which no further information exists and which will not be further considered here. These wells are not listed in the online Minnesota Well Index (MWI) database, maintained by the Minnesota Department of Health, apparently because they predate the index, which began in 1974. A query of MWI for all wells within a 1.6 km radius of Schieks Cave found no temperature data for wells terminating in the glacial drift, Platteville Limestone, or St. Peter Sandstone (A. J. Retzler, personal communication, 2021).

On 15 April 2007, I made another trip to Schieks Cave during which I conducted an extended thermometric survey of groundwater in the cave and underlying tunnel system, using the same thermometer as before.

LMF was measured in the same spot as before and the groundwater showering from OAW was measured in a concrete basin built into the floor of the cave. The water in this shallow basin undergoes rapid turnover. The same day, seepages in the Washington Avenue tunnel, as well as flowing nonwet weather storm water where tributary tunnels entered the main tunnel (for comparison purposes) were measured. The tunnel forms a 1.4 km transect from northwest to southeast at an average depth of 30 m (Fig. 10.2). The deep-water table in the St. Peter Sandstone was accessed where it seeped through breaks in the concrete tunnel lining, usually leaving reddish staining on the walls (Fig. 10.5).

10.4. RESULTS

The water temperatures from the 2000 and 2007 visits are listed in Table 10.1 along with descriptions of sampling points. The data include temperature measurements

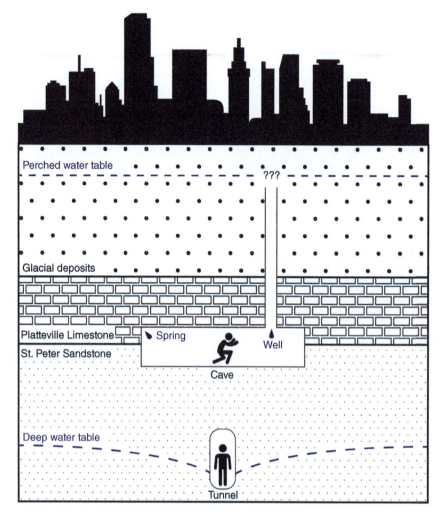

Figure 10.7 Geological cross section of Minneapolis at Schieks Cave (not to scale). LMF (spring) and OAW (well) are shown inside cave. A 1 m layer of Glenwood Shale between the Platteville and St. Peter has been omitted for clarity. The tunnel is 30 m below ground surface (artwork by Jessica Rogge).

Table 10.1 Thermometric survey of cave and tunnels, 15 April 2007*

Origin	Description/location of temperature readings	Temp °C
Platteville Limestone groundwater in Schieks Cave.	C1: Old Artesian Well (OAW) concrete basin (rapid turnover)	17.9
	C2: Little Minnehaha Falls (LMF) ceiling spring at bedding plane*	17.1
	C3: Galvanized drip basin (south)	17.0
	C4: Galvanized drip basin (north)	16.9
	C5: Black Medusa formation, flowing like faucet	16.9
	C6: ML&T Co well hole, from still pool around steel casing	16.5
	C7: Seepage on flowstone in concrete drain	16.5
	C8: Black drip pool at base of ladder, seeping around steel shaft lining	16.0
St. Peter Sandstone seeps in storm drains via gaps in concrete lining.	T1: Red springs near floor, Washington Ave S, half way between 3rd & 4th Ave S	14.0
	T2: Red spring in floor of side passage, Washington Ave S/3rd Ave S	12.5
	T3: Leakage jetting from tunnel lining with red staining, Washington Ave S/Hennepin Ave	13.0
Non-wet weather flows in storm drains measured in tributary just before it joins flow in Washington Avenue tunnel.	T4: Hennepin Ave at Washington Ave S	13.0
	T5: Nicollet Mall at Washington Ave S	15.5
	T6: Marquette Ave at Washington Ave S	14.5
	T7: Side-passage half way between Marquette Ave & 2nd Ave S at Washington Ave S	13.0
	T8: 2nd Ave S at Washington Ave S	18.0
	T9: Iron Gate pool, Portland Ave/Washington Ave S	15.5
	T10: Chicago Ave Outfall at Mississippi River	13.5

*On 28 May 2000, the temperature of LMF was measured at 19°C. This was the only temperature measurement prior to the thermometric survey of 15 April 2007.

from (1) Platteville Limestone groundwater in Schieks Cave (C1–C8), which ranged from 16°C to 17.9°C; (2) St. Peter Sandstone seepage in the Washington Avenue tunnel (T1–T3), from 12.5°C to 14°C; and (3) storm water flow in the same tunnel (T4–T10), from 13°C to 18°C. Employing the terminology of Benz et al. (2017), the anthropogenic heat intensity (AHI) was 9.9K for this SUHI at the time of the 2007 survey. The AHI reports the difference between the local value and the average rural groundwater background temperature.

The Schieks Cave features and their temperature sampling points are shown in Fig. 10.3. LMF was 19°C in 2000 and 17.1°C in 2007, while groundwater captured in the OAW concrete basin was 17.9°C. The cave air temperature was 18°C, measured inside the concrete chamber containing LMF. The average surface air temperature for the month of April, 2007, recorded at the Minneapolis-St. Paul International Airport, was 8.44°C (NOAA, 2021).

Much of the nonwet weather baseflow of the Washington Avenue tunnel is derived from groundwater infiltration and each tributary tunnel (T4–T8) contributes its own separate temperature. The farthest downstream measuring point, the Chicago Avenue Outfall (T10), was 13.5°C, which presumably represents the mean temperature of all contributing flows in the storm drains mixed together. This temperature falls within the St. Peter seepage range, suggesting its ultimate origin as infiltration water.

10.5. DISCUSSION

The flows from LMF and OAW are documented inside Schieks Cave from the earliest map (1904) until today, more than a century, suggesting that the water does not originate from leaking pipes or water mains (cf. Lerner, 1986).

Other possible sources of heat to groundwater can be excluded. Tissen et al. (2019) considered potential natural causes of elevated temperatures for SUHIs, especially hot springs, which are not found in Minnesota. Tissen et al. (2019) also considered Acid Mine Drainage (AMD). While the Platteville Limestone contains pyrites, subject to oxidation (an exothermic reaction), monitoring of Platteville springs has never detected elevated pH among them (Minnesota Department of Natural Resources, 2021).

Groundwater entering Schieks Cave was elevated above background by 11K in 2000, larger than the 3K predicted by Taylor and Stefan (2008, 2009) for the Minneapolis latitude. As a conceptual model, basements and pavements are warming the shallow Quaternary aquifer by vertical conduction (Fig. 10.8) and this water finds its way via OAW and fissures down through the Platteville Limestone, into the cave.

LMF was 19°C in May 2000 and 17.1°C in April 2007, a 1.9K decrease during this 7 year interval. According to Taylor and Stefan (2008), seasonal temperature

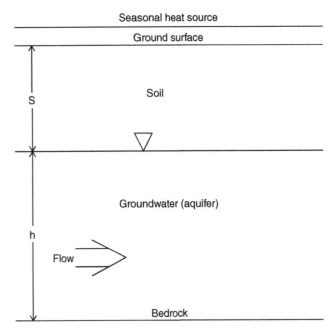

Figure 10.8 Schematic of aquifer of thickness h at depth S below the ground surface (adapted from Taylor & Stefan, 2009, p. 21). At Schieks Cave, S = 10.7 m and h = 5.5 m. The soil compartment represents the unsaturated zone.

fluctuations penetrate the ground to depths of 10 to 15 m, within the depth range of the shallow Quaternary aquifer at this location. Continuous, multiyear temperature monitoring would be necessary to determine the magnitude of fluctuations at LMF. Kasahara (2016) documented a systematic, seasonal fluctuation in the water temperature in Coldwater Spring of 2.3K (10.8 to 13.1°C), which lagged the surface air temperature by about 4 months. The change in temperature at LMF is within the range of documented seasonal fluctuation in the Platteville aquifer in Minneapolis.

Located within 23 m of the surface, LMF and OAW reveal elevated groundwater temperatures that are also within the range of depth for thermal anomalies measured by Taniguchi et al. (2007, p. 596) in boreholes in four Asian cities: "The depth of deviation from the regional geothermal gradient was deepest in Tokyo (140 m), followed by Osaka (80 m), Seoul (50 m), and Bangkok (50 m)."

LMF and OAW, located 27 m apart inside Schieks Cave, had temperatures within 1K of each other during the only event in which they were simultaneously measured, which would be expected if they are derived from the same aquifer.

At OAW, the temperature went from feeling "icy" (Longnecker, 1907), and thus presumably normal groundwater temperature of 8°C, to 17.9°C in the 100 years from 1907 to 2007. While the first observation is qualitative and the second is quantitative, the century-long trend is unambiguous. This is strongest signal of anthropogenic groundwater warming in the state of Minnesota.

Where seepage waters were encountered in the St. Peter Sandstone below the Platteville Limestone, they were several degrees cooler, yet still about 5K above background groundwater temperatures. The tunnel seepages (T1–T3) cannot be easily revisited because the Central City Tunnel System project (2020–2023) plans to eliminate infiltration and the proposed construction of relief sewers would likely disturb the hydrology of the setting (CDM Smith Inc., 2018).

The warm groundwater pool of the SUHI thus appears to be perched in the strata above Schieks Cave, most likely (given vertical heat conduction considerations) in the shallow Quaternary aquifer but also the Platteville Limestone.

Attard et al. (2016) defined a thermally affected zone (TAZ) around urban structures as where the groundwater temperature is elevated 0.5K above expected values. Menberg et al. (2013b) found hotspots below German cities elevated as much as 20K. Menberg et al. (2013c) concluded:

> By modeling the anthropogenic heat flux into the subsurface of the city of Karlsruhe, Germany, in 1977 and 2011, we evaluate long-term trends in the heat flux processes. It revealed that elevated GST (ground surface temperature) and heat loss from basements are dominant factors in the heat anomalies. (p. 9747)

In the case of Schieks Cave, the groundwater direction in the shallow Quaternary aquifer indicates flow from the west through the heavily commercialized Nicollet Mall and the densest skyscraper cluster toward the cave (Kanivetsky, 1989). The source of the heat is likely due to heated buildings as was shown by Krcmar et al. (2020), who reported groundwater temperatures had risen 3.2K by flowing past a building in Bratislava, Slovakia.

Minneapolis is rare among SUHIs in that while the literature is almost exclusively based on data from observation wells, this example involves a cave functioning as a collection gallery for groundwater deep below the city surface. Schieks Cave thus affords a parallel with the historic cellar of the Paris Observatory, which lies at a comparable depth (28 m) and has been used for temperature measurements for centuries (Dettwiller, 1970).

Scott Alexander (personal communication, 2021), hydrogeologist at the Earth Sciences Department at the University of Minnesota, stated that several shallow campus wells in the Platteville Limestone (used for teaching purposes) have had a long-term temperature of about 25°C. His opinion is that these elevated temperatures are due to the campus steam tunnel system, which is carved in the underlying St. Peter Sandstone. If true, this campus SUHI would be the inverse of the Schieks thermal anomaly in that the greater heat source is from below, rather than above.

The twin city of Minneapolis is the neighboring city of St. Paul, capital of the state of Minnesota. Unlike its twin, however, St. Paul features a multilevel utility tunnel system carved within the St. Peter Sandstone, on a larger scale than the university campus (Brick, 2009, pp. 179–188). One of the levels contains a district heating system employing hot water. The conjectural St. Paul SUHI would likely involve both the shallow (perched) and deep water tables. This SUHI could be more intense than that of Minneapolis, even though St. Paul has a smaller population, challenging some fundamental assumptions about urban heat islands based on population modeling (e.g., Oke, 1973).

10.6. CONCLUSIONS

Anthropogenic subsurface urban heat islands (SUHIs) in groundwater under cities are known worldwide. A subsurface urban heat island (SUHI) under the center of the built-up area of Minneapolis, Minnesota, USA, was first detected at a spring inside Schieks Cave located 23 m below street level. SUHIs are potentially threats to springs because much spring biota, like trout, amphipods, and rare plants, is cold stenothermal.

Minneapolis is rare among SUHIs in that while the literature is almost exclusively based on data from observation wells, this case involves a cave functioning as a collection gallery for groundwater deep below the city surface.

A thermometric survey of the cave and tunnel seepages in 2007 revealed a temperature trend extending from 1907 to 2007, showing rising temperature from normal groundwater temperatures for this latitude (8°C) to 17.9°C in one century. This is strongest signal of anthropogenic groundwater warming in the state of Minnesota and is attributed to vertical heat conduction from basements and pavements.

Future studies of the spring and well in Schieks Cave should involve multilevel sampling, data loggers, lengthy time series spanning several years to detect possible seasonal trends, and MODFLOW visualization of data. Any future trip should take water samples as well as additional temperature readings of this feverish spring.

ACKNOWLEDGMENTS

The author would like to thank Andrew Retzler (Minnesota Geological Survey) and Scott Alexander (University of Minnesota) for assistance with groundwater temperature data. The recommendations of two reviewers, E. Calvin Alexander Jr. (University of Minnesota) and Kathrin Menberg (Karlsruher Institut für Technologie), made immense improvements in the manuscript.

REFERENCES

Alexander, E. C., Jr., & Brick, G. A. (2021). Minnesota caves and karst. In G. A. Brick & E. C. Alexander Jr. (2021). *Caves and karst of the upper midwest, USA*. Switzerland: Springer International Publishing. https://doi.org/10.1007/978-3-030-54633-5

Attard, G., Rossier, Y., Winiarski, T., & Eisenlohr, L. (2016). Deterministic modeling of the impact of underground structures on urban groundwater temperature. *Science of the Total Environment, 572*, 986–994.

Bayer, P., Attard, G., Blum, P., & Menberg, K. (2019). The geothermal potential of cities. *Renewable and Sustainable Energy Reviews, 106*, 17–30.

Benz, S. A., Bayer, P., & Blum, P. (2017). Identifying anthropogenic anomalies in air, surface and groundwater temperatures in Germany. *Science of the Total Environment, 584–585*, 145–153.

Brick, G. A. (1993). *The Platteville Springs of Minneapolis*. Term Paper, GEO 5613, Fall Quarter 1993. Minneapolis: University of Minnesota.

Brick, G. A. (1997). Along the Great Wall: Mapping the springs of the Twin Cities. *Minnesota Ground Water Association Newsletter, 16*(1), 1–7. https://www.mgwa.org/newsletter/mgwa1997-1.pdf

Brick, G. A. (2009). *Subterranean Twin Cities*. Minneapolis: University of Minnesota Press.

Brick, G. A. (2014). Thermal pollution of groundwater under Minneapolis. MGWA Spring Conference. https://www.mgwa.org/meetings/2014_spring/2014_spring_abstracts.pdf

Brick, G. A. (2017a). A third kind of cave in the world: Anthropogenic. *Proceedings of the 17th International Congress of Speleology, Sydney, Australia, II*, 248.

Brick, G. A. (2017b). *Minnesota spring inventory guidance document*. Minnesota Department of Natural Resources, Ecological and Water Resources, County Geologic Atlas Project, St. Paul. https://files.dnr.state.mn.us/waters/groundwater_section/mapping/msi/MSI_GuideDoc.pdf

Brick, G. A. (2021). "This strange voyage": Significance of the 1907 Longnecker narrative for underground Minneapolis. *Minnesota Speleology Monthly, 52*(1), 3–11.

CDM Smith Inc. (2018). Preliminary design report, Central City Tunnel System, City of Minneapolis. https://www.minneapolismn.gov/media/-www-content-assets/documents/wcmsp-214802.pdf

Dettwiller, J. (1970). Deep soil temperature trends and urban effects at Paris. *Journal of Applied Meteorology, 9*(1), 178–180.

Dornberg, D. (1939). Camera safari explores "Lost World" under loop. *Minneapolis Journal*, April 16, pp. 1, 8.

Epting, J., Scheidler, S., Affolter, A., Borer, P., Mueller, M. H., Egli, L., et al. (2017). The thermal impact of subsurface building structures on urban groundwater resources: A paradigmatic example. *Science of the Total Environment, 596–597*, 87–96.

Ferguson, G., & Woodbury, A. D. (2004). Subsurface heat flow in an urban environment. *Journal of Geophysical Research, 109*, B02402.

Ferguson, G., & Woodbury, A. D. (2007). Urban heat island in the subsurface. *Geophysical Research Letters, 34*, L23713–L23713.

Fitzsimmons, R. J. (1931). Burrowing workers risk lives for public's health. *Minneapolis Tribune*, August 23, p. 1.

Hemmerle, H., Hale, S., Dressel, I., Benz, S, A., Attard, G., Blum, P., et al. (2019). Estimation of groundwater temperatures in Paris, France. *Geofluids,* Article ID 5246307. https://doi.org/10.1155/2019/5246307

Kanivetsky, R. (1989). Quaternary hydrogeology. In N. H. Balaban (Ed.), *Geologic atlas of Hennepin County, Minnesota*. County Atlas C-4. St. Paul: Minnesota Geological Survey, Plate 5. https://hdl.handle.net/11299/58491

Kasahara, S. (2016). *A hydrogeological study of Coldwater Spring, Minneapolis, MN*. Honors Thesis, University of Minnesota. https://hdl.handle.net/11299/182336

Krcmar, D., Flakova, R., Ondrejkova, I., Hodasova, K., Rusnakova, D., Zenisova, Z., et al. (2020). Assessing the impact of a heated basement on groundwater temperatures in Bratislava, Slovakia. *Groundwater, 58*(3), 406–412.

Kress, A., & Alexander, E. C., Jr. (1980). Farmers and Mechanics Bank cave. In E. C. Alexander Jr. (Ed.), *An introduction to caves of Minnesota, Iowa, and Wisconsin* (pp. 59–67). Guidebook for the 1980 National Speleological Society Convention, Huntsville, AL.

Lerner, D. N. (1986). Leaking pipes recharge groundwater. *Groundwater, 24*(5), 654–662.

Longnecker, J. (1907). In caverns of eternal night. *Minneapolis Tribune*, July 7.

Meersman, T. (2012). The comeback of urban trout: Brook trout dart in Ike's Creek in Bloomington, a rehabilitated sliver of nature not far from mall parking ramps and jets flying low overhead. *Minneapolis Star Tribune*, June 5, B1, B5.

Menberg, K., Bayer, P., & Blum, P. (2013a). Elevated temperatures beneath cities: An enhanced geothermal resource. In *Proceedings of the European Geothermal Congress, Pisa, Italy*.

Menberg, K., Bayer, P., Zosseder, K., Rumohr, S., & Blum, P. (2013b). Subsurface urban heat islands in German cities. *Science of the Total Environment, 442*, 123–133.

Menberg, K., Blum, P., Schaffitel, A., & Bayer, P. (2013c). Long-term evolution of anthropogenic heat fluxes into a subsurface urban heat island. *Environmental Science & Technology, 47*(17), 9747–9755.

Minnesota Department of Natural Resources (2021). Minnesota spring inventory. https://www.dnr.state.mn.us/waters/groundwater_section/springs/msi.html

Nicollet, J. N. (1845). *Report intended to illustrate a map of the hydrographical basin of the Upper Mississippi River*. Document of the U.S. House of Representatives, 28th Congress, 2nd Session, No. 52. Washington, D.C.: Blair and Rives. http://www.worldcat.org/title/report-intended-to-illustrate-amap-of-the-hydrographical-basin-of-the-upper-mississippiriver/oclc/166586906?referer=di&ht=edition

NOAA National Centers for Environmental Information (2021). *April 2007, local climatological data, Minneapolis, MN*. https://www.ncdc.noaa.gov/cdo-web/datasets/LCD/stations/WBAN:14922/detail

Oke, T. R. (1973). City size and the urban heat island. *Atmospheric Environment, 7*(8), 769–779.

Retzler, A. J. (2018). Bedrock geology, Plate 2. In J. R. Steenberg et al. (Eds.), *Geologic atlas of Hennepin County, Minnesota, County atlas series C-45, Part A*. Minnesota Geological Survey.

Retrieved from the University of Minnesota Digital Conservancy. https://hdl.handle.net/11299/200919

Steenberg, J., Runkel, A., & Tipping, R. (2011). Hydrostratigraphy of a fractured, urban aquitard: The Platteville formation in the Twin Cities metropolitan area. *MGWA Newsletter, 30*(4), 23–26.

Taniguchi, M., Shimada, J., Fukuda, Y., Yamano, M., Onodera, S. I., Kaneko, S., et al. (2009). Anthropogenic effects on the subsurface thermal and groundwater environments in Osaka, Japan, and Bangkok, Thailand. *Science of the Total Environment, 407*(9), 3153–3164.

Taniguchi, M., Takeshi, U., & Karen, J. (2007). Combined effects of urbanization and global warming on subsurface temperature in four Asian cities. *Vadose Zone Journal, 6*(3), 591–596.

Taniguchi, M., & Uemura, T. (2005). Effects of urbanization and groundwater flow on the subsurface temperature in Osaka, Japan. *Physics of the Earth and Planetary Interiors, 152*(4), 305–313.

Taylor, C. A., & Stefan, H. G. (2008). *Shallow groundwater temperature response to urbanization and climate change in the Twin Cities metropolitan area: Analysis of vertical heat convection effects from the ground surface.* SAFL Report No. 504. http://home.safl.umn.edu/bmackay/pub/pr/pr504.pdf

Taylor, C. A., & Stefan, H. G. (2009). Shallow groundwater temperature response to climate change and urbanization. *Journal of Hydrology, 375*(3–4), 601–612. https://doi.org/10.1016/j.jhydrol.2009.07.009

Tissen, C., Benz, S. A., Menberg, K., Bayer, P., & Blum, P. (2019). Groundwater temperature anomalies in central Europe. *Environmental Research Letters, 14*(10), 104012.

Yalcin, T., & Yetemen, O. (2009). Local warming of groundwaters caused by the urban heat island effect in Istanbul, Turkey. *Hydrogeology Journal, 17*(5), 1247–1255.

Zalusky, J. W. (1953a). The Loop Cave continued. *Hennepin County History, 13*(2, April), 2–3, 5.

Zalusky, J. W. (1953b). What about the Loop Cave? *Hennepin County History, 13*(1; January), 5–6.

Zhu, K., Blum, P., Ferguson, G., Balke, K. D., & Bayer, P. (2010). The geothermal potential of urban heat islands. *Environmental Research Letters, 5*(4), 044002.

Part III
Policy and Governance Approaches for the Protection of Springs

11

Community-Based Water Resource Management: Pathway to Rural Water Security in Timor-Leste?

Tanja Rosenqvist[1], George Goddard[2], Jack Nugent[3], Nick Brown[1], Eugenio Lemos[4], Elsa Ximenes[5], and Aleixo Santos[6]

ABSTRACT

Achieving water security is a growing challenge for many rural communities in Timor-Leste. Rural communities typically rely on natural springs for their daily water needs, but fluctuations in the quantity and quality of water are impacting water security. Some communities have started exploring community-based water resource management (CBWRM) as a pathway to improve water resilience. They have participated in training by the Timorese NGO Permatil and learned to construct retention basins and revegetate catchments to improve spring recharge. In this chapter, we explore whether rural communities, many who already struggle to maintain water infrastructure under a community-based management model (CBM), are willing to take on further water responsibilities after such training and whether CBWRM is likely to improve water resilience. We draw on qualitative interviews conducted with community members from six rural communities, and apply socioecological-technical systems (SETS) lens to explore water resilience under CBWRM. This study is timely as the government of Timor-Leste is in the process of implementing integrated water resource management (IWRM) at a national level. The study provides suggestions for the role communities and CBWRM may play in IWRM in Timor-Leste going forward.

11.1. INTRODUCTION

Achieving water security is a growing challenge for many rural communities in Timor-Leste who rely on natural springs for both irrigation and domestic use. Water security is defined by UN-Water (2013) as "the capacity of a population to safeguard sustainable access to adequate quantities of acceptable quality water." While 80.5% of rural residents have access to a so-called basic drinking water service (WHO and UNICEF, 2021a), water supply is not always reliable. Basic water supply is defined as: "drinking water from an improved source, provided collection time is not more than 30 minutes for a roundtrip including queuing." An improved source is accessible on premises, available when needed, free from contamination (WHO and UNICEF, 2021b) Many communities experience fluctuations in the quantity and quality of spring water (Saikal, 2019). This is caused by a range of shocks and ongoing stresses, including natural weather patterns, changing water use practices, population growth, challenges maintaining infrastructure, the widespread use of slash-and-burn agriculture, deforestation, disasters, and climate change.

[1] Royal Melbourne Institute of Technology, Melbourne, Australia
[2] Engineers Without Borders Australia, Brisbane, Australia
[3] Engineers Without Borders Australia, Melbourne, Australia
[4] Permatil, Dili, Timor-Leste
[5] Engineers Without Borders Australia, Dili, Timor-Leste
[6] Plan International, Dili, Timor-Leste

Threats to Springs in a Changing World: Science and Policies for Protection, Geophysical Monograph 275, First Edition.
Edited by Matthew J. Currell and Brian G. Katz.
© 2023 American Geophysical Union. Published 2023 by John Wiley & Sons, Inc.
DOI:10.1002/9781119818625.ch11

Over the last couple of years, some communities in Timor-Leste have started experimenting with community-based water resource management (CBWRM) to improve water security. They are supported by the Timorese nongovernment organization (NGO) Permaculture Timor Lorosa'e (Permatil), which provides training on how to protect and rejuvenate local springs. This training typically includes understanding geographical and topographical features, the construction of retention basins, Earth mounds and check dams, revegetating catchments, and changing community behavior. Members of more than 70 communities have joined Permatil training, to date.

Rural communities are already, by law, responsible for operating and maintaining water infrastructure (i.e., spring boxes, pipes, and tap stands) through community-based management (CBM), and several studies have shown that many experience significant challenges doing so (Hamel, 2009; Bond, 2009; Willetts, 2012; Saikal, 2019; World Bank, 2018). This raises the questions, if communities are already struggling to maintain water infrastructure, are they then willing to take on further water responsibilities through CBWRM, and if so, is it likely that CBWRM will improve water security? In this chapter, we answer these two questions. We take the stance that improved water security is achieved by improving the resilience of rural water systems (Rodina, 2018) and draw on a socioecological-technical systems (SETS) framework introduced by Markolf et al. (2018) to explore rural water system resilience in Timor-Leste under CBWRM.

Exploring community willingness and ability to implement CBWRM in Timor-Leste is timely. The government of Timor-Leste (GoTL) is in the process of implementing integrated water resource management (IWRM) and is setting up new departments and drafting legislation to support this. As the final governance structures are still uncertain, it is an opportune time to consider what role communities can play in the move toward IWRM and whether CBWRM may be one.

This chapter has eight sections. We start with a short introduction to community-managed rural water supply in Timor-Leste, followed by a section outlining the growing challenge of water security. In the next section, we describe current government and community initiatives to improve rural water security through IWRM and CBWRM. This leads to an introduction to the SETS framework used to analyze the data. This is followed by a methodology section and a findings section in which we answer the two questions posed above. Finally, in the recommendation section, we propose some pathways toward more resilient rural water supply through IWRM and CBWRM. The chapter ends with a conclusion.

11.2. COMMUNITY-MANAGED RURAL WATER INFRASTRUCTURE

To date, the main focus of the Timorese water sector has been to reestablish and expand water infrastructure. This is partly in response to the destruction of large parts of the nation's water systems during the withdrawal of Indonesia and the fight for independence in 2002 (GoTL, 2019).

Despite the need to rebuild infrastructure, progress toward the Sustainable Development Goal (SDG) 6 (Ensuring availability and sustainable management of water for all) (United Nations, 2015), has been impressive. Data from the WHO and UNICEF Joint Monitoring Program (JMP) show that 85.5% of the national population had access to a basic water service in 2020, up from just 55.7% in 2002 (WHO & UNICEF, 2021a). This is remarkable progress, especially considering some of the practical challenges faced; for example, 70% of Timor-Leste's 1.2 million people live in rural areas (Direcção Geral de Estatiística, 2015), many of which are highly mountainous, with 44% of the country having a slope greater than 40% (Wallace et al., 2012).

The rapid expansion of water infrastructure has partly been a result of the government of Timor-Leste (GoTL) giving it high priority. A key aim of the Timor-Leste Strategic Development Plan 2011–2030 is to ensure "a safe pipe 24-hour water supply to all households by 2030" (GoTL 2010, p.78). This promise has led to significant government investment in infrastructure (Willetts, 2012) and to the development of extensive water supply legislation. A set of laws, developed over the past two decades, lays out the division of roles and responsibilities for the implementation and maintenance of water systems between various stakeholders from governments ministries to communities. This includes Decree Law 04/2004 (Regime of water distribution for public consumption) and Government Resolution 43/2020 (National policy for public water supply) (GoTL, 2004, 2020a). As will become apparent later, the governance of water resources (rather than infrastructure) is not as clear.

Rural communities have also played an essential role in the rapid expansion of water systems. With the support from nongovernment organizations (NGOs), international donors and/or the GoTL, rural communities have designed, built, and funded gravity-fed water systems drawing from natural springs. These systems typically collect spring water in spring boxes and, utilizing the mountainous topography, transport the water through simple piped distribution systems either directly to households and/or to shared standpipes. Figure 11.1a to c shows examples of typical spring boxes found in rural communities in Timor-Leste. Figure 11.1d and e shows typical community water collection points and tap stands fed by the spring boxes. Figure 11.1f and g

Figure 11.1 Examples of water infrastructure in rural Timor-Leste: (a), (b), and (c) show spring boxes; (d) and (e) show water collection points; (f) shows a typical distribution system; (g) shows a spring box and water tank.

shows other typical components of rural water systems including plastic tanks and open guttering.

Rural water systems are community-managed in Timor-Leste, which means that after implementation, the system is handed over to a community-based organization (CBO), a so-called *grupu maneja fasilidade* (GMF) or, in English, a facility management group (GoTL, 2010). The GMF is responsible for collecting water-user tariffs and minor rehabilitation, while Municipal Water, Sanitation and Environment Services (SMASA), with support from the National Suco Development Program (PNDS), is responsible for major rehabilitation and system implementation. This management model is inscribed into Decree Law No. 4/2004 (GoTL, 2004), which requires water systems servicing less than 1,000 people to be owned and operated by a GMF. A GMF is typically made up of three volunteer community members: a group leader, a treasurer, and a technician. The GMF is elected by the community and formally registered with SMASA. It receives ongoing support from a *facilitador postu administrativu* (FPA) or subdistrict facilitator, who is believed to play an essential role in the success of community-managed water service delivery (Saikal, 2019). GMFs also receive support from the National Directorate for Water Services (DNSA) through the development of guidelines for infrastructure maintenance and through funding for water system rehabilitation (Willetts, 2012).

11.3. GROWING WATER SECURITY CHALLENGES

While the reliance on CBM has allowed for rapid implementation of water systems with limited resources, many rural communities in Timor-Leste are now dealing with water security challenges. In fact, several studies have shown that the functionality of rural water systems in Timor-Leste, a measure of both technical functionality, spring water availability, and water quality, often decreases over time. In 2009, Hamel (2009) surveyed all water systems in rural Aileu and Lautem districts and found that only 30% were fully functioning just one year after implementation and that only 15% were fully functioning 6 years after. (Systems were considered "fully functioning" if there was: an appropriate quantity of water less than 100 m from 80% of *aldeia* (subvillage, administrative division), a permanent source of water through the year, no major problem of water quality, and no or limited number of missing or broken elements (Hamel, 2009).) The more recent, 2019 Sustainable Water in Timor-Leste Municipalities (SWiM) study (referenced as Saikal, 2019) explored the functionality of 220 systems in Aileu and Ainaro districts. (The functionality of water systems was calculated based on (1) the structural integrity of the water system, its resilience against disasters, and whether measures were in place to minimize contamination; (2) the number of tapstands flowing compared to the total number of tapstands; and (3) seasonal variations in water availability. Each factor was given a one-third weighting in the overall functionality score (Saikal, 2019).) The study found that 72% were either nonfunctional (15%), required major repairs (20%), or were semifunctional (38%) (Saikal, 2019). The remaining 28% were considered mostly functional (Saikal, 2019).

Next we outline some of the key challenges influencing the functionality of rural water systems in Timor-Leste, which in turn affect rural water security. This includes challenges related to CBM, water quantity, and water quality. As will become apparent later, these challenges are deeply interrelated and mutually reinforcing.

11.3.1. Challenges of Community-Based Management

The functionality of water systems is strongly influenced by the challenges communities experience with CBM. Over the past decade, GoTL and NGOs have sought to improve CBM through better community participatory planning processes and by training GMF members in Operation and Maintenance (O&M) and financial literacy (Willetts, 2012). However, despite these efforts, recent studies confirm that community management remains a significant challenge in the majority of rural communities (Hamel, 2009; Bond, 2009; Willetts, 2012; Saikal, 2019; World Bank, 2018). In fact Hamel (2009) argues that the main cause of water system failure is the failure of CBM. A lack of active GMFs (Hamel, 2009) and low acceptance around paying water fees (Bond, 2009) are two of the primary challenges. Support from FPAs may also be insufficient. A study across five municipalities indicated that FPAs were not performing the recommended three visits a year per community to provide support to the GMFs and were providing limited operation and maintenance support in the communities surveyed (WaterAid, 2016). Other challenges include social issues in communities including conflict, theft, and vandalism, environmental issues such as landslides, as well as a lack of institutional support (Willetts, 2012).

It is worth noting that challenges of CBM described here are not unique to Timor-Leste. Rural water supply has been found to be unsustainable under CBM in many parts of the world (see, e.g., Berner & Phillips, 2005; Chowns, 2015; Hope, 2015; van den Broek & Brown, 2015).

11.3.2. Challenges of Water Quantity

In addition to the challenges of CBM, the functionality of water systems is also impacted by limited quantities of spring water. The 2019 SWiM study, found that the quantity of spring water was insufficient to meet the needs of the majority of the 220 sampled communities year round (Saikal, 2019). Only 34% of surveyed aldeia (villages, administrative unit in Timor-Leste) were achieving the highest bracket of supply (9–12 months of adequate spring water supply a year), while 52% had access only 6–9 months of the year, with the remaining 14% having less than 6 months of access (Saikal, 2019). The study confirmed that communities experienced the biggest decrease in the dry season. These findings resonate with a similar study of eight communities in Liquica and Manufahi districts by WaterAid (2016), which found that none of the water points surveyed produced enough water year round to meet the national minimum standard of 30 L of water per person per day.

There are many possible causes for the limited quantities of spring water. Most obviously, rain patterns naturally change across wet and dry seasons, which, combined with the ENSO cycle (El Niño and La Niña), lead to natural fluctuations in rainfall across the year and from year to year, impacting groundwater and aquifer recharge. However, widespread use of slash and burn agricultural practices and logging has also exposed top soils and led to high levels of erosion, flash flooding, and landslides, which is further impacting spring recharge (World Bank, 2018). Water demand has also increased due to population growth, increased usage per capita, and additional nonagricultural uses (Powell et al., 2006; World Bank, 2018; Rofi & Saragih, 2019). Evidence further suggests that climate change has already impacted and likely will further impact groundwater supply in Timor-Leste in the decades to come. The first major study of the hydrogeology of Timor-Leste was completed by Geoscience Australia in 2012 under the project Vulnerability Assessment of Climate Change Impacts on Groundwater Resources in Timor-Leste, cited as Wallace et al. (2012). The study found that the decrease in groundwater quantity is likely due to reduced aquifer recharge caused by increased temperatures (increased evaporation) and reduced rainfall (Wallace et al., 2012). Based on the recent IPCC report, rainfall is expected to decrease further in the region and with fewer but more intense tropical cycles (IPCC, 2021).

11.3.3. Challenges of Water Quality

The quality of water is also impacting water system functionality. The 2019 SWiM study identified *Escherichia coli* contamination across all 220 sampled communities (Saikal, 2019). (*E. coli* contamination was measured using portable Aquagenx CBT (Compartment Bag Test) *E. coli* kits; see https://www.aquagenx.com/wp-content/uploads/2020/05/MPNCBT-ECTC-Instructions-DrinkingWater-May2020.pdf.). *E. coli* was detected using

Table 11.1 Proportion of tested water systems recording some level of *E.coli* contamination

Province	Administrative post	Systems tested	Systems with contaminated springs	Systems with contaminated distributions	Systems with contaminated springs (%)	Systems with contaminated distributions (%)
Aileu	Liquidoi	8	4	7	50	88
	Laulara	15	14	15	93	100
	Aileu Vila	41	35	40	85	98
	Total	64	53	62	83	97
Ainaro	Maubise	35	27	30	77	86
	Hatudo	5	5	5	100	100
	Ainaro Vila	26	21	26	81	100
	Hatubulico	16	7	14	44	88
	Total	82	60	75	73	92

Source: Adapted from Saikal (2019) with permission from Plan International.

a most probable number (MPN) score used as a proxy for contamination. While all systems were contaminated, at the spring, in the distribution system, or both, not all fell into the unsafe category as defined by the World Health Organization (2011). The study found that, in most cases, water taken directly from the spring was contaminated, which suggests contamination takes place in the catchment. In the majority of communities, additional contamination took place in the distribution system. Further details are found in Table 11.1.

There are many possible sources of spring contamination in communities in rural Timor-Leste. This includes free roaming animals, proximity to sanitation systems, unprotected and broken spring boxes, and open or broken pipes. According to Wallace et al. (2012), the reduction in spring water quality in Timor-Leste could also be driven by saltwater intrusion caused by sea level rise, which may modify flows leading to greater contamination from sources such as solid waste and sewage.

11.4. TOWARD IMPROVED WATER RESOURCE MANAGEMENT: GOVERNMENT AND COMMUNITY APPROACHES

The water security challenges experienced by rural communities in Timor-Leste point toward a growing need to shift attention away from infrastructure implementation toward resilient water service provision. As we have described here, this will require (1) improvements in the management of rural water systems (to overcome current challenges of CBM), and (2) improved water resource management to increase the quantity and quality of spring water year round.

The focus of this chapter is water resource management. Although, as will become apparent, the challenges of CBM cannot be separated.

11.4.1. Toward Integrated Water Resource Management

The urgent need for water infrastructure in Timor-Leste after independence left little opportunity to fund, legislate, or build institutional capacity and coordination for water resource management (World Bank, 2018). This is now changing, with GoTL in the process of implementing integrated water resource management (IWRM). This is crucial, as IWRM has been promoted as the most sustainable approach to managing water resources since the first UNESCO International Conference on Water in 1977 (Jeffrey & Gearey 2006) and is required for reaching SDG 6. IWRM is inscribed into SDG 6 through target 6.5.1: Degree of integrated water resource management implementation (UN-Water, 2021). In the most recent monitoring report on SDG 6.5.1, Timor-Leste was the only country in the Southeast Asia region to score very low on IWRM implementation (GWP & UNEP-DHI, 2021).

According to the Global Water Partnership, Technical Advisory Committee (GWP-TAC), IWRM can be seen as "a process which promotes the co-ordinated development and management of water, land and related resources in order to maximize the resultant economic and social welfare in an equitable manner without compromising the sustainability of vital ecosystems" (GWP-TAC, 2000, p. 68). This strongly aligns with the main objective of water resource management in Timor-Leste, which is to promote the coordinated planning, development, management, and protection of the nation's water resources, to optimize, social, economic, and cultural benefits without compromising the sustainability of essential water-dependent ecosystems and the environmental benefits that those ecosystems provide for people (GoTL, 2020a).

To kick-start IWRM implementation, the GoTL is putting water resource management legislation in place. The National Water Resources Management Policy (Government Resolution 42/2020; GoTL, 2020b) was recently approved by the Council of Ministers, while the Water Resource Management Decree Law has remained in draft format since 2016 and is still awaiting approval from the Council of Ministers. It is unknown when this will happen.

New institutions have also been put in place, among other things, to support IWRM. In October 2020, the National Water and Sanitation Authority (ANAS), which will play a pivotal role in IWRM, was established. According to National Water Resources Management Policy resolution No. 38/2020 (GoTL, 2020d) ANAS's mandate is to (1) propose, monitor and ensure the implementation of national policy on water resources, in order to ensure their sustainable and integrated management; (2) support the work of the Integrated Water Resources Management Coordination Council; and (3) promote efficient water use and water use planning through Water Resources Management Planning.

In January 2021, Timor-Leste, furthermore, began transitioning water supply governance to management by a public utility Be'e Timor-Leste E. P. (BTL). While ANAS's mandate is strongly related to IWRM (GoTL, 2020d), BTL's mandate is more focused on the provision of water through implementation and operation and maintenance of infrastructure (GoTL, 2020c). (BTL will be responsible for ensuring the public supply of water through the design, construction, operation and management of water supply systems. BTL will support the government in defining the policy for water supply and raising financial resources for the sector. It will also promote the efficient use of water and a continuous improvement of water quality (GoTL, 2020c). Finally, it will be responsible for ensuring the quality of public water supplies, including testing and treatment (GoTL, 2020e).) A period of establishment and institutional development is required for these new institutions to become fully operational. It is interesting that in the initial months of operation, BTL has shown significant institutional capacity and leadership, backed with more financial and human resources, while ANAS has been established at a slower rate.

There is still a lot of work to be done to fully implement IWRM in Timor-Leste. The role and responsibilities of governing institutions apart from ANAS and BTL, including municipal, suco (larger village), and aldeia (smaller village) levels, remains unclear. As an example, Decree Law 03/2016, a law related to decentralization, has a section outlining water, sanitation, and hygiene (WASH) responsibilities (article 11) for municipal administration. How this decentralization of authority will interact with ANAS's and BTL's mandates has not yet been defined. The role of communities and GMFs, and how water resource management will interact with CBM of water systems, is equally unclear at this point.

Furthermore, according to the new National Water Resources Management Policy (Government Resolution 42/2020; GoTL, 2020b), ANAS will support work of the Integrated Water Resources Management Coordination Council (IWRM-CC). To the best of our knowledge, this council has not yet been formed or is still inactive.

11.4.2. Toward Community-Based Water Resource Management

While IWRM is widely applied globally, it has also been critiqued, in part for not sufficiently including community perspectives and capabilities (Day, 2009). IWRM promotes citizen participation in decision making, but is typically not very people centered in practice (Butterworth et al., 2010). To avoid ignoring existing community practices and knowledge, Day (2009) suggests community-based water resource management (CBWRM) should be a component of IWRM.

CBWRM is a type of community-based natural resource management (CBNRM) focused on water. It is an alternative to centralized management of resources and involves transferring some degree of resource management and decision-making authority to communities (Armitage, 2005). CBWRM activities may include water conservation, land use changes, or water rationing. Day (2009) argues that including CBWRM as part of IWRM may have many benefits, including taking advantage of community members' detailed knowledge of local water resources, their needs and historical changes in water use, their ability to monitor water use on a daily basis, and the mechanisms for conflict resolution that many communities have already established (Day, 2009). However, CBNRM, and by extension CBWRM, is not always successful, often at least partly due to complex community dynamics and heterogeneity (Leach et al., 1999; Armitage, 2005; Kumar, 2005).

While GoTL works toward implementing IWRM, many rural communities are already working with Permatil to improve water security through CBWRM. Since 2019, Permatil's approach to CBWRM has involved training communities to conserve water and control erosion using permaculture principles. According to Permatil, the primary objective is to protect and rejuvenate springs, with the secondary objectives to reduce land erosion, create healthier and more sustainable ecosystems, and improve income and food production opportunities for rural communities.

Permatil works closely with communities to deliver hands-on implementation-focused training. They take a train the trainer approach to build community ownership of interventions and to achieve scalable impact with minimal resources. Training equips participants with the ability to identify important geographical and

Figure 11.2 (a) Retention basin; (b) tree saplings ready for planting.

topographical features and implement earth mounds and check dams, retention basins (see Fig. 11.2a), drainage channels, and so on, with the aim of reducing surface water velocity and increasing infiltration to groundwater. Other actions, include strategic tree planting (see Fig. 11,2b) or removal of certain types of vegetation, community behavior change (e.g., ceasing cut and burn practices), or direct protection of spring sources. These efforts are in line with recommendations from the recent SWiM study, which highlights the need for improving catchment management and suggests planting deep-rooting trees and installing terraces with appropriate vegetation and/or Earth dams to prevent runoff (Saikal, 2019).

Considering the significant interest in Permatil's work with communities, there may be support for the use of CBWRM as part of IWRM in Timor-Leste. Community interest in, and demand for, Permatil training has grown significantly in recent years as anecdotal evidence from communities has shown positive impact. Training in Aileu in October 2020 attracted approximately 400 volunteers and unpaid participants. According to Permatil, around 90% of trainees have since implemented solutions within their home communities, and around 70 water conservation sites have been established across Timor-Leste to date. The GoTL has also shown interest in Permatil's training programs. In 2019, Permatil's water conservation activities were financially supported by the Timor-Leste Ministry of Public Works through the General Directorate of Water and Sanitation (DGAS) and the General Director of DGAS has attended several Permatil events.

The growing interest in CBWRM and positive anecdotal evidence of impact suggest CBWRM could play an important role in IWRM in Timor-Leste. The question, however, remains whether communities are willing to engage with CBWRM and whether CBWRM efforts will improve ongoing water security.

11.5. WATER SECURITY REQUIRES RESILIENT SOCIOECOLOGICAL-TECHNICAL WATER SYSTEMS

Achieving water security requires water systems that are resilient to a broad range of shocks and stresses. In rural communities in Timor-Leste this may include stresses like changing water use practices, population growth, and climate change, as well as acute shocks like landslides and flooding. A resilient water system is able to "cope with risk, shocks and stresses...while continuing to maintain certain key functions and structures" (Rodina, 2018, p. 2).

To analyze the resilience of rural water systems under CBWRM and the vulnerabilities that might develop over time, we explore such systems as complex socioecological-technical systems (SETS). Drawing in Markolf et al. (2018), we take the stance that traditional ways of adapting to external shocks and stresses, which tend to be highly technocentric (i.e., increasing robustness of infrastructure), do not result in resilient infrastructure systems. Stronger pipes, as an example, will not make a rural water system in Timor-Leste resilient. The availability of spring water may still be impacted by limited spring

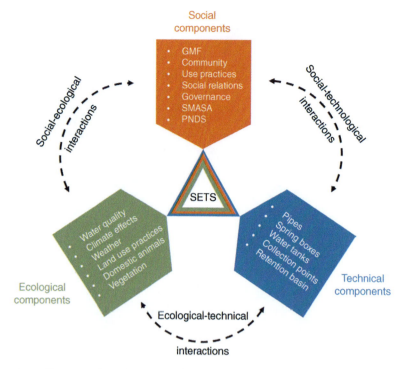

Figure 11.3 Overview of key social, ecological, and technical components of rural water systems in Timor-Leste (adapted from Markolf et al., 2018).

recharge (ecological), and challenges of CBM (social) may still result in poorly maintained infrastructure.

From a SETS perspective, there are strong interconnections, and at times significant overlaps, between social, ecological, and technical components (Markolf et al., 2018). Examples of SETS in this context of rural water supply in Timor-Leste are outlined in Figure 11.3. Social components constitute a broad category, which includes both human actors, their roles and activities. Things like rules and regulation, governance models, moral values and beliefs, behaviors, water system users and operators, education and communication, are all considered social components. Ecological components include, for example, natural resources such as water, weather, and climate, land use practices, ecological behavior, and environmental degradation. Technical components, in this case, are simply water infrastructure components such as pipes, water tanks, and tapstands. Importantly, many things do not sit neatly within one category. A retention basin, as an example, can be considered both a technical and an ecological component, which has been created through social components (community member's actions).

In this chapter we view rural water systems as SETSs and explore whether and how CBWRM impacts system resilience. As Markolf et al. (2018) argue, a SETS lens allows us to identify both vulnerabilities emerging from the interactions between social, ecological, and technical components as well as what appropriate and effective adaptation strategies to avoid vulnerabilities.

11.6. METHODOLOGY

In this chapter, the willingness of community members to implement CBWRM and the SETS resilience of rural water systems under CBWRM is explored by drawing findings from 19 semistructured interviews conducted with 24 members of six rural communities in Aileu District, Timor-Leste. This method was applied to gain in-depth insight into the complex social, ecological, and technical components influencing water system resilience.

The six communities were selected based on data from the 2019 SWiM survey and a list of communities who had received training from Permatil in 2019 and 2020. Selected communities were all engaged with Permatil and represented in the SWiM survey (see Table 11.3). All communities furthermore used gravity-fed water systems implemented between 2008 and 2014, with the support from either Red Cross, World Vision, Plan International, GoTL, or the World Bank. The communities were purposefully selected to be at different stages of CBWRM implementation. Members of four communities had received training twice over 2 years, while members of two communities had only received their first training

Table 11.2 Overview of the number of community members engaged in interviews in each of the six communities

Interviewees	Community 1	Community 2	Community 3	Community 4	Community 5	Community 6
Xefe	1	1	1	1	1	1
GMF	1	1	1	1	1	1
Trainees	2	2	2	2	1	1
Women's group	0	0	0	0	0	2
Total	4	4	4	4	3	5

3 months prior. The first four had implemented various CBWRM interventions, while the last two had not yet implemented intervention based on the training. Table 11.3 provides an overview of SWiM study data for each of the six communities and lists the CBWRM interventions implemented at the time of the interviews. The identities of the specific communities have been removed to preserve anonymity. Within each community, three to five community members were interviewed (see Table 11.2). Participants in each community included the xefe of the aldeia (community leader), a member of the GMF, and one or two community members who participated in the Permatil training (trainees). These participants were chosen to get a comprehensive perspective on water system and water resource management from a community leader, someone engaged in CBM and trainees engaged in CBWRM. To avoid participants influencing each other's responses, the interviews were conducted with each group separately, that is, the xefe was not present during interviews with the GMF member or trainees. In one community, two members of the women's group were also interviewed, since none of the other interviewees were women. GMF members, trainees, and members of the women's group were identified with the support from the xefe. Interviews took place in community meeting places or other community locations as suggested and requested by research participants.

During each community visit, the xefe took researchers on a tour to see water systems and any CBWRM interventions. Photographs were captured of the various sites with the permission of the xefe. Some of these pictures feature in this chapter.

Table 11.3 Key data from SWiM survey from each of the six communities

	Pop.	Water quality[1]	Months of flow	System functionality rating[2]	GMF functionality rating[3]	Estimated cost of repairs	Permatil training	CBWRM
1	376	Unsafe	6–9 months	41.70%	0%	$3,300	2019 2020	Retention basins, trees, fence, tara bandu
2	268	Low risk/safe	6–9 months	73.90%	61.10%	$4,288	2019 2020	Retention basins, fence
6	270	Low risk/safe	6–9 months	22.20%	16.70%	$8,357	2019 2020	Retention basin, trees, tara bandu
4[4]	224	na	6–9 months	na	na	na	2019 2020	Retention basins, tara bandu
3	441	Unsafe	9–12 months	69.40%	27.80%	$3,300	2020	Tara bandu
5	142	Unsafe	3–6 months	70.40%	38.90%	-	2020	Tara bandu

Source: Adapted from Saikal (2019).
Note: GFS = Gravity-flow scheme.
[1] A measure of water quality as assessed using an Aquagenx Compartment Bag Test Kit to test for *E.coli* levels. Results were analyzed using the World Health Organizations "Guidelines for Drinking Water Quality": Safe = 0/100 mL; Low risk = 1–10 /100 mL; Intermediate risk = 11–100/100 mL; Unsafe = >100/100 mL (Saikal, 2019).
[2] A measure of how well a rural water system is functioning. The rating is largely based on technical functionality and is calculated based on structural integrity of infrastructure and resilience against disasters, whether measures are in place to minimize contamination, the number of tapstands flowing, and seasonal variations in water availability (Saikal, 2019).
[3] A measure of how well a GMF is functioning. The rating is developed based on four proxy indicators: level of organization, regular meetings and involved women (out of 4); record of conducting technical maintenance (out of 3); community manages their water resources and is not wasteful (out of 3); record of managing funds to perform maintenance (out of 5); GMF relationship with community leaders (out of 3). The total was combined into a weighted GMF functionality percentage score (Saikal, 2019).
[4] The full survey was not conducted for community 4, hence the missing fields.

Structured interview guidelines, questions, and participant information sheets were developed in accordance with standard Australian and Timorese ethical procedures. (Ethics approval was processed through RMIT University (approval no. 2020-23410-11808) in Australia, and from the Human Research Ethics Committee at the Cabinet of Health Research and Development in Timor-Leste.) Timor-Leste-based team members conducted, transcribed, and translated all interviews. COVID-19-related travel restrictions prevented broader team participation in field visits.

Interviews were analyzed using a coding framework implemented through Excel to ensure accessibility for the whole research team without access to more powerful tools such as NVivo. An initial deductive coding framework was developed based on the SETS lens, described above. This coding framework prompted the team to explore the social, ecological, and technical complexity of rural water systems leading to vulnerabilities under CBWRM. While deductively coding interviews from the first community, the coding framework was further inductively expanded, as the richness of the data in multiple key areas of interest became evident.

In early April 2021, major flash flooding hit Timor-Leste and resulted in the local team prioritizing preservation of life, disaster response, and rebuilding. (The flash floods significantly damaged water systems in Timor-Leste. In Dili, it was estimated that 30% of the centralized water supply system was destroyed. Poor land management practices appear to have played a key role in the severity of consequences from the rainfall event. Cutting and burning practices have left steep and unstable slopes vulnerable to collapse and high velocity runoff has left many landscapes deeply scarred while filling natural and human-made drainage systems with sediment over time. This event highlighted the fragility of Timor-Leste's water systems to sudden shocks.). As a result the coding was completed by Australia-based team members. A confound was identified that there could be misinterpretation as the interviews were conducted by one half of the research team, and were analyzed by the other half. To minimize possible distortions of interview data during analysis, insights from the coding were shared with and sense checked by Timor-Leste-based team members, who have also been involved in the write up and provided clarification when required.

11.7. COMMUNITY-BASED WATER RESOURCE MANAGEMENT: A PATHWAY TO RURAL WATER SECURITY IN TIMOR-LESTE?

In this section, we share research findings. We first explore community willingness to take on CBWRM and then explore whether CBWRM is likely to lead to resilient rural water systems.

11.7.1. Willingness to Use Community-Based Water Resource Management

Interviewees in all six communities showed willingness to engage in water resource management. Interviewees from communities where CBWRM interventions had been implemented, showed interest in extending the interventions and in one of the communities they had already done so. In one community, where CBWRM interventions had not yet been implemented, a trainee simply stated, "We are ready to support." In another community without CBWRM interventions, a trainee explained that they had already implemented part of the training in his own garden, demonstrating their personal interest.

The willingness to engage in CBWRM seems driven by positive outcomes. Interviewees from communities where CBWRM interventions had been implemented described how the spring flow had improved in the 2020 dry season compared with previous years. "It helped, because the spring used to lack water. Even dry," one xefe noted. In a separate interview, a member of the GMF confirmed: "During the rainy season the retention basin dug by Permatil, now the small quantity springs in previous years have been re-filled with water." The xefe in another community also noted positive outcomes: "[It's] good because it conserved [water]. Until August the water [is] still available and continues flowing." He appeared confident that the positive impacts would continue in the 2021 dry season: "The water will be enough. Because conserving the water is not only in a hole, but a big retention basin." This aligns with anecdotal evidence Permatil has received from many rural communities to date, however, long-term spring flow and water quality data are still needed to quantify improvements.

In communities where CBWRM interventions had not yet been implemented, willingness seemed driven by expectations of positive benefits and interviewees appeared optimistic about a future with CBWRM. Trainees were particularly positive, arguing that CBWRM interventions will help improve water security. One trainee mentioned, "I attended for two days and it was about digging retention basins and I felt so happy, because we can use the water to plant bananas, flowers and vegetation in the dry season." Interviewees mentioned a range of future opportunities that CBWRM could offer their communities, like raising more cattle, engaging in aquaculture, increasing rice and vegetable irrigation, extending farms, engaging in agroforestry, as well as planting more trees to make the community greener.

11.7.2. Water System Resilience Under CBWRM

While interviewees showed willingness to engage in CBWRM, this has not necessarily translated to resilient water systems and in turn ongoing water security. From a SETS perspective, CBWRM does not remove all existing water system vulnerabilities and new ones may emerge from the ongoing complexities of interactions between social, ecological, and technical components. Table 11.4 provides an overview of the key SETS factors mentioned by interviewees, which may result in system vulnerabilities. These are further explored in the sections below.

Social

While interviewees showed willingness to engage in CBWRM, many argued this was not the case for the broader community. Trainees in one community noted that other community members did not believe Permatil and, therefore, were not willing to offer their support. In another community, interviewees described fellow community members as unsupportive and said they had demanded payment for their work. In yet another community, a trainee noted, "Some individuals have interest to support, some just speak, some just [go] away. So we do not feel good about it." This poses a significant vulnerability. As one of the xefes stated: "The challenge about the community, [it] depends on individuals. To call for work, we should not force, [just] call each other, [and] go and work."

Ongoing social conflicts, caused by or further exaggerated by, may make it even more challenging to gain wide community support. The xefe in one community for example noted how some community members do not control their water use: "[they do] not close the tap, so it causes anger…when you see they open the tap just remind them by saying 'please close the tap after use.'" In another community water insecurity had resulted in large-scale conflict between community members. As one interviewee described: "Even some people almost kill each other just because of water…. It is like when those who control the water and the others just wait and use it, and then causes emotions…arguing …hit each other."

Community conflicts can exacerbate water insecurities. Several interviewees described how some community members, faced with a lack of water, had modified water infrastructure. "They use the piece of wood and put [it] into the pipe, so we need to cut the pipe and remove the wood to ensure the supply of water," one community member described. In another community, several interviewees described how people kept breaking pipes: "There are always problem about the pipe, we need to replace frequently because [they are] broken by people." The SWiM study similarly highlights a range of challenges associated with illegal connections and modifications of water systems (Saikal, 2019). This not only causes immediate challenges to water supply, but also poses challenges to ongoing water security as it "affect(s) the community's willingness to contribute tariffs and is how many GMFs become dysfunctional, further exacerbating overall supply system functionality" (Saikal, 2019, p.10).

Other social factors can lead to vulnerabilities in water systems under CBWRM. In one community, interviewees noted it was difficult to secure permission from the land owner to implement CBWRM interventions. This was part of the reason why nothing had been implemented yet. The SWiM report also notes such potential social sensitivities around spring ownership (Saikal, 2019). External and unforeseen factors like COVID-19 can also impact CBWRM. As one of the trainees noted: "I attended [Permatil training], but to implement in our village, we informed the chiefs of the village and sub-villages and discussed about retention basin, but, unfortunately, because of coronavirus, it has been canceled."

Once CBWRM interventions are implemented, the complexity of social systems can lead to ongoing vulnerabilities. It was, for example, unclear across communities who will be responsible for maintaining retention basins or vegetation after implementation. In fact, a range of different maintenance models where proposed, ranging from trainees being responsible for maintenance

Table 11.4 Overview of social, ecological, and technical (SETS) factors impacting the resilience of rural water systems

Social	Ecological	Technical
Lack of broader community trust in Permatil	Hard soil makes it difficult to build retention basins.	Broken pipes
Lack of broader community willingness to support CBWRM	Animals destroy retention basins and eat tree saplings.	Broken water tanks
Sensitive nature of spring ownership	Animals are a source of spring water pollution.	Pipes too big
Lack of clear roles and responsibilities within communities regarding ongoing CBWRM	Animals break water pipes.	Pipes too small
Existing challenges with CBM	Landslides break water pipes and destroy retention basins.	Broken spring boxes
Ongoing water conflicts	Falling trees break water pipes.	No tools
Community members diverting spring flow leading to exacerbated water insecurity	Rubbish and sand block pipes after flooding events.	
	No tree saplings	

(suggested by trainees), to those who have gardens in the area being responsible (suggested by a GMF member), or the xefe aldeia and the GMF taking responsibility (suggested by a GMF member). In one community, the xefe also suggested that Permatil should be responsible. There was no consistency in the suggestions across communities or within communities.

While social components pose a range of vulnerabilities, communities have social mechanisms available that may improve water system resilience as well. Tara bandu, a type of customary law in Timor-Leste, which provides protection for natural resources (Miyazawa, 2013), is widely implemented by rural communities. In fact, five out of the six case study communities had implemented tara bandu or wanted to do it. In the words of one xefe: "For the future we can motivate them to not let their animals pick up the small saplings. We can ban them to not cut the trees and around and also not to burn the grass near the pipes. If some of them are not obey the rule, we can apply the sanction in our village: tara bandu is a sanction from the village to condemn them." Tara bandu can be a good start toward building resilience through local governance structures, and aligns with Markolf et al.'s (2018) argument that "strong and just governance is a key element within the social domain of SETS."

Ecological

Ecological CBWRM interventions, like retention basins and revegetation, may minimize some vulnerabilities in rural water systems, although they do not completely disappear. Interview participants noted that retention basins, like other water infrastructure, are vulnerable to ecological components such as heavy rainfall, flooding, storms, and landslides. In the words of one trainee, "Recently we did the maintenance and conserve the water, but [it was] destroyed by the rain and the water in [the] basin also flowed to the forest." These vulnerabilities are likely to increase with the effects of climate change and ongoing population growth.

Ecological CBWRM interventions also introduce new vulnerabilities to rural water systems. Trees planted as part of CBWRM may fall and break pipes, water tanks, or tap stands, and retention basins may attract animals like goats, cattle, and buffalos, which can create a potential new source of water contamination. Free roaming animals may also destroy retention basins by stepping on the edges. In two communities, their recent efforts to plant tree saplings had, furthermore, been negated by loose animals eating the saplings.

Technical

CBWRM does not minimize the vulnerabilities of rural water infrastructure. Permatil training focuses on intervening in ecological and, to a more limited extent, social systems (behavior change), and will, therefore, not fix the many ongoing technical challenges rural communities face managing water infrastructure. These challenges were top of mind for many interviewees. They showed great concern for the functionality of water infrastructure, and outlined a wide range of issues: pipes being too small for a high water pressure from the spring or too large for low water pressure, sand and leaves blocking pipes, water tanks and reservoirs cracking, and pipes breaking.

The technical functionality of water infrastructure is a more urgent need than CBWRM. If water infrastructure does not function, water will not be available now. If water resource management is not functioning, problems with water security may be experienced down the track. In other words, CBWRM is a long-term fix for a long-term problem, while the functionality of water infrastructure is a very immediate problem. Unsurprisingly, CBWRM, therefore, appears to be a secondary priority.

11.7.3. Pathways Toward Resilient Rural Water Systems in Timor-Leste: Between IWRM and CBWRM

From a SETS perspective, CBWRM may not lead to resilient rural water systems. Interactions between social, ecological, and technical systems, as seen above, are complex and vulnerabilities may continue to exist while new ones may emerge under CBWRM.

It seems that especially socioecological and sociotechnical interactions can lead to vulnerabilities when social components are fragile. While some community members are willing to engage in CBWRM, this is not necessarily the case for the entire community, and trainees may find it challenging to mobilize community members, especially if not in community leadership positions. Ongoing community conflicts resulting from water insecurity may pose further barriers for CBWRM implementation and further exacerbate CBM challenges as well. These findings align with existing CBWRM and CBNRM literature, which has highlighted the inherent heterogeneity of community as a challenge (Leach et al., 1999; Armitage, 2005; Kumar, 2005). As Kumar (2005, p. 280) argues, "communities are not necessarily clearly bounded social or geographic units, nor are they likely to be homogeneous entities with single or agreed interests." Such heterogeneity can make it challenging to agree on how to manage water resources. As Day (2009, p. 52) argues: "The inevitable challenge of community-based water resource management' is to bring together heterogeneous community groups to agree on sharing available water resources for the benefit of all."

While CBWRM, at least as currently trialed in Timor-Leste, may not lead to resilient rural water systems, the critique of IWRM highlights the importance of involving communities in water resource management in some way. The question is therefore, how may communities engage in IWRM in a way that improves rather than hinders resilience? This is an important question to consider as GoTL works toward finalizing IWRM legislation and institutional arrangements. Drawing on the findings presented in this chapter, we offer the following suggestions:

1. Quantitatively confirm improvements in water quantity and/or quality across all 70 communities engaged in Permatil training to date and develop a long-term strategy for data collection, storage, and use.

2. Conduct further research across all 70 communities to understand community willingness and ability to engage in water resource management and what types of support they may want.

3. Consider expanding water resource management training to engage entire communities (instead of train the trainer approach), and utilize community leaders who are in a position of authority and are well placed to motivate broader community engagement.

4. Use training as a way to settle community water conflicts and engage community members in developing shared community-led water resource management plans.

5. Strengthen CBM before, or as part of, CBWRM, for example, by increasing the engagement and financing of FPAs.

6. Develop legislation outlining community roles and responsibilities in water resource management and what is beyond the remit of communities.

7. Offer ongoing support to communities engaging in water resource management, for example by extending the role of FPAs, and provide transparent high level support from ANAS and the Integrated Water Resources Management Coordination Council, including community funding and guidance.

Ongoing support to communities engaging in water resource management seems particularly important. The role of FPAs has been shown to significantly improve CBM (Saikal, 2019) and our research also suggests some community members wish and hope for support. In one community, a GMF member noted: "The solutions the people can do, but not enough. It has to be supported by the government Mana. Mana also work in an NGO, [and] can also help us. Help us so we can do [it]. We alone cannot do anything. Talk is easy, but it is difficult sometimes." External support can come in many forms. Interviewees suggested support to restart user fee collection, purchasing tools, food during building of retention basins and revegetation, ongoing training and guidance, help purchasing seeds and saplings, as well as water quality testing.

11.8. CONCLUSION

Rural communities in Timor-Leste may play an important role in the current national move toward IWRM. However, CBWRM, as currently trialed, is unlikely to lead to ongoing water security at scale. Findings from interviews with 24 community members across six rural communities suggest that while some community members are willing to take on water resource management responsibilities, this may not be the case for everyone. As shown in this chapter, CBWRM, furthermore, can lead to a range of vulnerabilities that may impact the resilience of rural water systems. The inherent heterogeneity of communities and ongoing community conflicts resulting from water insecurity are significant ones. This chapter offers a list of recommendations for how communities may play a role in IWRM calling for changes to legislation, institutional arrangements, and for further research at scale. Research should be performed after the COVID-19 pandemic to ensure communities have had more opportunity to implement CBWRM interventions.

ACKNOWLEDGMENTS

The authors thank interviewees for their time and willingness to participate in the research and for sharing their knowledge and experiences. We also thank Francisco Guterres dos Reis, Elsa Maria Claudia Ximenes, Chandra Monteiro Ximenes, Maria Manuela da Costa Franco, Jose dos Reis, Berjita Desiana Bernadino da Costa, Nelvia Agostinha Correia, Abel Fernandes, and Luis Pereira for their work arranging, conducting, transcribing, and translating interviews. This research would not have been possible in the COVID era without their commitment and perseverance.

REFERENCES

Armitage, D. (2005). Adaptive capacity and community-based natural resource management. *Environmental Management*, *35*(6), 703–715.

Berner, E., & Phillips, B. (2005). Left to their own devices? Community self-help between alternative development and neo-liberalism. *Community Development Journal*, *40*(1), 17–29. https://doi.org/10.1093/cdj/bsi003

Bond, M. (2009). *Support to decentralising service delivery assigning concrete functions to the municipalities preparation for decentralising WATSAN functions to municipalities in East Timor mission report*. April 2009.

Butterworth, J., Warner, J., Moriarty, P., Smits, S., & Batchelor, C. (2010). Finding practical approaches to Integrated Water Resources Management. *Water Alternatives*, *3*(1), 68–81.

Chowns, E. (2015). Is community management an efficient and effective model of public service delivery? Lessons from the rural water supply sector in Malawi. *Public Administration and Development*, *35*(4), 263–276. https://doi.org/10.1002/pad.1737

Day, S. J. (2009). Community-based water resources management. *Waterlines*, 47–62.

Direcção Geral de Estatística (2015). Timor-Leste population and housing census, data sheet. https://www.statistics.gov.tl/wp-content/uploads/2016/11/Wall-Chart-Poster-Landscape-Final-English-rev.pdf

GoTL (2004). *Decree Law No. 4/2004: On water supply for public consumption*. http://extwprlegs1.fao.org/docs/pdf/tim63487.pdf

GoTL (2010). *Timor-Leste Strategic Development Plan 2011–2030*. https://www.adb.org/sites/default/files/linked-documents/cobp-tim-2014-2016-sd-02.pdf

GoTL (2019). *Report on the implementation of the sustainable development goals: From ashes to reconciliation, reconstruction and sustainable development*. Voluntary National Review of Timor-Leste 2019.

GoTL (2020a). *Government Resolution No. 43/2020: National Policy for Public Water Supply*.

GoTL (2020b). *Government Resolution No. 42/2020 on the National Water Resources Management Policy*.

GoTL (2020c). *Government Resolution No. 41/2020: Creating the Bee Timor-Leste public company and approving the related statute*.

GoTL (2020d). *Government Resolution No. 38/2020: Creating the National Authority for Water and Sanitation and approving the related Statute*.

GoTL (2020e). *Decree Law No. 31/2020:* Monitoring the quality of water for human consumption

GWP and UNEP-DHI (2021). *Progress on integrated water resources management (IWRM) in the Asia-Pacific region 2021: Learning exchange on monitoring and implementation toward SDG 6.5.1*.

GWP-TAC (2000). *Integrated water resources management*. TAC Background Papers No. 4, GWP, Stockholm, Sweden.

Hamel, S. (2009). *Baseline survey of Aileu and Lautem district rural water supply and sanitation coverage, Plan, Timor-Leste*.

Hope, R. (2015). Is community water management the community's choice? Implications for water and development policy in Africa. *Water Policy*, 17(4), 664–678. https://doi.org/10.2166/wp.2014.170

IPCC (2021). *Sixth assessment report: Regional fact sheet, Asia*. Working Group 1, The Physical Science Basis. https://www.ipcc.ch/report/ar6/wg1/downloads/factsheets/IPCC_AR6_WGI_Regional_Fact_Sheet_Asia.pdf

Jeffrey, P., & Gearey, M. (2006). Integrated water resources management: Lost on the road from ambition to realization? *Water Science and Technology*, 53(1), 1–8.

Kumar, C. (2005). Revisiting "community" in community-based natural resource management. *Community Development Journal*, 40(3), 275–285.

Leach, M., Mearns, R., & Scoones, I. (1999). Environmental entitlements: Dynamics and institutions in community-based natural resource management. *World Development*, 27(2), 225–247.

Markolf, S. A., Chester, M. V., Eisenberg, D. A., Iwaniec, D. M., Davidson, C. I., Zimmerman, et al. (2018). Interdependent infrastructure as linked social, ecological, and technological systems (SETSs) to address lock-in and enhance resilience. *Earth's Future*, 6(12), 1638–1659. https://doi.org/10.1029/2018EF000926

Miyazawa, N. (2013). Customary law and community-based natural resource management in post-conflict Timor-Leste. In J. Unruh & R. C. Williams (Eds.), *Land and post-conflict peacebuilding*. London: Earthscan.

Powell, B., Davies, P., Costin, G., Wairiu, M. & Ross, H. (2006). *Interdisciplinary approaches to catchment risk management: Case studies of Timor-Leste and the Solomon Islands*. International Riversymposium.

Rodina, L. (2018). Defining "water resilience": Debates, concepts, approaches, and gaps. *Wiley Interdisciplinary Reviews: Water*, 6(2), e1334.

Rofi, A., & Saragih, S. (2019). Protect the water sources through community participation on the micro-watershed management in the Region of Oecusse Timor-Leste. In *IOP conference series: Earth and environmental science* (vol. 256, no. 1, p. 012006). IOP Publishing.

Saikal, A. (2019). *Sustainable water in Timor-Leste municipalities (SWiM) technical report*. Dili: Plan International Timor-Leste.

United Nations (2015). *Transforming our world: The 2030 agenda for sustainable development*. http://www.un.org/ga/search/view_doc.asp?symbol=A/RES/70/1&Lang=E

UN-Water (2013). *Water security & the global water agenda: A UN-Water Analytical Brief*. https://www.unwater.org/publications/water-security-global-water-agenda/

UN-Water (2021). Indicator 6.5.1, Degree of integrated water resources management implementation (0-100). https://www.sdg6monitoring.org/indicator-651/

van den Broek, M., & Brown, J. (2015). Blueprint for breakdown? Community based management of rural groundwater in Uganda. *Geoforum*, 67, 51–63. https://doi.org/10.1016/j.geoforum.2015.10.009

Wallace, L., Sundaram, B. Brodie, R. S., Marshall S., Dawson, S., Jaycock, J., et al. (2012). *Vulnerability assessment of climate change impacts on groundwater resources in Timor-Leste*. Record 2012/55. Canberra: Geoscience Australia.

WaterAid (2016). *Social audit for rural water supply services in Timor-Leste*. https://www.dfat.gov.au/about-us/publications/Pages/timor-leste-social-audit-rural-water-supply-services-report

WHO and UNICEF (2021a). https://washdata.org/data/downloads#TLS

WHO and UNICEF (2021b). https://washdata.org/monitoring/drinking-water

Willetts, J. (2012). *A service delivery approach for rural water supply in Timor-Leste: Institutional options and strategy*. Prepared by Institute for Sustainable Futures, University of Technology Sydney for BESIK (Timor-Leste Rural Water Supply and Sanitation Program), March 2012.

World Bank (2018). *Timor-Leste water sector assessment and roadmap*. Washington, D.C.

World Health Organization (2011). *Guidelines for drinking-water quality*, Fourth ed. Geneva.

12

Setting Benthic Algal Abundance Targets to Protect Florida Spring Ecosystems

Robert A. Mattson

ABSTRACT

The establishment of quantitative targets for algal abundance in lakes (usually expressed as water column chlorophyll *a* concentration in µg/L) is well-ensconced in lake management. Setting targets for abundance of benthic/attached algae in streams has not garnered as much attention in stream ecology and management. Establishment of targets for benthic algae in streams is in part dependent upon the particular attribute that is to be protected, including water withdrawal/water supply, aesthetics, recreation, or ecosystem protection. Abundance targets may be expressed as % cover, chlorophyll *a* density (as mg Chl a per unit area), or standing crop (g dry weight or ash-free dry weight per unit area). This chapter will review benthic algal abundance targets proposed in the stream literature and compare them with epiphytic and macroalgal abundance measured in a 2015 study of 14 spring-run streams in Florida. Various investigators have proposed quantitative targets based on either mean/median or maximum algal abundance. Efforts in temperate streams have suggested macroalgal targets of 20%–40% cover, 100–150 mg/m^2 chlorophyll *a* density, or 40 g/m^2 ash-free dry weight (AFDW). These targets were mainly based on aesthetics and recreational issues and may or may not be relevant for Florida spring-run streams, but they may be adopted as a starting point. Given the changes in benthic algal abundance seen in Florida spring-run streams over the past few decades, establishing targets for algal abundance to guide restoration attempts appears to be warranted.

12.1. INTRODUCTION

Florida's karst geology is the basis for the existence of over 1,000 springs in the state. These aquatic ecosystems are a signature feature of the state and they attracted visitors and tourists before Florida became known for its beaches (Florida Springs Task Force, 2000; Scott et al., 2004). Two broad categories of Florida springs have been identified (Copeland, 2003): (1) seep springs, fed by seepage of water from shallow, surficial (water table) aquifers; and (2) vent springs, fed by artesian flow from the deeper Floridan Aquifer. The latter are what most people think of as a Florida spring and is the focus of this chapter. Springs have been classified by their mean annual discharge or flow using the system first developed in Meinzer (1927); ranging from First magnitude (≥ 100 ft^3/second mean annual flow) to Eighth magnitude (<1 pint/minute). Florida springs have also been classified by base water chemistry (Slack & Rosenau, 1979; Woodruff, 1993), ranging from low ion (soft water) springs with minimal concentrations of dissolved salts and minerals to salt springs with high concentrations. One of the most noticeable characteristics of Florida springs and the streams that they create (spring-run streams) is the exceptional clarity of the water, allowing for abundant penetration of

St. Johns River Water Management District, Palatka, Florida, USA

Threats to Springs in a Changing World: Science and Policies for Protection, Geophysical Monograph 275, First Edition.
Edited by Matthew J. Currell and Brian G. Katz.
© 2023 American Geophysical Union. Published 2023 by John Wiley & Sons, Inc.
DOI:10.1002/9781119818625.ch12

light energy, which sustains the growth of dense beds of submerged aquatic vegetation (SAV), including rooted macrophytes and attached and free-living algae (Mattson et al., 2019; Odum, 1957). These SAV communities support diverse food webs of invertebrates, fish, turtles, and other fauna (Odum, 1957).

Over the past several decades, Florida springs have exhibited significant changes in hydrology, water quality, and biological integrity, as have many other aquatic ecosystems in the state (both freshwater and marine). Many of these changes are driven by the substantial growth in the state's human population over this period of time. The principal change in water quality has been significant increases in dissolved inorganic nitrogen as nitrate, typically measured as nitrate-nitrite nitrogen or NOx-N (Copeland et al., 2009; Katz, 2020). Associated with these increases in NOx-N have been changes in the SAV communities in springs and spring-run streams (hereafter "springs" for brevity), including replacement of rooted macrophytes by benthic filamentous algal mats and greater abundance of epiphytic, attached algae on the macrophytes (Florida Springs Task Force, 2000; Mattson et al., 2019; Stevenson et al., 2007). There is evidence that increased NOx-N may be responsible for these vegetation shifts (Cowell & Dawes, 2004; Stevenson et al., 2007), but there is also evidence that nitrogen is not a factor in these changes (Brown et al., 2008; Cohen et al., 2017; Heffernan et al., 2010). Other drivers influencing these vegetative shifts may be changes in current regimes due to alterations in spring flow (King, 2014; Reaver et al., 2019) or changes in grazer/herbivore populations (Liebowitz et al., 2014). Whatever is driving these vegetation changes, they result in alterations to the food webs in springs (Camp et al., 2014; Mattson, 2009) and may legitimately be considered degradation of spring ecology.

The objectives of the analysis discussed in this chapter are (1) review the existing stream literature for quantitative benthic algal targets proposed to protect stream ecology and/or human use, (2) compare these targets to quantitative algal data collected in a relatively recent (2015) study of 14 Florida springs to evaluate whether and how extensively algal abundance in these streams exceeds these targets, and (3) provide suggestions on adoption of these targets for possible future use in management of Florida springs.

12.2. DERIVATION OF BENTHIC ALGAL TARGETS IN STREAMS: REVIEW OF THE LITERATURE

In the Florida water quality standards (Chapter 62-302.530(48)(a) & (b), Florida Administrative Code (FAC)), the criterion for nutrients, referring to human-induced enrichment of nitrogen, N, and phosphorus, P, is a narrative standard, reading in part, "In no case shall nutrient concentrations of a body of water be altered so as to cause an imbalance in natural populations of aquatic fauna and flora." This includes nutrient enrichment causing algal blooms. In 2012, Florida adopted numeric nutrient criteria for inland waters as Chapter 62-302.531, FAC, (Numeric Interpretation of Narrative Nutrient Criteria), adding quantitative criteria to the narrative standard. This included adoption of a nitrate criterion of 0.35 mg/L NOx-N for springs based on data showing quantitative relationships between NOx-N and algal abundance and taxonomic composition (FDEP, 2012).

It has been argued that chemical criteria, while important, are not enough to fully protect surface waters (Davis & Simon, 1995); these must be supplemented by biological criteria and metrics. Algae have been used for decades as biological indicators of water quality (Bellinger & Sigee, 2010) including the use of algal abundance thresholds that constitute a "nuisance" or undesirable condition. In lake management, the use of algal abundance thresholds or targets (i.e., water column chlorophyll *a* concentrations) is well established, but development of algal abundance thresholds or targets for streams has not been done as much (Dodds et al., 1998).

Attached and benthic algae, also referred to as "periphyton," are known to be useful as biological indicators of water quality (Rosen, 1995), and these are the dominant type of algal growth in most streams. As noted above, establishment of abundance targets for benthic algae has rarely been done in stream management. As a management tool, the importance of developing and adopting benthic algal/periphyton abundance targets may be becoming more critical, as the proliferation of nuisance blooms of benthic, filamentous algae appears to be an increasing global phenomenon (Hudon et al., 2014; Paerl & Otten, 2013; Stevenson et al., 2007). When taxa of filamentous forms in the Cyanobacteria, Chlorophyta, or Xanthophyta become dominant in the periphyton, macroalgal mats may result from their proliferation. The choice of a target for algal abundance (periphyton or macroalgal mats) is dependent upon the specific resource value in a stream that is to be protected (Biggs, 1996). This could include water withdrawal (e.g., macroalgal mats clogging intakes), aesthetics (macroalgal mats look unsightly and produce objectionable odors), recreation (algal accumulation interfering with wading, swimming, or paddling watercraft), or ecological impacts (aquatic life impairment and food web or habitat disruption).

When deriving targets based on biological conditions, an important foundation is establishing a reference condition (Davies & Jackson, 2006; Stoddard et al., 2006). The concept of "reference condition" means the state or status of an ecosystem used to evaluate the effects of human impact (Stoddard et al., 2006). In general, this equates to a natural background condition (in terms of ecosystem structure, composition, function, etc.) in the absence of human impact. This can be based on historic condition or on the condition of minimally impacted ecosystems. There is a lack of historic quantitative data on algal abundance in springs, which makes determining a reference condition difficult. Some data on epiphytic algal abundance have been reported in Silver Springs (Odum, 1957), and this will be discussed in more detail below, but no data appear to exist for macroalgal mat abundance.

Algal abundance measures that could be considered in setting targets (for macroalgal mats, periphyton, or epiphytic algae) include percent (%) cover, chlorophyll a density (as mass per unit area), or dry weight (DW), or ash-free dry weight (AFDW) as mass per unit area. Horner et al. (1983) suggested a nuisance chlorophyll a value of 150 mg/m^2 of periphyton, based on experiments in artificial stream channels and a review of maximum periphytic algal abundance levels reported in the stream literature (including spring-run streams in Missouri and Florida). Studies conducted in rocky northwestern U.S. streams also concluded that a threshold of 150 mg/m^2 chlorophyll a constituted an aesthetic nuisance level of macroalgal growth, mainly by interfering with recreational fishing (Welch et al., 1988). This equated to a macroalgal coverage of about 20%. Biggs (1996), citing his own and others' work in streams worldwide, suggested thresholds of 40% maximum cover of benthic filamentous algae, and/or maximum benthic algal standing crop of 100 mg/m^2 chlorophyll a or 40 g/m^2 AFDW. The basis for these targets was broad and included constraining recreational use and aquatic life impairment.

Ecological impacts or aquatic life impairment may be the most compelling resource value for deriving stream periphyton quantitative targets. Chlorophyll a criteria in lakes can be tied to ecological impacts, including light attenuation (effects on rooted, submerged macrophytes) and accumulation of dead algal cells on the lake bottom, resulting in anoxic or hypoxic conditions from decomposition (Brönmark & Hansson, 1998). Excessive growth of epiphytic algae on the leaves of submerged macrophytes in springs can reduce the amount of light energy reaching the host plant, with consequent negative effects, and there have been specific studies of this in Florida springs. Szafraniec (2014) found that epiphyte growth on the leaves of *Sagittaria kurziana* could intercept 20%–75% of the incident light reaching the blades in two spring-run streams in west-central Florida. This reduction is in addition to any water column attenuation of light, potentially resulting in substantial cumulative reductions of light energy available for macrophyte growth and proliferation. Additionally, Szafraniec's work found that the epiphytes were particularly efficient at intercepting light in the blue range of the spectrum, a wavelength that also appeared important for the growth and persistence of *Sagittaria*, indicating that both quantity and quality of light are important.

Guan et al. (2020a) found that excessive accumulation of epiphytic algal growth on blades of *Vallisneria americana* in the Chassahowitzka River, Florida, resulted in reductions in light reaching the leaves of 12.8%–95.5%, similar to the results of Szafraniec. In subsequent work, Guan et al. (2020b) documented changes in growth and physiology of *Vallisneria* due to excessive epiphytic algal accumulation on the leaves. Chen et al. (2007) documented changes in physiology and reduced production due to epiphyte accumulation on leaves of the submerged macrophyte *Potamogeton crispus* in mesocosm experiments but did not propose any epiphyte targets to prevent impacts. Guan et al. (2020b) based on their results, proposed a maximum epiphyte burden of 4–5 mg dry weight (DW)/cm^2 of leaf surface to be an unacceptable level of algal growth in locations receiving full sunlight, and in locations more heavily shaded by the riparian tree canopy thresholds for algal abundance were an order of magnitude lower: 0.4–0.8 mg DW/cm^2 of leaf area.

Quantitative criteria for periphyton may also be developed based on impacts on fauna and/or food webs. A review of the literature and existing data relating benthic macroinvertebrate community characteristics to benthic algal abundance in Florida springs concluded that while some invertebrate community characteristics (taxa richness and abundance) were increased by more abundant algae, others (diversity, evenness, and the % dominance index) declined (Mattson, 2009). Camp et al. (2014) saw similar trends (higher density but reduced diversity) in small fish and invertebrate communities in macroalgal mats versus macrophyte beds in the spring-fed Homosassa and Chassahowitzka rivers in Florida. Threshold targets for algal abundance could not be derived from these algal/invertebrate relationships, but these could be useful for algal target development in conjunction with other thresholds, such as those proposed earlier.

Changes in both algal and macroinvertebrate species composition may also be important in deriving quantitative targets for periphyton. Mattson (2009) found that more sensitive benthic invertebrate groups, such as the EPT taxa (Ephemeroptera, Plecoptera, and Trichoptera; mayflies, stoneflies, and caddisflies, respectively), generally dropped out of the invertebrate community with increasing relative abundance of Cyanobacteria and Chlorophyta in the benthic algal community in Florida springs (as indicated by the EPT score). These two algal Divisions contain many of the filamentous taxa that can form extensive mats of macroalgae. In contrast, the EPT score increased with increasing relative abundance of diatoms in the periphyton community in springs (Mattson, 2009). Power (1990) also found that benthic macroalgal mats supported fewer taxa of mayflies and stoneflies and were more dominated by chironomid midges, reflecting that macroalgal mats may only be preferred as habitat by a few, more tolerant invertebrate groups.

12.3. COMPARISON OF ALGAL TARGETS TO SELECTED FLORIDA SPRING-RUN STREAMS

While algae have always been an important component of the SAV (primary producer) communities in Florida springs (Whitford, 1956), the proliferation of benthic algae seen in recent decades in these springs (Stevenson et al., 2007) may justify the development and adoption of benthic algal targets for springs management. To evaluate the abundance of benthic algae in Florida springs (both macroalgal mats and epiphytes attached to macrophytes) and assess whether these constitute a nuisance level, algal data from a recent study of 14 springs by the St. Johns River Water Management District (SJRWMD) in north and central Florida (Table 12.1) were compared with some of the benthic algal targets discussed in the prior section. The SJRWMD study used quantitative methods to sample algal cover (%), chlorophyll a (mg/m^2), and AFDW (g/m^2). Sampling was conducted in the spring and fall of 2015 (AMEC, 2016; Mattson et al., 2019). For macroalgal mats (large accumulations of filamentous taxa), maximum algal abundance data from the 2015 study were compared with the targets Biggs (1996) proposed (Table 12.2).

Table 12.1 Spring-run streams and sampling transect identifiers in the SJRWMD 2015 study

Stream system	Upstream transect	Downstream transect	Spring magnitude	Spring chemistry
Alexander Springs Creek	ALE1	ALE2	1	Mixed
Volusia Blue Spring Run	VOL1	--	1	Salt
Gum Slough	GUM1	GUM2	2	Calcium/bicarb
Ichetucknee River	ICH1	ICH2	1	Calcium/bicarb
Juniper Creek	JUN1	JUN2	2	Low ion
Manatee Spring Run	MAN1	--	1	Calcium/bicarb
Rainbow River	RAI1	RAI2	1	Calcium/bicarb
Rock Springs Run	ROC1	ROC2	2	Calcium/bicarb
Silver River	SIL1	SIL2, SIL3	1	Calcium/bicarb
Silver Glen Spring Run	SLG1/SILG1	--	1	Salt
Wacissa River	WAC1	WAC2	1	Calcium/bicarb
Wakulla River	WAK1	WAK2	1	Calcium/bicarb
Weeki Wachee River	WEE1	WEE2	1	Calcium/bicarb
Wekiva River	WEK1	WEK2	2	Calcium/bicarb

Source: Adapted from Mattson et al. (2019) for description and site location coordinates.

Table 12.2 Comparison of maximum macroalgal abundance measures of nine springs from the 2015 survey with maximum benthic algal abundance targets proposed

	Max cover (%) spring/fall	Max Chl a (mg/m^2) spring/fall	Max AFDW (g/m^2) spring/Fall
Alexander Springs Creek	**78** / **60**	**1,509.0** / **400.0**	**262.5 / 65.6**
Volusia Blue Spring Run	**85** / **90**	58.2 / 55.2	**62.5 / 46.9**
Gum Slough	0 / **85**	0 / **145.2**	0 / 39.1
Manatee Spring Run	**100 / 100**	94.4 / **226.4**	27.5 / 21.4
Rainbow River	**89** / 15	**890.0 / 250.2**	**118.8** / 35.27
Rock Springs Run	2.5 / 0	**145** / 0	**56.3** / 0
Silver Glen Spring Run	**98** / 0	**1,127.0** / 0	**262.5** / 0
Wakulla River	**99 / 55**	**168.2 / 1,008.1**	22.5 / **332.8**
Weeki Wachee River	30 / **45**	**323.7 / 1,123.0**	**181.3 / 131.3**
Biggs criteria	<40% Max seasonal cover	<100 mg/m^2 to protect other aquatic life	<40 g/m^2 AFDW

Source: Adapted from Biggs 1996.
Note: Bold text indicates exceedance of Biggs criteria at one or more sampling transects; shaded cell indicates exceedance in both seasons.

The Biggs targets were selected because they were specific for macroalgae and were based on a broad range of stream types and protection of multiple resource values. Information on the algal sampling methods used was not available. Of the 14 springs sampled, 9 had macroalgal mats present at the transect sites; the other 5 springs did not have macroalgal mats present at the sampling transects, but that does not mean macroalgal mats were not present in the spring. All 9 springs exceeded at least one of the Biggs targets in at least one season (spring or fall 2015). Six of the 9 springs exceeded at least one of the Biggs targets in both seasons (Table 12.2). Mats were dominated by filamentous taxa in the Cyanobacteria (primarily *Microseira* [formerly *Lyngbya*] *wollei*), Chlorophyta (*Rhizoclonium heiroglyphicum*, *Dicotomosiphon* sp., *Cladophora glomerata*, or *Hydrodictyon* sp.), and/or Xanthophyta (*Vaucheria* sp.). Many of these taxa are regarded as nuisance taxa in streams due to their propensity to form extensive mats (Biggs, 1996; Hudon et al., 2014; Stevenson et al., 2007). This comparison indicates that macroalgal abundance in these springs does exceed one or more thresholds that may constitute a nuisance condition that may result in negative impacts on spring ecology (macroinvertebrate or fish communities) or spring uses (recreation).

Given that macroalgal abundance exceeds potential nuisance thresholds in these nine springs, what constitutes a reference or natural background level of macroalgae, and how far above this are these springs? Almost no historic data exist to evaluate this in Florida springs. In Silver Springs, Odum (1957) references the occurrence of macroalgal mats, but he indicates that they were uncommon/low abundance. Currently, extensive areas of macroalgal mat do occur in the Silver Springs main headspring; these were sampled by Quinlan et al. (2008). They reported a mean macroalgal dry weight of 381 g/m^2 in winter 2003 and 641 g/m^2 in summer 2004. Using a conversion factor from DW to AFDW of 0.35 (Mattson, unpublished data), this equates to 133.4 and 224.4 g/m^2 AFDW, both of which exceed the Biggs criteria and appear to be substantially higher than what Odum measured in the 1950s (although he did not report macroalgal abundance data). Silver Springs was sampled in the 2015 study but the transects were dominated by rooted macrophytes with attached epiphytes and macroalgal mats were not present at the sampling transects.

Whitford (1956) mentions the occurrence of mats of filamentous Chlorophyta and the cyanobacterium *M. wollei* in the Florida springs he sampled but does not mention their abundance or frequency of occurrence. The temporal and spatial variability in algal abundance in springs is a complicating factor in comparing historic versus current conditions and even currently comparing among springs. For example, no macroalgal mats were present at the Gum Slough sample transects in spring but they were there in fall (Table 12.2), and the transects at Rock Springs Run and Silver Glen Run did not have mats in the fall but did support fairly extensive mats in spring 2015 (Table 12.2).

Of the 14 springs sampled in 2015 (Table 12.1), sampling transects at 12 of these supported rooted submerged macrophytes with attached epiphytic algae. Comparison of mean epiphyte abundance data with the targets proposed by Guan et al. (2020b) is shown in Figure 12.1. Their DW data were converted to AFDW using a conversion factor of 0.35 (Mattson, unpublished data) and then converted to g/m^2. In spring, one transect each in Gum Slough, Ichetucknee River, and Rainbow River exceeded the targets for sites in full sunlight (14–17.5 g/m^2 AFDW). Almost all the transects exceeded the targets for shaded sites in spring (1.4–2.8 g/m^2 AFDW, Fig. 12.1). In the fall, epiphyte abundance was generally reduced at many transects (Fig. 12.1), and one transect (the same Gum Slough transect) exceeded the full sun targets. Many transects still exceeded the shaded site targets in the fall (Fig. 12.1). As for macroalgae, considerable spatial and temporal variability in epiphytic algal abundance is seen in these data, but many of these springs exceed epiphytic algal thresholds, which could potentially be harmful to the macrophytes.

Epiphyte AFDW data from the 2015 study were compared with similar AFDW data collected in an earlier study (GreenWater Labs, 2010) from a similar location in Alexander Springs Creek, just downstream of the headspring (Fig.12.2). This site is open and would be a "full sun" site after Guan et al. (2020b). AFDW varied widely, but rarely reached or exceeded the algal targets for full sun (Fig. 12.2). Conversely, most of the time AFDW did equal or exceed the targets for shaded sites (Fig. 12.2). Again, considerable temporal variation is seen in these data. Also of note, N and P concentrations in Alexander Spring are very low (reflecting natural background conditions in Florida springs), yet algal abundance exceeds some of the targets examined.

Also, like the macroalgal data, there are few historic data for comparison to indicate a background or reference condition for epiphytic algae. Odum (1957) did report epiphyte dry weight data for Silver Springs. Some calculations had to be done to make his data comparable to what we report here; after these adjustments, Odum's estimate was 2.7 g/m^2 AFDW. This was the only value provided; the replicate data were not available, although he did sample at multiple locations in the upper reach of Silver Springs. The 2015 study measured from 2.08–12.89 g/m^2 (mean 6.05 g/m^2) AFDW in the same area Odum sampled in Silver Spring, so these data would suggest that epiphyte AFDW currently is higher than the historic

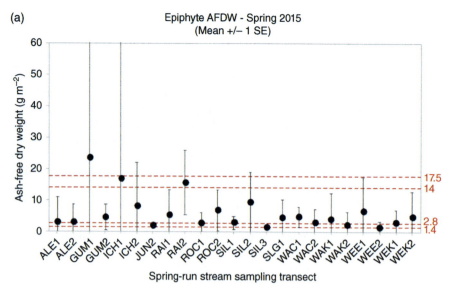

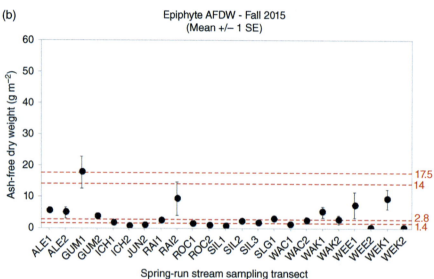

Figure 12.1 Plots (a) and (b) showing mean epiphytic algal AFDW at transects in the 2015 survey compared with the abundance thresholds adapted from Guan et al. (2020b). Note that their DW values were converted to AFDW.

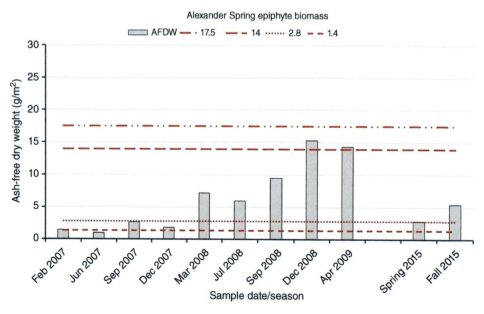

Figure 12.2 Mean epiphytic algal AFDW at a site on Alexander Springs Creek over time compared with the thresholds derived by Guan et al. (2020b: 14–17.5 in sunlight, 1.4–2.8 in shade). 2007–2009 data from GreenWater Labs (2010).

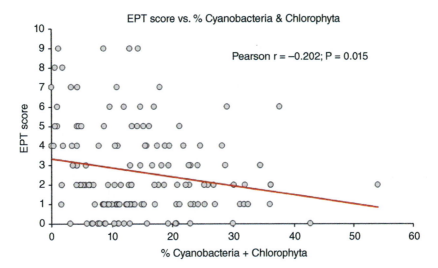

Figure 12.3 Correlation comparison of the EPT Index score and % Cyanobacteria and Chlorophyta in the periphyton community. Data are collected in springs and are from the FDEP bioassessment program (2000–2006).

condition, and a reference standing crop for algal epiphytes of <3 g/m^2 AFDW is proposed here for consideration. Based on the targets proposed in Guan et al. (2020b), this would be well below the full sun-site targets, but slightly above the shaded-site targets. The reach sampled in Silver Spring (both Odum and the 2015 study) was partially shaded in some areas and full sun in others

In contrast to looking at measures of algal abundance, the relative composition of the algal community could provide some quantitative criteria for benthic algae in springs related to impacts on fauna. Mattson (2009) obtained macroinvertebrate and algal composition data from the Florida Department of Environmental Protection (FDEP) bioassessment program collected in springs between 2000 and 2006. The macroinvertebrate data were collected using their Stream Condition Index methodology and the algal data were qualitative samples collected from multiple substrates and examined in the laboratory. There was a significant negative correlation between the relative abundance of Cyanobacteria and Chlorophyta in the algal community and the invertebrate EPT Index (Fig. 12.3). While the variability in this relationship indicates there are other factors influencing the EPT Index score, thresholds can be seen, with most of the higher scores (≥6) dropping out above 20%, and the EPT taxa essentially gone from the community above 40% Cyanobacteria and Chlorophyta (Fig. 12.3).

12.4. DISCUSSION AND CONCLUSIONS

This chapter presented benthic algal abundance targets (for macroalgae and epiphytic algae) proposed in the stream literature and compared some of these to a data set collected from 14 springs in Florida to evaluate their applicability for setting algal abundance targets in Florida springs. In some cases, the proposed targets were developed for temperate, rocky-bottom streams (Welch et al., 1988), or from streams in more northern climates (Biggs, 1996), but some of these targets were developed specifically in Florida spring systems (Guan et al., 2020b), or included data collected in Florida springs (Horner et al., 1983).

Florida does have guidelines for thresholds of benthic algal abundance in streams, which can be applied to springs. The FDEP has developed an assessment methodology for stream algae known as the Rapid Periphyton Survey (RPS). This methodology measures cover and thickness of benthic and attached algae (periphyton, epiphytes, etc.) at 99 sampling points in a 100 m reach of stream. Algal thickness is ranked, with higher ranks indicating higher algal abundance, and FDEP has determined that a rank of 4–6 in >25% of the RPS samples constitutes an "imbalance" as stated in the narrative nutrient standard (FDEP, 2013). FDEP staff members are currently evaluating the need for more quantitative monitoring of epiphytic and benthic algae in springs to support water quality restoration efforts and comparison of the RPS methodology with the methods used in the 2015 SJRWMD springs study and other survey methods (Mary Paulic, Florida Department of Environmental Protection, personal communication, 27 April 2021). The analysis in this chapter may provide some additional guidance in terms of desired future conditions and targets for epiphyte and macroalgal abundance in springs.

The epiphytic algal targets developed by Guan et al. (2020b) are directly applicable to Florida springs, since they were developed in one, and are considered a good target to use for epiphyte abundance thresholds on submerged macrophytes in spring-run streams. Setting targets for macroalgal abundance is still going to need additional work and consideration. The percent cover, chlorophyll *a*, and AFDW targets suggested by Biggs (1996; see Table 12.2) seem to have broad applicability to streams, as these were developed using a broad range of streams worldwide and considered a variety of environmental values to be protected. These might be considered as interim targets for macroalgal abundance in Florida springs. Highest macroinvertebrate diversity (Shannon Index) and evenness were seen at a macroalgal abundance <40 g/m^2 AFDW on snags in the Ichetucknee River (Mattson, 2009), suggesting that the Biggs targets may be useful/applicable for Florida springs. Supplemental thresholds could include the relative abundance of the algal taxa generally regarded as nuisance taxa (Cyanobacteria, Chlorophyta, and/or Xanthophyta) with thresholds of 20%–40% composition of the algal community (Fig. 12.3 in this chapter; Mattson, 2009).

It could be argued that biological criteria or targets should be tied to a cause/effect relationship with specific physical/chemical criteria subject to management, for example, nutrient criteria or water quality standards, or stream flow criteria. As discussed earlier in this chapter, the evidence linking benthic algal abundance in springs to nutrient concentrations, particularly nitrogen as NOx-N, is mixed (Brown et al., 2008; Heffernan et al., 2010; Stevenson et al., 2007). However, numerous physical, chemical, and biological drivers affect biotic communities and ecosystems, including algal communities in Florida springs. Biological targets such as those suggested here for benthic algae provide additional management tools to assess the "physical, chemical, and biological integrity" of surface waters as set forth in the U.S. Clean Water Act (Davis & Simon, 1995). While the specific environmental drivers affecting benthic algal abundance in Florida springs need additional investigation, benthic algal abundance criteria provide a benchmark or target for management that can help with assessment of current ecological condition. Establishing and using benthic algal abundance thresholds or targets can be combined with other management tools, including instream flow standards, water quality criteria, and continued research on cause/effect as part of an adaptive management approach to protection and restoration of Florida springs (Walters & Holling, 1990).

Some of the springs sampled in the 2015 study are considered reference springs in regard to their water quality; their NOx-N concentration is at or near the natural background of 0.05 mg/L NOx-N defined by the Florida Geological Survey (Scott et al., 2004); these are Juniper Creek, Alexander Springs Creek, and Silver Glen Spring Run. Wacissa River also has relatively low NOx-N concentrations. Despite this good water quality, algal abundance (epiphytes and/or macroalgae) in these springs exceeded one or more of the Biggs or Guan targets at many of the transects in the 2015 study (Table 12.2, Figs. 12.1, 12.2). These springs may not be good candidates for a biological reference condition because they receive a considerable amount of recreational impact or have been physically modified to some extent. They also differ in their base water chemistry, with Juniper Spring a low-ion spring, Alexander Spring a mixed spring, and Silver Glen Spring a salt spring (Woodruff, 1993). The SAV community in Silver Glen Spring Run regularly contains extensive mats of macroalgae, Alexander Springs Creek occasionally exhibits blooms of macroalgal mats, and Juniper Creek rarely develops macroalgal mats. Studies evaluating algal abundance in relation to base chemistry have not yet been done.

Additional research and long-term monitoring of SAV communities in springs are needed to help refine these algal targets. The SJRWMD 2015 springs study did not use the FDEP Rapid Periphyton Survey method, and, as noted earlier, a pilot study is being planned to compare the RPS results with the types of data collected in the 2015 study and other ongoing springs SAV monitoring. Additional evaluation of the effects of recreational and other physical disturbance on algal proliferation, and the influence of spring base chemistry on algal abundance also needs to be done to further evaluate these suggested targets or develop more appropriate targets.

As noted earlier, proliferation of nuisance benthic algal blooms in freshwater ecosystems, including Florida springs, is an emerging global issue (Hudon et al., 2014; Paerl & Otten, 2013; Stevenson et al., 2007). Whatever environmental drivers are promoting this trend, targets for algal abundance need to be adopted that result in reduced or minimal impacts to human use and ecological integrity. Additional research is also needed on the relationships between algal proliferation and other aquatic ecosystem characteristics (faunal communities, ecosystem productivity, etc.) to develop appropriate algal targets. These biological criteria can supplement and help support other criteria, such as water quality standards (concentration targets) or hydrologic/instream flow standards.

ACKNOWLEDGMENTS

I am grateful to the St. Johns River Water Management District for supporting this springs work. L. McCloud, E. Marzolf, and D. Dobberfuhl provided helpful reviews prior to submission of this manuscript to the editors, and the contributions of two anonymous reviewers were also very helpful.

REFERENCES

AMEC (2016). Synoptic biological monitoring of springs data collection, final report. (Special Publication SJ2019-SP1). Palatka, FL: St. Johns River Water Management District. https://www.sjrwmd.com/documents/technical-reports/special-publications/2019-2/

Bellinger, E. G., & Sigee, D. C. (2010). *Freshwater algae: Identification and use as bioindicators*. West Sussex, UK: Wiley-Blackwell.

Biggs, B. J. F. (1996). Patterns in benthic algae in streams. In R. J. Stevenson, M. L. Bothwell, & R. L. Lowe (Eds.), *Algal ecology. Freshwater benthic ecosystems* (pp. 31–56). New York, NY: Academic Press.

Brönmark, C., & Hansson, L. (1998). *The biology of lakes and ponds*. Oxford, UK: Oxford University Press.

Brown, M. T., Reiss, K. C., Cohen, M. J., Evans, J. M., Reddy, K. R., Inglett, P. W., et al. (2008). *Summary and synthesis of the available literature on the effects of nutrients on spring organisms and systems*. Water Institute Report. Gainesville, FL: University of Florida. https://floridadep.gov/dear/water-quality-evaluation-tmdl/documents/summary-and-synthesis-available-literature-effects

Camp, E. V., Staudhammer, C. L., Pine III, W. E., Tetzlaff, J. C., & Frazer, T. K. (2014). Replacement of rooted macrophytes by filamentous macroalgae: Effects on small fishes and macroinvertebrates. *Hydrobiologia, 722*, 159–170. https://doi.org/10.1007/s10750-013-1694-3

Chen, C., Yin, D., Yu, B., & Zhu, H. (2007). Effects of epiphytic algae on photosynthetic function of *Potamogeton crispus*. *Journal of Freshwater Ecology, 22*(3), 411–420.

Cohen, M. J., Coveney, M. F., & Slater, J. (2017). Physicochemistry: Nitrogen dynamics and metabolism. In K. R. Reddy et al. (Eds.), *Collaborative research initiative on sustainability and protection of springs (CRISPS)*. Water Institute Report. Gainesville, FL: University of Florida. https://www.sjrwmd.com/static/waterways/springs-science/CRISPS_Final_Report-All_Sections.pdf

Copeland, R. (2003). *Florida spring classification system*. Special Publication No. 52. Tallahassee, FL: Florida Geological Survey.

Copeland, R., Doran, N. A., White, A. J., & Upchurch, S. B. (2009). *Regional and statewide trends in Florida's spring and well groundwater quality (1991–2003)*. Bulletin No. 69. Tallahassee, FL: Florida Geological Survey.

Cowell, B. C., & Dawes, C. J. (2004). Growth and nitrate-nitrogen uptake by the cyanobacterium *Lyngbya wollei*. *Journal of Aquatic Plant Management, 42*, 69–71.

Davies, S. P., & Jackson, S. K. (2006). The biological condition gradient: A descriptive model for interpreting change in aquatic ecosystems. *Ecological Applications, 16*(4), 1251–1266.

Davis, W. S., & Simon, T. P., Eds. (1995). *Biological assessment and criteria: Tools for water resource planning and decision making*. Boca Raton, FL: Lewis Publishers.

Dodds, W. K., Jones, J. R., & Welch, E. B. (1998). Suggested classification of stream trophic state: Distributions of temperate stream types by chlorophyll, total nitrogen, and phosphorus. *Water Research, 32*(5), 1455–1462.

FDEP (2012). *Technical support document: Development of numeric nutrient criteria for Florida lakes, spring vents, and streams*. Report by the Standards and Assessment Section. Tallahassee, FL: Florida Department of Environmental Protection.

FDEP (2013). *Implementation of Florida's numeric nutrient standards*. Report by the Florida Department of Environmental Protection submitted to the U.S. Environmental Protection Agency. Tallahassee, FL: Florida Department of Environmental Protection. https://floridadep.gov/dear/water-quality-standards/documents/nnc-implementation

Florida Springs Task Force (2000). *Florida's springs: Strategies for protection and restoration*. Report by the Florida Springs Task Force. Tallahassee, FL: Florida Department of Environmental Protection. https://floridadep.gov/dear/water-quality-evaluation-tmdl/documents/florida%E2%80%99s-springs-report

GreenWater Labs (2010). *Monitoring of algal communities in the Wekiva River, Rock Springs Run, Juniper Creek, Alexander Spring Creek, and Silver Glen Run*. Final Report. Palatka, FL: St. Johns River Water Management District.

Guan, J., Jacoby, C. A., & Frazer, T. K. (2020a). Light attenuation by periphyton on *Vallisneria americana*. *Ecological Indicators, 116*. https://doi.org/10.1016/j.ecolind.2020.106498

Guan, J., Jacoby, C. A., & Frazer, T. K. (2020b). In-situ assessment of the effects of periphyton on the growth of *Vallisneria americana*. *Ecological Indicators, 119*. https://doi.org/10.1016/j.ecolind.2020.106775

Heffernan, J. B., Liebowitz, D. M., Frazer, T. K., Evans, J. M., & Cohen, M. J. (2010). Algal blooms and the nitrogen-enrichment hypothesis in Florida springs: Evidence, alternatives, and adaptive management. *Ecological Applications, 20*(3), 816–829.

Horner, R. R., Welch, E. B., & Veenstra, R. B. (1983). Development of nuisance periphytic algae in laboratory streams in relation to enrichment and velocity. In R. G. Wetzel (Ed.), *Periphyton of freshwater ecosystems* (pp. 121–134). The Hague: Dr. W. Junk Publishers.

Hudon, C., De Séve, M., & Cattaneo, A. (2014). Increasing occurrence of the benthic filamentous cyanobacterium *Lyngbya wollei*: A symptom of freshwater ecosystem degradation. *Freshwater Science, 33*(2), 606–618. http://www.bioone.org/doi/full/10.1086/675932

Katz, B. G. (2020). *Nitrogen overload: Environmental degradation, ramifications, and economic cost* (Geophysical Monograph 250). American Geophysical Union and J. Wiley & Sons.

King, S. A. (2014). Hydrodynamic control of filamentous macroalgae in a sub-tropical spring-fed river in Florida, USA. *Hydrobiologia, 734*, 27–37. https://doi.org/10.1007/s10750-014-1860-2

Liebowitz, D. M., Cohen, M. J., Heffernan, J. B., Korhnak, L. V., & Frazer, T. K. (2014). Environmentally mediated consumer control of algal proliferation in Florida springs. *Freshwater Biology, 59*, 2009–2023. https://doi.org/10.1111/fwb.12403

Mattson, R. A. (2009). *Relationships between benthic algae and benthic macroinvertebrate communities in Florida spring-run streams*. Professional Paper SJ2009-PP1. Palatka, FL: St. Johns River Water Management District. https://www.sjrwmd.com/documents/technical-reports/professional-papers/

Mattson, R. A., Hall, D. L., Szafraniec, M. L., & Guyette, M. Q. (2019). *Synoptic biological survey of 14 spring-run streams in north and central Florida: I. Submerged aquatic vegetation communities-macrophytes*. Technical Publication SJ2019-2. Palatka, FL: St. Johns River Water Management District. https://www.sjrwmd.com/documents/technical-reports/technical-publications/2019-2009/

Meinzer, O. E. (1927). *Large springs in the United States*. Water-Supply Paper 557. Washington, DC: US Geological Survey.

Odum, H. T. (1957). Trophic structure and productivity of Silver Springs, Florida. *Ecological Monographs*, 27(1), 55–112.

Paerl, H. W., & Otten, T. G. (2013). Harmful cyanobacterial blooms: Causes, consequences, and controls. *Microbial Ecology*, 65, 995–1010. https://doi.org/10.1007/s00248-012-0159-y

Power, M. E. (1990). Benthic turfs versus floating mats of algae in river food webs. *Oikos*, 58, 67–79.

Quinlan, E. L., Phlips, E. J., Donnelly, K. A., Jett, C. H, Sleszynski, P., & Keller, S. (2008). Primary producers and nutrient loading in Silver Springs, FL, USA. *Aquatic Botany*, 88, 247–255.

Reaver, N. G. F., Kaplan, D. A., Mattson, R. A., Carter, E., Sucsy, P. V., & Frazer, T. K. (2019). Hydrodynamic controls on primary producer communities in spring-fed rivers. *Geophysical Research Letters*, 46(9), 4715–4725. https://doi.org/10.1029/2019GL082571

Rosen, B. H. (1995). Use of periphyton in the development of biocriteria. In W. S. Davis & T. P. Simon (Eds), *Biological assessment and criteria: Tools for water resource planning and decision making* (pp. 209–215). Boca Raton, FL: Lewis Publishers.

Scott, T. M., Means, G. H., Meegan, R. P., Means, R. C., Upchurch, S. B., Copeland, R. E., et al. (2004). *Springs of Florida*. Bulletin No. 66. Tallahassee, FL: Florida Geological Survey.

Slack, L. J., & Rosenau, J. C. (1979). *Water quality of Florida springs*. Map Series 96. Tallahassee, FL: Florida Geological Survey.

Stevenson, R. J., Pinowska, A., Albertin, A., & Sickman, J. O. (2007). *Ecological condition of algae and nutrients in Florida springs: The synthesis report*. Technical Report. Tallahassee, FL: Florida Department of Environmental Protection.

Stoddard, J. L., Larsen, D. P., Hawkins, C. P., Johnson, R. K., & Norris, R. H. (2006). Setting expectations for the ecological condition of streams: The concept of reference condition. *Ecological Applications*, 16(4), 1267–1276.

Szafraniec, M. L. (2014). *Spectral distribution of light in Florida spring ecosystems: Factors affecting the quantity and quality of light available for primary producers*. Doctoral dissertation. Gainesville, FL: University of Florida.

Walters, C. J., & Holling, C. S. (1990). Large-scale management experiments and learning by doing. *Ecology*, 71(6), 2060–2068.

Welch, E. B., Jacoby, J. M., Horner, R. R., & Seeley, M. R. (1988). Nuisance biomass levels of periphytic algae in streams. *Hydrobiologia*, 157, 161–168.

Whitford, L. A. (1956). The communities of algae in the springs and spring streams of Florida. *Ecology*, 37(3), 433–442.

Woodruff, A. (1993). *Florida springs chemical classification and aquatic biological communities*. Master's Thesis. Gainesville, FL: University of Florida.

13

Protecting Springs in the Southwest Great Artesian Basin, Australia

Mark Keppel[1], Anne Jensen[2], Melissa Horgan[3], Aaron Smith[3], and Simone Stewart[1]

ABSTRACT

The Great Artesian Basin (GAB) springs are unique environments of great ecological and cultural value, while the supporting groundwater is also an important economic resource. A high concentration of springs occurs within the far north of South Australia (SA). Historically, management and protection of springs, community, and industry needs have not necessarily aligned; water extraction led to decline and extinction of many springs. Concerns over these impacts led to significant research to understand hydrogeological and environmental processes associated with the springs, enabling the development of management measures and tools. This chapter demonstrates the use of evidence-based policies to manage ongoing threats to GAB spring environments. The primary policy and management tool implemented in SA to mitigate spring impacts is the legal declaration of the Far North Prescribed Wells Area (FNPWA), under which groundwater is monitored and regulated via the FNPWA Water Allocation Plan (FNWAP). This is designed to promote integrated and sustainable management of water resources, through licensing of groundwater extractions, ensuring judicious usage, and safeguarding pressures near springs. The GAB Springs Adaptive Management Plan (GABSAMP) is a recent, nonlegal, cross-jurisdictional initiative that collates current evidence on spring condition and threats and proposes objective methods to assess and manage identified risks. The GABSAMP also recommends a national interactive spring database and a coordinated, GAB-wide monitoring program. The FNWAP and the GABSAMP collectively promote evidence-based management actions that will maintain sufficient artesian pressure to sustain spring flows and encourage sensitive land use practices in and around springs.

13.1. INTRODUCTION

Groundwater plays an integral part in the ongoing sustenance of the natural environment, as well as supporting the traditional and modern societies and economies of the arid and semiarid region of Australia. For these reasons, the Great Artesian Basin (GAB) is an important groundwater resource in Australia, covering a large tract of the Australian states of Queensland (QLD), New South Wales (NSW) and South Australia (SA), and the Northern Territory (NT) (Fig. 13.1). The groundwater of the GAB supports springs and the associated spring wetland communities, considered some of the most revered, endemic and fragile groundwater-dependent natural environments in Australia (Ponder, 1986; Fatchen, 2000; Arthington et al., 2020). Groundwater that supplies GAB springs in SA is generally understood to enter recharge beds located on the eastern and western margins of the basin where it flows toward the discharge springs located in SA and NSW (Habermehl, 1980; Habermehl, 2020b) (Fig. 13.1).

[1] Department for Environment and Water, Government of South Australia, Adelaide, Australia
[2] Environmental Consultant, Adelaide, Australia
[3] South Australia Arid Lands Landscape Board, Port Augusta, Australia

182 THREATS TO SPRINGS IN A CHANGING WORLD

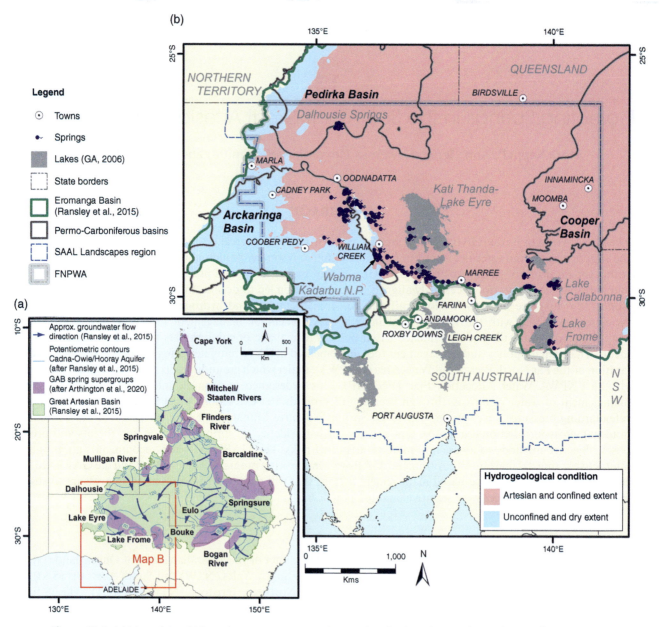

Figure 13.1 (a) Map of the GAB, spring super groups and generalized inferred groundwater flow in the main GAB aquifer (adapted from Ransley et al., 2015); (b) location map of springs in the FNPWA of SA (DEW, 2021a) hydraulic conditions of the GAB in SA (adapted from Love et al., 2013). Permo-Carboniferous basins from Wohling et al. (2013), Keppel et al. (2015), and GA (2013).

After recent field and remote sensing-based survey work by Gotch (2013), Gotch et al. (2016), and Keppel et al. (2016), the total number of recorded individual spring vents in SA is now over 5,000 (DEW, 2021a). These springs are contained within 23 spring complexes and spread over a length of approximately 800 km (Fig. 13.1b), extending from Lake Frome in the southeast in a northwest direction to the largest spring complex per discharge volume found in SA, Dalhousie Springs (Fig. 13.2a), located near the SA border with the NT (Zeidler & Ponder, 1989).

Images of springs like Thirrka (Blanche Cup) and Pirdalinha (The Bubbler) (Fig. 13.2b), two of the most famous limestone springs found in the Wabma Kadarbu National Park, are iconic as visual representations of the GAB. Further, there is a wide variety of environments found within the SA GAB springs. As well as the distinctive limestone mounds that endure in the landscape

Figure 13.2 Springs of the South Australian portion of the GAB: (a) Irrwanyere (Main Spring) at Witjira-Dalhousie Springs. Tourists are permitted to swim in this spring-fed pool. (b) Pirdalinha (The Bubbler), Wabma Kadarbu National Park, one of the most famous springs in the GAB. An inselberg, formed from a now extinct mound spring complex, can be seen in the background top right. (c) Yurilja (Margaret Spring), is a classic limestone GAB mound spring (Keppel et al., 2011/with permission of ELSEVIER). (d) Mud mound spring found in the Mulligan Spring complex (Kurnuwardna), near Lake Callabonna. There is evidence to suggest water supply is not restricted to the GAB (Keppel et al., 2020b). (e) Waralanha (McLachlin Spring), near Kati Thanda-Lake Eyre (South) formed in a dunal environment. (f) Springs on the surface of Kati Thanda-Lake Eyre (North). Little is known about these springs due to their remote and inaccessible location.

(Fig. 13.2b,c), there are also springs that form structures out of other substrates, such as mud (Fig. 13.2d), sand (Fig. 13.2e), or salts (Fig. 13.2f), that are more susceptible to erosion and ablation.

The GAB springs have supported occupation by Indigenous Australians for tens of thousands of years and have deep cultural significance for Indigenous Australians (Ah Chee, 2002; Arthington et al., 2020; Moggridge, 2020). Important cultural stories and practices are linked to springs and associated topographical features (Hercus & Sutton, 1985). Archaeology in and around spring sites reflects the importance of these permanent water sources in the otherwise dry landscapes (Florek, 1993). Many springs determined important trade and communication corridors for Indigenous People (Ah Chee, 2002; Harris, 2002). These connections continue today with communities maintaining cultural, social, and spiritual connections with basin springs and their associated ecological communities and landscapes (Moggridge, 2020). More recently, European exploration and settlement relied on the springs for water sources. In SA, the route of the Overland Telegraph and associated transport routes, such as the Oodnadatta Track and the narrow-gauge Ghan railway to Alice Springs, were governed by the springs (Harris, 1981). More than 120 towns and settlements in the basin currently rely on GAB water (GABCC, 2018), as well as the pastoral, mining, energy, and tourism industries.

The GAB springs are home to some of the most unique and fragile ecosystems in Australia. For example, in their documentation of the ecological significance of SA GAB springs, Gotch (2013) highlighted several key findings, including up to 42 significant evolutionary units within the aquatic invertebrate fauna investigated, most of which are interpreted to be short-range spring endemics. The same study also found that the abundance and richness of endemic flora was correlated with high flow rates and springs supportive of flowing water; both findings emphasize such spring-supported environments' vulnerability to flow reductions or other rapid changes. This propensity to support endemic fauna and flora is due in part to their persistence in an arid environment, which not only isolates these communities but also allows endemism to develop. The magnitude of this persistence over time has only recently been recognized: for example, Murphy et al. (2009) used mitochondrial DNA and allozyme analysis of the freshwater amphipods of the family Chiltoniidae from the GAB springs of SA to suggest most evolutionary lineages originated in the late Miocene.

The communities of native species that depend on the natural discharge of groundwater have been protected through a declaration as an endangered ecological community under the Australian government's Environment Protection and Biodiversity Conservation Act (EPBC)(1999), a key piece of national environmental protection legislation (DAWE, 2021). The overarching aim of the EPBC Act is to provide a legal framework for environment and heritage protection and biodiversity conservation that is focused on matters of national environmental significance (DAWE, 2021). While all GAB springs in SA are included under the EPBC Act, this does not include all springs with known connection to the GAB. Springs termed "recharge springs" are those found in recharge zones and represent overflow from aquifers (Ponder, 2002). Notably, this exclusion covers the entire Cape York supergroup (Fig. 13.1a). In addition, the EPBC Act does not include the Mitchell/Staaten rivers supergroup because it was identified as a separate supergroup only in 2007 (Fensham et al., 2007) (Fig. 13.1a).

All of SA's GAB springs are found within the South Australian Arid Lands (SAAL) Landscape Region (Fig. 13.1). This region is one of several administrative districts found within SA that are specifically declared for the purposes of supporting and enhancing sustainable development while recognizing and protecting the intrinsic values of the landscape. The SAAL Landscape Region is the largest of its type in SA, and water is its scarcest resource. It encapsulates the FNPWA, a special management zone within the SAAL Landscape Region under which water resources are highly regulated (Fig. 13.1).

As well as supporting a unique number of spring-dependent ecosystems, groundwater in the SAAL Landscape Region is also vital in supporting the SA economy. The region hosts several mining operations, the most significant of which is the Olympic Dam polymetallic mine. Olympic Dam is currently one of the largest copper, gold, silver, cobalt, and rare earths deposits and largest single uranium deposit in the world (Mudd & Jowitt, 2018). The region is also home to the SA portion of the Cooper Basin petroleum field, an important onshore petroleum and natural gas deposit (Fig. 13.3a), of which a considerable proportion of conventional production has been and continues to be sourced from the overlying Eromanga Basin (GAB) strata (Cotton et al., 2007; Schenk et al., 2016). Further, the region has a long established and historically important pastoral industry, and a growing tourism industry that is highly dependent upon the sustainability of the natural environment and the communities of the region. Currently, of the 178 Megalitres per day (ML/d) of groundwater licensed to be extracted from the FNPWA, 134 ML/d are sourced from the GAB (Fig. 13.4). Water for mining operations (44 ML/d), water extracted as a by-product of petroleum production (coproduced water; 57 ML/d), and stock water (16 ML/d) are the biggest collective approved water takes through license (LSA SAAL, 2021).

Consequently, a primary threat to the spring environments of the Far North of SA is pressure loss within the

Figure 13.3 Uses, impacts, and associated mitigation strategies: (a) Energy exploration rig in the Cooper Basin region (Keppel et al., 2019). Mining and energy industries are currently the largest licensed users of groundwater from the GAB in SA. (b) Badly degraded headworks at Montecollina Bore. Such degradation prevents bores from being shut off, causing wastage and pressure loss. This bore was decommissioned in 2019 under the IGABDR program. (c) Bore drain used to water cattle supplied by uncontrolled flow from an artesian bore. (d) Heavy compaction caused by cattle around an unfenced spring vent, leading to ecological degradation. (e) Watering trough that supplies water to cattle via a ball-float valve; a simple yet effective and robust water-saving delivery technique. (f) A modern metered artesian bore. Bore headworks are located within the concrete standpipe to the right of the shed.

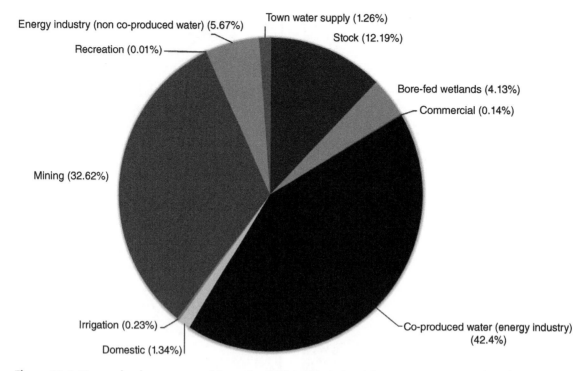

Figure 13.4 Licensed volumes sourced from the GAB in South Australia as a percentage of total (134 ML/d) presented by license purpose description. (Adapted from LSA SAAL, 2021).

supporting artesian aquifers via groundwater extraction. Not only does pressure loss directly impact spring flow, additionally, loss of confining pressure leaves groundwater increasingly vulnerable to ingress of different or poorer quality water from adjacent strata, particularly where such confining pressures adjacent unconfined conditions (Fig. 13.1a). Demand for water from industry inclusive of mining and extractive industries is expected to grow in SA (Adamo, 2021; DEW, 2021b). With respect to the GAB, while there have been recent requests from industry for temporary increases in water take (Government of South Australia, 2019), longer term, such water demands may lead to the development of further "cumulative drawdown impacts" (situations where groundwater drawdowns caused by multiple users merge to form a larger area of impact). Therefore, the concepts of judicious groundwater use and collective environmental stewardship are more important than ever with respect to the survival of the spring environments.

Two key initiatives driven by community and government have developed in response to these threats to spring wellbeing from changes wrought by development and groundwater exploitation. The first is the establishment of the FNPWA within the SAAL Landscape Region and the development of the accompanying FNWAP. More recently, a project team of researchers, managers, and individuals with springs knowledge and experience collaborated to develop the second response, the evidence-based GABSAMP, funded by the GAB state governments and the Australian government. While the FNWAP is a legislated requirement stipulated under the Landscapes South Australia Act (2019), the GABSAMP currently has no mandated requirement to be implemented. Consequently, GABSAMP adoption will be voluntary during initial roll out.

This chapter provides an overview of what the FNWAP and the GABSAMP are, the science and history pertinent to their development, and their context in the wider regulatory and management effort to protect GAB springs. In doing so, this chapter also provides an example of the use of evidence-based policies and adaptive management principles to address and manage ongoing threats to GAB spring environments.

13.2. THE GREAT ARTESIAN BASIN AND ITS ECONOMIC IMPORTANCE

The GAB is the hydrogeological name given to several interconnected sedimentary basins of Mesozoic age that covers an area of approximately 1.7 million km^2, or over 20% of the Australian continent (Krieg et al., 1995; Ransley & Smerdon, 2012; Habermehl, 2020b). The GAB is found over sizeable portions Qld, SA, NSW, and the NT (Fig. 13.1). The sedimentary basins include the Eromanga, Surat, Carpentaria, and Clarence-Morton,

Table 13.1 Volumes for oil, gas, and water production from the Cooper-Eromanga region

Basin	Oil ML (mbbl)	%	Gas (mmcf)	%	Water ML (mbbl)	%	Oil ML (mbbl)	%	Gas (mmcf)	%	Water ML (mbbl)	%
	Historic (Nov 1969–Feb 2021)						2020					
Eromanga (GAB)	34,488 (217)	83	75,655	1	333,216 (2,096)	92	1,569 (10)	89	2,450	2	21,496 (135)	98
Cooper	6,851 (43)	17	7,791,943	99	29,987 (189)	8	199 (1)	11	98,604	98	338 (2)	2
Total	41,339 (260)	100	7,867,598	100	363,203 (2,285)	100	1,768 (11)	100	101,054	100	21,834 (137)	100

Source: Data from Government of South Australia (2021).
Notes: mbbl: million barrels; mmcf: million cubic feet.

with each representing a depo-center within the larger GAB hydrogeological super-basin (Fig. 13.1). The largest of these sedimentary basins is the Eromanga Basin, which is synonymous with the GAB in all of SA, the NT, and much of southwest Qld and northwest NSW.

The most important aquifers of the GAB are composed of fine to coarse grained sandstone and interbedded finer grained units deposited in alluvial, fluvial, aeolian, lake, and coastal environments of late Jurassic to early Cretaceous age. Collectively this sequence of strata has been called the "J" aquifer (Habermehl, 1980) or the "Cadna-owie - Hooray" aquifer (Radke et al., 2000). Overlying this aquifer are a thick sequence of mudstone, siltstone, and shale collectively known as the Rolling Downs Group, which provides the confining layer that induces the artesian pressures in the aquifer below. Finally, overlying the Rolling Downs Group are less significant aquifers that nevertheless provide important sources of water in localized areas (DNRME, 2018).

The GAB's significance as a major regional water resource is magnified by its location: it underlies arid and semiarid landscapes to the west of the Great Dividing Range. (Fig. 13.1). Basin groundwater provides a climate-independent water supply in areas that often receive low or intermittent rainfall (GABCC, 2018). The GAB is still the only reliable source of water for stock and potable water supplies over much of the basin to support most human activity. This value as an aquifer imbues the GAB with great economic significance for both SA and Australia at large; consumptive use of GAB water was estimated in 2016 to return about $13 billion from production annually, including $4 billion in stock, $6 billion in mining, $2 billion in gas, and $1 billion from tourism (Frontier Economics, 2016).

• Underlying the GAB are also several Permo-Carboniferous sedimentary basins that are prospective source and reservoir rocks for oil, gas, and coal resources. While the Cooper Basin is the most important of these economically to SA, the Arckaringa and Pedirka basins further to the west are also considered prospective (DMITRE, 2012) (Fig. 13.1). The Cooper and Eromanga oil and gas province is Australia's largest inland oil and gas development (Core Energy Group, 2016). While conventional and unconventional gas production is predominantly from Cooper Basin formations, conventional oil production is largely from Eromanga Basin formations (Cotton et al., 2007; Schenk et al., 2016; Habermehl, 2020b). Petroleum industry data provided to and published by the government of South Australia under regulatory requirement stipulated under the SA Petroleum and Geothermal Energy Act (2003) (Government of South Australia, 2021) indicate that within the Cooper and Eromanga oil and gas province, historically 99% of gas is produced from Cooper Basin formations but 83% of oil and 92% of co-produced water were generated from Eromanga Basin formations. Additionally, after a period of declining production from a mid-1980's peak, discoveries made in the early 2000s have seen conventional oil production from the Eromanga Basin increase (Sales et al., 2015), with this proportional importance of the Eromanga Basin reflected in the 2020 full calendar year production figures (Table 13.1).

13.3. THREATS TO GAB SPRINGS THROUGH RECENT GROUNDWATER SCIENCE

The GAB springs are one of the few major artesian spring systems in the world that has not been entirely degraded by over-exploitation of the water-bearing aquifers and/or the impacts of land use in and around spring vents. However, in various parts of the GAB, spring extinction and wetland degradation have occurred and continue to occur to a notable extent. Across the GAB in its entirety, springs are under threat from two major processes: (1) reduction in artesian pressure from excess water extraction and uncontrolled bores that reduce flow to springs (Fig. 13.3b,c) and (2) physical disturbance of spring structures and dependent ecosystems resulting in loss of geological features and disruption of spring ecosystems through grazing, trampling, increased nutrients from animal waste, weed invasion, and excavation (Fig. 13.3d).

Further, since artesian groundwater was discovered in the GAB and its vulnerability recognized in the late nineteenth to early twentieth centuries, there has been a considerable number of scientific studies aimed at understanding the hydrodynamics of the groundwater systems and the groundwater-dependent ecosystems (GDEs) reliant on these. These studies have employed a wide variety of techniques and approaches, including hydraulic and hydrochemical analysis, geophysical investigations, and remote sensing. In recent times, several government-sponsored or endorsed research into the GAB resource and its connectivity to other surface and groundwater systems have occurred. For example, the Australian government's GAB Water Resources Assessment (Smerdon et al., 2012) and the Allocating Water and Maintaining Springs in the GAB (AWMSGAB) (NWC, 2013), were two complementary research undertakings delivered in 2013. Following this work, the SA Lake Eyre Basin Water Knowledge Project (LEBWKP) (DEW, 2015) and similar projects elsewhere in the GAB were undertaken in response to community concerns regarding the impact of coal seam gas and large coal mining developments targeting the Permo-Carboniferous strata underlying the GAB. These later projects were part of the Australian government sponsored Bioregional Assessments Program (CofA, 2020). There have been several publications providing an historical overview of scientific advancements, including Habermehl (2020a and 2020b). Some of the key findings pertinent to spring environment survival and management in SA from these studies that have been considered during development of the current FNWAP and GABSAMP included the following:

1. That the GAB is not in a natural steady state where recharge equals discharge. Recharge is subject to paleoclimatic-related change and the current state of drying in Australia means that recharge is currently estimated to be very much less than discharge (Love et al., 2000; Rousseau-Gueutin et al., 2013; Welsh, 2006). Further, Fu et al. (2020) presented evidence from more recent historical climatic data to suggest rainfall across the greater GAB region is in net decline, while evapotranspiration overall is increasing. These changes not only impact recharge but potentially the economic and environmental stressors impacting the GAB groundwater resource.

2. Where undeformed, the confining layers overlying the largest GAB aquifer are extremely tight, suggesting that previous estimates of natural water discharge via vertical leakage were overestimated (Harrington et al., 2013; Smerdon et al., 2014). Leakage through the main confining layer is now thought to be largely restricted to where structural deformation has caused preferential flow path development into adjacent strata or indeed the surface where groundwater may discharge via diffusion. Such structural deformation is also recognized as a key means through which spring conduits may also form in many instances. Diffuse leakage to surface similarly is restricted to the basin margins where the confining layer is thin and is also often associated with springs (Harrington et al., 2013; Costelloe et al., 2015; Matic et al., 2020).

3. The hydrostratigraphic and structural framework of the basin is more complex than previously conceptualized. The hydrogeological properties of aquifer and confining layers may change across the basin, potentially connecting the GAB to other aquifers or reducing the thickness of the previously understood aquifer. For instance, Ransley and Smerdon (2012) describe the Cadna-owie Formation within the depo-center of the Eromanga Basin to be more akin to a leaky aquitard than the historically understood conceptualization as a productive aquifer. Much of this complexity, including the formation of springs, is related to neotectonic deformation of the Australian continent (e.g., Karlstrom et al., 2013; Keppel et al., 2020a; Sandiford et al., 2020).

4. The quantitative relationship between spring flow and pressure is not simple. Spring conduit formation and the relationship to spring flow and supplying groundwater pressure can be conceptualized and is considered a key element in the assessment of spring environment vulnerability (Green & Berens, 2013; Flook et al., 2020). However, while there is some correlation between pressure head and spring flow as determined by wetland area (White & Lewis, 2011; Fatchen, 2000), the chemical, biological, and mechanical processes that either create or maintain flow within or at the opening of the spring conduit are also influential in controlling spring flow (Green & Berens, 2013; Keppel, 2013). Consequently, there is only a limited relationship observed between spring flow and artesian pressure head above the springs (called "excess head" by Green & Berens, 2013). By extension, changes in supplying pressure may not have a uniform or predictable impact on spring flow within spring complexes (Kinhill Engineers, 1997). With respect to risk assessment and management, Green and Berens (2013) surmised that understanding the relationship between spring flow and excess head was critical to understanding and assessing risk. They suggested that slight drawdowns where excess head was minimal may have more significant impacts than where excess head is larger, regardless of the spring flow rate. Therefore, an understanding of these processes and estimation of the excess head at individual spring complexes is necessary for sound springs management and risk assessment.

5. Spring flow has been found to have a key role in the risk to spring environments related to wetland hydrochemistry. Shand et al. (2016) suggested that a reduction in spring flow may lead to the oxidation of naturally occurring pyrite in spring wetland environments, leading to hypersulfidic soils and extreme acidification. While natural variation in spring flow and wetland dynamics may also result in acidification, Shand et al. (2016) suggested permanent loss of flow could cause such processes to accelerate with catastrophic consequences. Similarly, Green et al. (2013) argued that reduction of spring flow may also increase evapotranspiration within the spring wetland environment by increasing the residence time in the containing pool near the spring conduit mouth. Further, Green et al. (2013) quantified this potential salinity increase as a function of spring flow reduction using an analytical model approach, determining that springs that already have a high proportion of water loss via evapotranspiration within a captured spring pool are more susceptible to water quality impacts because of pressure loss than those that have a higher through flow.

6. Groundwater flow may be impacted by density forces related to the very high temperatures experienced in deeper parts of the basin. Consequently, any estimated flow and volumes using classical gravity-driven hydrogeological principles may be erroneous (Pestov, 2000; Rousseau-Gueutin et al., 2013).

7. In addition to the increased number of individual spring vents that were mapped, Gotch (2013), Gotch et al. (2016), and Keppel et al. (2016) also documented and summarized a contemporary snapshot of the ecology, hydrochemistry, and morphological setting of these springs, finding them to be highly variable, with levels of degradation due to modern land practices ranging from negligible to very high.

8. Finally, evidence has emerged to suggest that some GAB springs may have additional sources of groundwater beyond the largest GAB aquifer found in SA (Priestley et al., 2020), including the underlying Precambrian basement rocks (Halihan et al., 2020), Permo-Carboniferous Arckaringa (Keppel et al., 2015) and Pedirka (Wolaver et al., 2020) basins as well as aquifers in the overlying late Cretaceous, Cenozoic, and Quaternary strata not typically associated with SA's GAB springs (Keppel et al., 2020b). This finding emphasizes the importance of including all groundwater resources within management and regulatory approaches, rather than just those found in the largest GAB aquifer.

These findings and others summarized in Habermehl (2020a, 2020b) highlighted the inherent vulnerability of GAB springs to water take and land management practices. Revelations concerning the current rate of groundwater resource depletion heightened the need for change on both environmental and economic sustainability grounds.

Further, changes in and around springs associated with land-use practices that impact ecological, cultural, and geological heritage are also as important as those related to loss of artesian pressure. Both threatening processes require sustained action to ensure that springs are protected while the benefits that accrue from the use of GAB water and land continue.

13.4. THE NATIONAL REGULATORY RESPONSE TO SPRING THREATS

Governments recognized issues surrounding uncontrolled flowing bores as early as the late nineteenth and early twentieth century (Brake, 2020), but the threat to springs specifically by loss of pressure was first recognized around the mid-twentieth century (Brake, 2020; Habermehl, 2020b). In SA, the state government produced a report and accompanying bore-rehabilitation program in 1977 that rehabilitated 230 government-drilled bores (Brake, 2020; Habermehl, 2020b). Despite state-based measures to address such issues, Brake (2020) notes that prior to 1999 water pressures were still diminishing in many regions; springs and artesian bores continued to extinguish; and wastage was still prevalent. By the late 1990s, there were still more than 1,500 uncontrolled artesian bores feeding more than 34,000 km of open bore drains (Brake, 2020; DAWE, 2020b).

In response, the Great Artesian Basin Consultative Committee (GABCC) (in a later iteration called the Great Artesian Basin Coordination Committee) was formed to act as a facilitator between government, landholders, and other stakeholders to obtain a coordinated approach to capping and piping bores (Brake, 2020; Habermehl, 2020b). Further, the Australian government, with pivotal input from the GABCC, developed the first GAB Strategic Management Plan (GABSMP). The GABSMP has since provided a framework for governments, Indigenous Peoples, and other water users and stakeholders to "achieve economic, environmental, cultural and social outcomes for the GAB and its users" (DAWE, 2020a). Finally, the GABCC also had a role in helping improve landholder attitudes with respect to good on-ground management practice (Brake, 2020; Habermehl, 2020b).

Each GABSMP is developed in partnership with GAB-state governments and has a 15 year life, with the first GABSMP covering 2000–2015. Central to the GABSMP are the guiding principles for responsible groundwater and related natural resource management in the GAB, guiding governments, water users, and other stakeholders on policies, programs, and actions necessary for economic, environmental, and socially sustainable use of GAB groundwater resources. The guiding principles from the current GABSMP (DAWE, 2020a) include: (1) coordinated governance; (2) a healthy resource; (3) Indigenous values, cultural heritage, and other community values; (4) secure and managed access; (5) judicious use of groundwater; (6) information, knowledge, and understanding for management; and (7) communication and education.

Complementary to this, several investment programs pursuant to the overall strategic direction of the GABSMP have been implemented. At a federal level, the GAB Sustainability Initiative (GABSI) was initiated in 1999 to address the need for improved coordination and increased funding for bore and bore drain rehabilitation efforts, which had commenced in the 1980s. GABSI, which ran from 1999–2017, resulted in the rehabilitation of 759 bores, removal of over 21,000 km worth of bore drains, and an estimated water saving of 250,000 ML/y (Brake, 2020; DAWE, 2020b), largely through a program of piping, well capping, well rehabilitation, and, where necessary, well decommissioning and replacement. A further program, the Interim Great Artesian Basin Infrastructure Investment Program (IGABIIP) ran from 2019 to 2020. Sixteen infrastructure projects were funded under this initiative worth $1.842 million of Australian government funding and proved an estimated water saving of 3,050 ML/y, primarily in Qld (DAWE, 2020b). The current program Improving GAB Drought Resilience (IGABDR), commenced in 2019 and builds on the efforts under GABSI and IGABIIP though a further investment of $27.3 million to 2024 (DAWE, 2020b). At the commencement of this current program, Brake (2020) and DAWE (2020b) estimated that, across the entire GAB, 431 bores and over 5,000 km worth of bore drains remained that require bore capping and piping work for an estimated water saving of just over 116,000 ML/y.

13.5. THE SOUTH AUSTRALIAN REGULATORY RESPONSE: THE FAR NORTH WATER ALLOCATION PLAN

In SA, the initial response to the unfolding water security predicament, both in the GAB and elsewhere in the state, was the adoption of first the Water Resources Act (1976) (now repealed), followed by its later replacements in the Water Resources Act (1990) (repealed), Water Resources Act (1997) (repealed), Natural Resources Management Act (2004) (repealed), and Landscape South Australia Act (2019). Key to these pieces of legislation with respect to GAB groundwater management and regulation in SA is the declaration of groundwater resources of economic and ecological significance as Prescribed Wells Areas. The groundwater resources in such declared areas are subject to various government-managed functions including planning, management, regulation, licensing of take, and pricing within a water market framework. The main regulatory and management tool is a water allocation plan (WAP), a legal document that sets out the rules for the management of take and use of water resources from a Prescribed Wells Area. WAPs are collaboratively developed by the relevant landscape board, supported by the state government Department for Environment and Water (DEW), other government entities, and the broader community of water users. The FNPWA was declared on 27 March 2003 and the inaugural FNWAP was adopted on 16 February 2009 (SAAL NRM, 2009). The FNWAP was reviewed and updated in 2020 and formally adopted in February 2021 (LSA SAAL, 2021). The FNPWA is the largest in SA, covering approximately 315,000 km^2 in the northeast corner of the state (Figs. 13.1 and 13.4).

A founding principle within the FNWAP and the declaration of the FNPWA was to maintain sufficient artesian pressure to ensure the survival of GAB springs, through the capping of water take and encouragement of judicious use of what is taken. Such survival requires sufficient pressure to maintain flow and to maintain water quality (by preventing leakage of poorer quality water into the GAB aquifers from other, saltier sources). Central to the implementation of this principle were the following requirements for any new groundwater affecting activity (SAAL NRM, 2009; LSA SAAL, 2021): (1) no exceedance of a 1 meter cumulative drawdown at the boundaries of a buffer zone, the Southwest Springs Zone (2009 WAP), or equivalent 50 km radial buffer zone (Zone B) around each spring (2021 WAP) (Fig. 13.5); (2) no exceedance of a 0.5 meter cumulative drawdown at the boundary of a 5 km radial exclusion zone around a spring (Zone A) (Fig. 13.5). The use of a 5 km buffer zone for drawdown assessment of springs provides for the potential of direct monitoring and measurement of drawdown around springs to either confirm or calibrate drawdown estimates without risk of causing detrimental surficial impacts to spring wetland environments; and (3) no extraction within the 5 km radial buffer around springs (2009 WAP) or demonstration that the water taken will not detrimentally impact upon the ecology of the spring (2021 WAP).

If a deliberately conservative technical assessment shows that a proposed activity could not satisfy these requirements, proponents are required either to undertake further assessment or to modify their activity to ensure that their activity would/does not adversely impact springs. Further, the emphasis on cumulative drawdown, rather than just drawdowns only attributable to individual projects, addresses a deficiency in the EPBC Act (Pointon & Rossini, 2020). While the first FNWAP (2009), in combination with the GABSI, GABSMP, and IGABDR programs, provided a necessary first step, it contained several deficiencies that meant the full benefits of

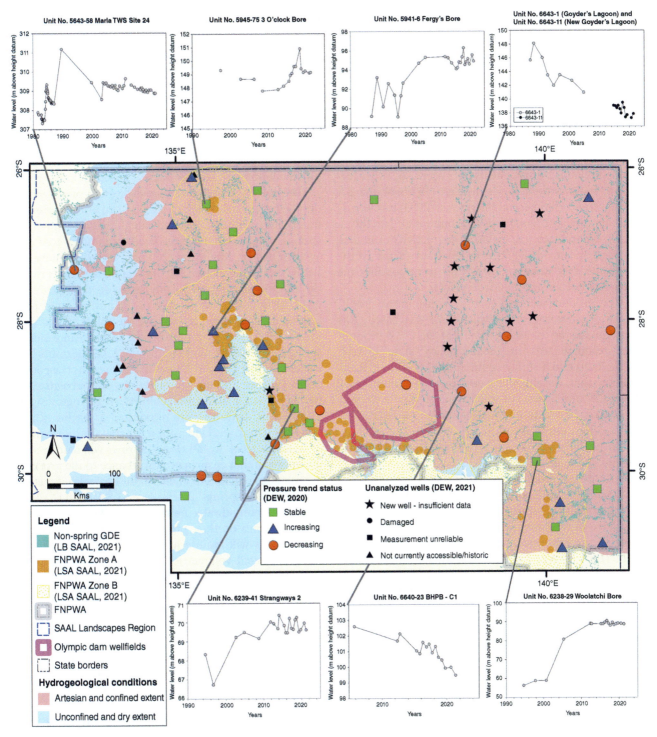

Figure 13.5 FNPWA monitoring network. Note pressure status trends are based on the previous 5 years. Such trends may fit within longer term trend. New wells are wells that have insufficient data, due to their inclusion in the network less than 5 years ago.

regulating the groundwater resources were not met (Lewis & Harris, 2020):

1. A lack of investment in monitoring, compliance, and enforcement measures, such as metering and inspection. For example, the systematic rollout of bore metering has been hampered by many issues, including the remoteness of many wells; vulnerability of meters to the elements; expense associated with rollout, maintenance, and monitoring; and the previous lack of clear direction and funding to overcome such issues (Fig. 13.3f).

2. Compliance by proponents not covered by other legislation, such as the SA Mining Act (1971) or the SA Petroleum and Geothermal Energy Act (2003) was largely voluntary and associated targets aspirational. Notably, a lack of compliance and enforcement activity was also identified by Hawke (2009) as a deficiency of the EPBC Act (Pointon & Rossini, 2020), thus compounding the problem at both a state and federal level.

3. The groundwater volumetric "consumptive pool," upon which licensed groundwater take was based, was determined using an assumption that the GAB groundwater resource was in pseudo-steady state (SAAL NRM, 2009) and a spring count and flow-volume determination that was highly uncertain. The consumptive pool is the total volume of water that, from time to time, will be taken to constitute the groundwater resource available for allocation (LSA SAAL, 2021).

4. Purpose-based allocation rules engendered a sense of inequality among users.

5. Insufficient recognition of non-GAB spring GDEs or the water rights of Indigenous Peoples. For the FNWAP, non-GAB spring GDEs are locations where it is considered likely, based on available data, that the taking of shallow groundwater has the potential to impact upon GDEs (LSA SAAL, 2021). Non-GAB spring GDEs are often associated with riparian vegetation (Fig. 13.5).

The current WAP (adopted 27 February 2021) addresses a number of these issues (LSA SAAL, 2021). For instance, the SAAL Landscapes Board will be investigating modern, recent, or innovative technologies with respect to metering and water accounting and implementing these solutions progressively to address this gap. Volumetric caps, of which the determination was recognized as highly uncertain, have been replaced with an increased importance on the principle of "management by pressure." Although this principle was in the 2009 WAP, it has taken on added importance as practice has demonstrated it to be a more reliable method with which to ensure adequate artesian pressure surrounding springs. Further, a new numerical groundwater model is being constructed that reflects new scientific knowledge concerning the hydrodynamics and framework of the GAB to aid groundwater management and compliance (Keppel et al., 2019). It is intended that this model can be used by government and proponents to assess the cumulative impacts of current and any proposed groundwater affecting activity and to estimate the amount of excess head at spring complexes for springs management and risk assessment purposes. Additionally, the current FNWAP is moving in the direction of an even-handed approach to the concept of "judicious use" by the removal of purpose-based allocation rules, including examining the reinjection of coproduced water by the energy industry and reinforcement of the pastoral industry to adopt closed-water delivery systems that use piping to distribute water and reduce losses (Fig. 13.3e).

The protection of the surface expression of springs, as opposed to the maintenance of sufficient artesian groundwater pressure and groundwater quality supporting springs, can at best be considered an indirect goal of a WAP. Further, McGrath (2005) notes that under the EPBC Act, all impacting activity that commenced prior to the act commencing on 16 July 2020 was deemed authorized, excluding any expansion or intensification of prior activity. In an attempt to improve the ability of the WAP to assist in the task of protecting the surface expression of springs, a system of works and water take approvals built around the recognition of Priority Springs has been developed. Priority Springs are those that are recognized as having particular ecological, cultural, or geological significance, or if disturbed, are recognized as having a good potential to recover. Under these criteria, the direct taking of water for stock from Priority Springs will require a site use approval that ensures no disturbance to the processes or structures required to maintain the ecological functionality of these springs takes place (LSA SAAL, 2021). Further, if groundwater users are unwilling or unable to protect springs on their property from livestock ingress, such springs will require a water license and accompanying estimated volumetric take from the spring, which contributes to proponents' total water take against their water allocation. Prior to this, such water from springs was unaccounted and therefore was treated as "free use." Priority Springs will be identified in consultation with community in line with the WAP implementation plan. Finally, application to take from a spring must ensure the taking of water will not damage, disturb, or interfere with any site of cultural significance (LSA SAAL, 2021).

13.6. TRACKING PROGRESS THROUGH MONITORING

In SA, determining the impact of groundwater extractions, bore rehabilitation, and water saving measures regulated under the FNWAP is reliant on regular monitoring of the FNPWA monitoring network, which covers 87 upgraded and adapted pastoral, monitoring, and former petroleum bores spread over the GAB in SA. The current network was commissioned in 2009/2010 using an older state government network as a basis (Sibenaler, 2010). It is currently monitored once to twice annually, with data and associated status reports uploaded to the publicly accessible online portal Waterconnect (DEW, 2020; DEW, 2021a).

Water resource trends are reported upon and discussed on a regular basis using a 5 year trend analysis. The most recent discussion of water resource trends presented by DEW (2020) suggest that, using the previous 5 years, just under half of monitoring bores in the network have a stable trend, whereas the remaining bores of approximately equal number either have a statistically notable rising or declining trend (Fig. 13.5). Bores with rising trends tend to be clustered around an area coincident with springs located near the central, northern, and southeast portions of the FNPWA, whereas bores with declining trends are found in the eastern and central portions to the east and north of the springs (Fig. 13.5). However, such 5 year trends may fit within longer term trends of stability, recovery, or decline (Fig. 13.5) and may therefore be more reflective of shorter-term resource demands. Water quality is stable where there is sufficient data to determine trends (DEW, 2020).

While the network has value in providing a regional snapshot of pressure and water quality conditions, it has several limitations. The heavy reliance on water supply bores compromises the ability to discriminate between cumulative regional or localized drawdown impacts (Bekesi et al., 2012). This reliance is necessitated by the great difficulty and expense typically required to construct bores within the artesian portion of the GAB. Further, there is currently no comparable government-managed monitoring of springs regularly undertaken to correlate between pressure change and spring flow. Finally, the lack of temporal consistency across the network limits trend analysis in certain areas, such as the Cooper Basin region (Fig. 13.5).

Outside of the FNPWA monitoring network, there are several privately maintained monitoring networks designed to assess impacts related to specific groundwater affecting activities. The largest of these is the one associated with the water supply wellfields of Olympic Dam mine. This large multimetallic mining operation located to the south of the GAB has assured water supply under the Roxby Downs Indenture Ratification Act (1982). However, one of the stipulations of this indenture is the requirement to undertake regular groundwater pressure monitoring, as well as spring-flow monitoring near the wellfields and to report results annually (BHP, 2020). Monitoring prior to 1997 indicated a significant decline in several spring complexes located near the original wellfield associated with increases in water abstraction related to growth of the mining operation (BHP, 2020; Kinhill Engineers, 1997). These declines have stabilized after the commissioning of a second wellfield located farther away from the springs (WMC, 1997), capping of abstraction from the original wellfield, and adoption of target management protocols (Kinhill Engineers, 1997; BHP, 2020). Nevertheless, flows at many affected springs remain below pre-mining levels, albeit within a predicted range (BHP, 2020).

13.7. GAB SPRINGS ADAPTIVE MANAGEMENT PLAN AND TEMPLATE (GABSAMP)

While the threat of pressure loss and water quality degradation is the subject of the FNWAP, degradation of the surficial environment due to land-use practice requires additional recourse, as such land-use practices are not considered the direct remit of a WAP. Land-management practices and their associated impacts are of equal importance to groundwater pressure loss (Lewis & Harris, 2020). To address these threatening processes, the GAB Springs Adaptive Management Plan and template (GABSAMP) was developed using evidence-based methodologies to identify, assess, and manage risks to spring groups across the GAB (Brake et al., 2020; Jensen et al., 2020). The GABSAMP seeks to emphasize the importance of minimizing impacts on flow rates through pressure loss and/or physical disturbance of spring vents. The project team that developed the GABSAMP had extensive knowledge and experience of the SA GAB springs and sought to apply this knowledge to the development of a national framework for sustainable management of GAB springs.

The GABSAMP is a basin-wide initiative that collates current evidence on spring condition and threats and proposes objective methods to assess and manage identified risks to springs. It aims to ensure maintenance of artesian pressures that sustain spring flows and encourages sensitive land-use practices in and around springs to protect spring geology and ecology, while minimizing disruption to current users of basin water resources. The focus is on spring environments, including spring-dependent ecosystems, rather than just on flow volumes.

The GABSAMP and template provides a decision framework for assessing the risks to springs and determining appropriate management actions to address those risks. This framework is drawn from existing legislation and regulatory plans such as the FNWAP, as well as evidence from recent research. The GABSAMP incorporates the concepts of "judicious use" of water resources and "willing compliance" of resource users, as outlined in the GABSMP, and identifies further information that is needed. The adaptive management tool deliberately emphasizes the need to assess the context and features of individual springs or spring complexes in the process of selecting the appropriate management tools.

The template consists of two stages, with the initial spring situational analysis undertaken to determine the relevant parameters and issues of the site being evaluated,

Table 13.2 Summary of key elements of GABSAMP template

Springs situation analysis: Assess current status of spring complex under the following categories:	Adaptive management response: Negotiated response based on springs situation analysis and coordinated strategies and options under the following categories:
1. Spring complex characteristics, including values and typology 2. Legislation and regulation, including tenure and relevant governance 3. Current condition and possible threatening processes, including threats from water extraction and land uses 4. Evidence base for decision making and development of management approach	1. Engagement for negotiating response with relevant parties 2. Risk assessment (pressure loss and land uses causing surface disturbance) 3. Create a culture of willing compliance 4. Risk management strategies to maintain spring flows and to protect surface structures, GDEs, and cultural values 5. Development of agreed on-ground actions between all parties 6. Implementation, funding, and maintenance strategy 7. Monitoring, evaluation, and reporting 8. Review and adaptive management actions

Source: Adapted from Brake et al. (2020).

followed by the adaptive management response that selects from various management options as appropriate. The key stages for using evidence-based methodologies for managing risks to spring values are summarized in Table 13.2 with more detail provided in Tables 1 and 2 in Brake et al. (2020). These options encourage proactive behavior from landholders and water users.

Effective implementation of the GABSAMP template requires a robust, comprehensive, and interactive basin-wide database that combines all available information on spring characteristics, condition, trends, values, groundwater-dependent ecosystems, risk factors, and their impacts. Access to this information will be critical for achieving effective situational analyses for target springs and development of appropriate management actions to maintain spring values. An objective, rigorous, and cost-effective basin-wide monitoring program linked to the database will be essential to assess the condition of assets and the effectiveness of management actions. Recommendations have been put forward for a collaborative project to develop the online portal and associated interactive database, based on work already under way by Qld and SA government agencies.

There is also a key recommendation for a coordinated, basin-wide monitoring program to track spring condition and pressure status. Further, like the FNWAP, the GABSAMP sets out options for the judicious use and efficient delivery of water to reduce pressure losses for GAB springs. While the GABSAMP currently has no legal authority, it has been recommended that the GABSAMP and template be adopted as an endorsed implementation strategy as part of the implementation plan for the updated national GABSMP (2018–2033). This step will ensure that these measures are incorporated into the longer-term governance framework for sustainable management of GAB springs.

13.8. CLOSING REMARKS

The most recent FNWAP and the GABSAMP both represent examples of evidence-based policies and management plans that provide the governance framework to address and manage ongoing threats to GAB spring environments. Each recognize and contain strategies and actions to address the two broad issues that threaten both the existence and viability of GAB spring wetlands as significant cultural and ecological hot spots, namely, the issues of reduced flow, degraded flow, or flow extinction resulting from groundwater pressure loss or environmental degradation because of land management practices. Further, both the current FNWAP and the GABSAMP reflect iterative changes to our understanding of the SA GAB springs and the risks to their survival, as well as knowledge and recommendations obtained from past regulatory and management plans. However, the success of each plan in terms of achieving an improvement in spring health is highly dependent on both the adoption and compliance by the communities and stakeholders that utilize the GAB and appropriate planning and development protocols, monitoring of progress, and compliance to assess progress. A recommendation of the GABSAMP that will aid in this task is the construction of a Web-based portal and associated online tools and databases of relevant geographical, hydrological, and ecological information. This portal will serve as a means of educating the community about springs and the relevant ways they can contribute to their protection from a land management perspective. With respect to the FNWAP, appropriate planning, monitoring, and compliance measures are the key tool to assess progress and therefore the risk of further pressure loss. On this latter point, the planning protocols of the FNWAP have been effective because they are easy to understand and only compliant plans for groundwater affecting activities may receive a permit or license. As discussed above, the success of such planning protocols and principles concerning bores have now been enhanced and expanded with respect to water take from and around springs and the concept of identifying Priority Springs at a pre-new-development stage. For compliance, the key issue concerning the lack of bore metering and water accounting measures will be progressively addressed in response to recommendations made during the WAP consultation process. Moving forward, monitoring and compliance would be aided by a targeted presentation and synthesis of the rehabilitative impacts such measures have on the spring environments. Currently, government-related monitoring focuses on pressure, while regular spring flow monitoring is focused on assessing impacts related to an active wellfield and is undertaken by the impacting industry.

Finally, and importantly, a long-term recommendation for both the FNWAP and the GABSAMP and the tools developed to aid their implementation, such as the Far North Numerical Groundwater Model, are that they be subject to constant iteration and improvement as new evidence emerges. This reflects a deficiency highlighted by Pointon and Rossini (2020) that progressing development approval within the current Australian legislative system is often undertaken within a data-deficient environment and therefore requires constant adaption. Further, this recognizes that while there have been many advances in our understanding of the GAB springs and related hydrogeology, a great deal of uncertainty remains while the demands on the groundwater resource are in a constant state of flux. Both plans will need to reflect these changes over time and continue to adapt as new knowledge and new risks emerge.

ACKNOWLEDGMENTS

The authors would like to thank Christopher Wilson and three anonymous reviewers for their reviews of the manuscript. This chapter is dedicated to the memory of Lynn Brake, champion of GAB springs and lead author of the GABSAMP, whose leadership, vision, and energy have done so much to protect and raise awareness of the springs of the GAB.

REFERENCES

Adamo, E. (2021). Hunt for water in the Flinders begins. *The Transcontinental, Port Augusta.* http://www.transcontinental.com.au/story/7389782/searc-is-on-for-new-water-source/

Ah Chee, D. (2002). *Indigenous people's connection with Kwatye (water) in the Great Artesian Basin, GAB FEST 2002: A resource under pressure.* Great Artesian Basin Coordinating Committee, Toowoomba, Queensland.

Arthington, A. H., Jackson, S. H., Tomlinson, M., Walton, C. S., Rossini, R. A., & Flook, S. C. (2020). Springs of the Great Artesian Basin: Oases of life in Australia's arid and semi-arid interior. *Proceedings of the Royal Society of Queensland, Springs of the Great Artesian Basin Special Issue, 126,* 1–10.

Bekesi, G., Tyler, M., & Waterhouse, J. (2012). *Conclusions from using an active production well for mine drawdown compliance.* International Mine Water Association Annual Conference, September, 2012, Bunbury, Australia. https://www.imwa.info/docs/imwa_2012/IMWA2012_Bekesi_073.pdf

BHP (2020). *Olympic Dam Great Artesian Basin Wellfields Report 1 July 2019–30 June 2020.* https://www.bhp.com/-/media/bhp/regulatory-information-media/copper/olympic-dam/0000/annual-environment-reports/fy20-great-artesian-basin-wellfields-report.pdf

Brake, L. (2020). Development, management and rehabilitation of water bores in the Great Artesian Basin, 1878–2020. *Proceedings of the Royal Society of Queensland, Springs of the Great Artesian Basin, Special Issue, 126,* 153–176.

Brake, L., Harris, C., Jensen, A., Keppel, M., Lewis, M. & Lewis, S. (2020). *Great Artesian Basin Springs: A plan for the future. Evidence-based methodologies for managing risks to spring values.* Prepared for the Australian Government Department of Agriculture, South Australian Department for Environment and Water, Queensland Department of Natural Resources, Mines and Energy, New South Wales Department of Planning, Industry and Environment, and the Northern Territory Department of Environment and Natural Resources.

CofA (2020). *Bioregional assessments.* https://www.bioregionalassessments.gov.au/

Core Energy Group (2016). *Cooper-Eromanga Basin outlook, 2035.* https://www.petroleum.sa.gov.au/__data/assets/pdf_file/0007/270826/Core_Energy_-_Cooper-Eromanga_Basin_Outlook_-_Final_-_Oct2016v1.pdf

Cotton, T. B., Scardigno, M. F., & Hibburt, J. E. (Eds.) (2007). *The petroleum geology of South Australia, vol. 2: Eromanga Basin, 2nd ed.* South Australia, Department of Primary Industries and Resources. Petroleum Geology of South Australia Series.

Costelloe, J. F., Matic, V., Western, A. W., Walker, J. P., & Tyler, M. (2015). *Estimating leakage around the southwestern margin of the Great Artesian Basin.* Final report for the Australian Research Council (ARC) Linkage Project LP0774814: Quantifying near-surface diffuse discharge from the southwest Great Artesian Basin. https://www.researchgate.net/publication/280240319_Estimating_leakage_around_the_southwestern_margin_of_the_Great_Artesian_Basin

DAWE (2020a). *Great Artesian Basin strategic management plan.* https://www.agriculture.gov.au/water/national/great-artesian-basin/strategic-management-plan

DAWE (2020b). *Great Artesian Basin.* https://www.agriculture.gov.au/water/national/great-artesian-basin

DAWE (2021). *The community of native species dependent on natural discharge of groundwater from the Great Artesian Basin in Community and Species Profile and Threats Database, Department of the Environment, Canberra.* http://www.environment.gov.au/sprat

DEW (2015). *Coal Seam Gas and Coal Mining Water Knowledge Program.* https://www.waterconnect.sa.gov.au/Industry-and-Mining/CSG-Coal-Mining/SitePages/Water%20Knowledge%20Program.aspx

DEW (2020). *Far North Prescribed Wells Area 2018–19 water resources assessment*. DEW Technical report 2020/33, Government of South Australia, Department for Environment and Water, Adelaide,. Government of South Australia.

DEW (2021a). *Watercconnect*. https://www.waterconnect.sa.gov.au/Pages/Home.aspx

DEW (2021b). *Water Security Statement 2021: Water for sustainable growth* (draft). https://www.environment.sa.gov.au/topics/water/water-security/water-security-statement

DMITRE (2012). *Roadmap for unconventional gas projects in South Australia*. http://www.energymining.sa.gov.au/__data/assets/pdf_file/0019/238033/Roadmap_Unconventional_Gas_Projects_SA_12-12-12_web.pdf

DNRME (2018). *Sediments above the Great Artesian Basin: Groundwater background paper July 2018*. https://www.mdba.gov.au/sites/default/files/pubs/qld-sediments-above-the-great-artesian-basin-%28GS57-GS58-GS59-GS60%29-groundwater-background-paper-2018_3.pdf

Fatchen, T. J. (2000). *A regional management plan for mound springs*. Proceedings of the 3rd Mound Spring Researchers Forum, Adelaide, 9 February 2000, Harris, CR (ed.), Department of Environment, Heritage and Aboriginal Affairs, Adelaide, pp. 32–36.

Fensham, R., Ponder, W., & Fairfax, R. (2007). *Recovery plan for the community of native species dependent on natural discharge groundwater from the Great Artesian Basin*. Report to the Department of the Environment, Water, Heritage and the Arts, Canberra, Queensland Parks and Wildlife Service, Brisbane. http://www.environment.gov.au/biodiversity/threatened/publications/pubs/gab.pdf

Flook, S. C., Fawcett, J., Erasmus, D., Singh, D., & Pandey, S. (2020). Evolution of knowledge on springs in the Surat and southern Bowen Basins: Survey, Conceptualization and Wetland Dynamics. *Proceedings of the Royal Society of Queensland, Springs of the Great Artesian Basin Special Issue, 126*, 47–64.

Florek, S. M. (1993). *Archaeology of the Mound spring campsites near Lake Eyre in South Australia*. Phd Thesis University of Sydney. https://core.ac.uk/download/pdf/212693498.pdf. Accessed 23rd September 2021

Frontier Economics (2016). *Economic output of groundwater dependent sectors in the Great Artesian Basin*. A report commissioned by the Australian government and Great Artesian Basin Jurisdictions based on advice from the Great Artesian Basin coordinating Committee. https://www.agriculture.gov.au/water/national/great-artesian-basin/economic-output-groundwater-dependent-sectors-great-artesian-basin

Fu, G., Zou, Y., Crosbie, R. S., & Barron, O. (2020). Climate change and variability in the Great Artesian Basin (Australia), future projections, and implications for groundwater management. *Hydrogeology Journal, 28*, 1–11. https://doi.org/10.1007/s10040-019-02107-8

GA (2006). *GEODATA TOPO 250K series 3: Bioregional assessment source dataset*. http://data.bioregionalassessments.gov.au/dataset/a0650f18-518a-4b99-a553-44f82f28bb5f

GA (2013). *Australian coal basins: Bioregional assessment source dataset*. http://data.bioregionalassessments.gov.au/dataset/9d5a8d74-a201-42bd-9d49-3c392244be16

Gotch, T. (Ed.) (2013). *Allocating water and maintaining springs in the Great Artesian Basin, vol. V: Groundwater-dependent ecosystems of the western Great Artesian Basin*. National Water Commission, Canberra.

Gotch, T., Keppel, M., Fels, K. & McKenzie J. (2016). *Lake Eyre Basin springs assessment: Ecohydrological conceptual models of springs in the western Lake Eyre Basin*. DEWNR Technical report 2016/02, Government of South Australia, Department of Environment, Water and Natural Resources, Adelaide.

Government of South Australia (2019). *The South Australian Government Gazette*. No. 42, Thursday 22 August 2019, pp. 3014–3015. http://governmentgazette.sa.gov.au/sites/default/files/public/documents/gazette/2019/August/2019_042.pdf.

Government of South Australia (2021). *Petroleum exploration and production system: South Australia (PEPS-SA)*. http://pepsa.sa.gov.au/home

GABCC (2018). *Great Artesian Basin Strategic Management Plan 2018 (draft consultation version)*. Report by the Great Artesian Basin Coordinating Committee, Canberra. https://haveyoursay.Awe.gov.au/37538/documents/85283

Green, G., & Berens, V. (2013). Relationship between aquifer pressure changes and spring discharge rates. In Love et al. (Eds.), *Allocating water and maintaining springs in the Great Artesian Basin, vol. III: Groundwater discharge of the western Great Artesian Basin*. National Water Commission, Canberra.

Green, G., White, M., Gotch, T., & Scholz, G. (2013). *Allocating water and maintaining springs in the Great Artesian Basin, vol. VI: Risk assessment process for evaluating water use impacts on the Great Artesian Basin Springs*. National Water Commission, Canberra.

Habermehl, M. A. (1980). The Great Artesian Basin, Australia. *BMR Journal of Geology and Geophysics, 5*. 9–37.

Habermehl, M. A. (2020a). Hydrogeological overview of springs in the Great Artesian Basin. *Proceedings of the Royal Society of Queensland, Springs of the Great Artesian Basin Special Issue*, 29–46. http://www.royalsocietyqld.org/2020-springs-special-issue-vol-126

Habermehl, M. A. (2020b). Review: The evolving understanding of the Great Artesian Basin (Australia), from discovery to current hydrogeological interpretations. *Hydrogeology Journal, 28*, 13–36. http://doi.org/10.1007/s10040-019-02036-6

Halihan, T., Love, A., Keppel, M., Dailey, M. K. M., Berens, V., & Wohling, D. (2020). Evidence for groundwater mixing at Freeling Spring Group, South Australia. *Hydrogeology Journal, 28*, 313–323. https://doi.org/10.1007/s10040-019-02069-x

Harrington, G. A., Smerdon, B. D., Gardner, P. W., Taylor, A. R., & Hendry, J. (2013). Diffuse discharge. In A J. Love et al. (Eds.), *Allocating water and maintaining springs in the Great Artesian Basin, vol. III: Groundwater discharge of the western Great Artesian Basin*. National Water Commission, Canberra.

Harris, C. (1981). Oases in the desert: The mound springs of northern South Australia. *Proceedings of the Royal Geographical Society of Australia (SA Branch), 81*, 28–39.

Harris, C. (2002). Culture and geography: South Australia's mound springs as trade and communication routes. *Historic Environment, 16*(2), 8–11. https://www.aicomos.com/wp-content/uploads/Culture-and-geography-South-Australia%E2%80%99s-mound-springs-as-trade-and-communication-routes1.pdf

Hawke, A. (2009). *Report of the independent review of the Environmental Protection and Biodiversity Act 1999.* Commonwealth Government of Australia

Hercus, L. A., & Sutton, P. (1985). *The assessment of Aboriginal cultural significance of mound springs in South Australia.* South Australian Department of Environment and Planning.

Jensen, A., Lewis, S. A., & Lewis, M. M. (2020). Development of an adaptive management plan and template for sustainable management of Great Artesian Basin Springs. *Proceedings of the Royal Society of Queensland, Springs of the Great Artesian Basin Special Issue, 126,* 289–304.

Karlstrom, K. E., Keppel, M. N., Love, A. J., & Crossey, L. (2013). Structural and tectonic History. In: Keppel, M. N., et al. (Eds.), *Hydrogeological framework of the western margin of the Great Artesian Basin, Australia.* National Water Commission, Canberra.

Keppel, M., Gotch, T., Inverarity, K., Niejalke, D. & Wohling, D. (2016). *A hydrogeological and ecological characterization of springs near Lake Blanche, Lake Eyre Basin, South Australia.* DEWNR Technical report 2016/03, Government of South Australia, Department of Environment, Water and Natural Resources, Adelaide.

Keppel, M., Sampson, L., Woods, J., & Osei-Bonsu, K. (2019). *The Far North Groundwater Model.* Paper presented at Australasian Groundwater Conference, 24–27 November, Brisbane, Australia.

Keppel, M., Wohling, D., Jensen-Schmidt, B., & Sampson, L. (2015). *A hydrogeological characterization of the Arckaringa Basin.* DEWNR Technical report 2015/03, Government of South Australia, Department of Environment, Water and Natural Resources, Adelaide. https://www.waterconnect.sa.gov.au/Content/Publications/DEW/DEWNR-TR-2015-03-Hydrogeol-characterisation-of-Arckaringa-Basin.pdf

Keppel, M., Wohling, D., Love, A., & Gotch, T. (2020b). Hydrochemistry highlights potential management issues for aquifers and springs in the Lake Blanche and Lake Callabonna region, South Australia. *Proceedings of the Royal Society of Queensland, Springs of the Great Artesian Basin Special Issue, 126,* 65–90. http://www.royalsocietyqld.org/2020-springs-special-issue-vol-126

Keppel, M. N. (2013). *The geology and hydrochemistry of calcareous mound spring wetland environments in the Lake Eyre South region, Great Artesian Basin, South Australia.* Ph.D Thesis. Flinders University, South Australia.

Keppel, M. N., Clarke, J. D. A., Halihan, T., Love, A. J., & Werner, A. D. (2011). Mound springs in the arid Lake Eyre South region of South Australia: A new depositional tufa model and its controls. *Sedimentary Geology, 240,* 55–70.

Keppel, M. N., Karlstrom, K., Crossey, L., Love, A. J., & Priestley, S. (2020a). Evidence for intra-plate seismicity from spring-carbonate mound springs in the Kati Thanda–Lake Eyre region, South Australia: Implications for groundwater discharge from the Great Artesian Basin. *Hydrogeology Journal, 28,* 297–311. https://doi.org/10.1007/s10040-019-02049-1

Kinhill Engineers (1997). *Olympic Dam Expansion Project, environmental impact statement.* Kinhill Engineers Pty Ltd., Adelaide.

Krieg, G. W., Alexander, E., & Rogers, P. A. (1995). Eromanga Basin. In J. F. Drexel & W. Preiss (Eds.), *The geology of South Australia* (pp. 101–105). Geological Survey of South Australia, Adelaide.

LB SAAL and LSA SAAL (2021). *Water allocation plan for the far north prescribed wells area.* 27 February 2021. https://cdn.environment.sa.gov.au/landscape/docs/saal/6087_wap_feb2021_final020321_002.pdf

Lewis, S., & Harris, C. (2020). Improving conservation outcomes for the Great Artesian Springs in South Australia. *Proceedings of the Royal Society of Queensland, Springs of the Great Artesian Basin Special Issue, 126,* 271–288. http://www.royalsocietyqld.org/2020-springs-special-issue-vol-126

Love, A. J., Herczeg, A. L., Sampson, L., Cresswell, R. G., & Fifield, L. K. (2000). Sources of chloride and implications for Cl 36 dating of old groundwater, southwestern Great Artesian Basin, Australia. *Water Resources Research, 36*(6), 1561–1574.

Love, A. J., Wohling, D., Fulton, S., Rousseau-Gueutin, p., & de ritter, s. (eds.) (2013). *Allocating water and maintaining springs in the Great Artesian Basin, vol. II: Groundwater recharge, hydrodynamics and hydrochemistry of the Western Great Artesian Basin.* National Water Commission, Canberra.

Matic, V., Costelloe, J. F., & Western, A. W. (2020). An integrated remote-sensing mapping method for groundwater dependent ecosystems associated with diffuse discharge in the Great Artesian Basin, Australia. *Hydrogeology Journal, 28,* 325–342. https://doi.org/10.1007/s10040-019-02062-4

McGrath, C. (2005). Key concepts of the Environmental Protection and Biodiversity Conservation Act (1999) (Cth). *Environmental and Planning Law Journal, 22*(1), 20–39.

Moggridge, B. J. (2020). Aboriginal People and groundwater. *Proceedings of the Royal Society of Queensland, Springs of the Great Artesian Basin Special Issue, 126.* 11–27.

Mudd, G. M., & Jowitt, S. M. (2018). Growing global copper resources, reserves and production: Discovery is not the only control on supply. *Economic Geology, 113*(6), 1235–1267.

Murphy, N. P., Adams, M., & Austin, A. D. (2009). Independent colonization and extensive cryptic speciation of freshwater amphipods in the isolated groundwater springs of Australia's Great Artesian Basin. *Molecular Ecology, 18*(1), 109–122.

NWC (2013). *Allocating water and maintaining springs in the Great Artesian Basin, vol. VII: Summary of findings for natural resource management of the Western Great Artesian Basin.* NWC, Canberra.

Pestov, I. (2000). Thermal convection in the Great Artesian Basin, Australia. *Water Resources Management, 14*(5), 391–403.

Pointon, R. K., & Rossini, R. A. (2020). Legal mechanisms to protect Great Artesian Basin springs: Successes and shortfalls. *Proceedings of the Royal Society of Queensland, Springs of the Great Artesian Basin Special Issue, 126,* 249–269.

Ponder, W. F. (1986). Mound springs of the Great Artesian Basin. In P. De Deckker & W. D. Williams (Eds.), *Limnology in Australia* (pp. 403–420). Commonwealth Scientific and Industrial Research Organization (CSIRO). Melbourne & Dordrecht: Dr. W. Junk Publishers.

Ponder, W. F. (2002). Desert springs of the Australian Great Artesian Basin. In D. W. Sada & S. E. Sharpe (Eds.), *Spring-fed wetlands: Important scientific and cultural resources of the intermountain region*. Conference Proceedings, Las Vegas, Nevada, May 7–9 2002, DHS Publication No. 41210.

Priestley, S. C., Shand, P., Love, A. J., Crossey, L. J., Karlstrom, K. E., Keppel, M. N., et al. (2020). Hydrochemical variations of groundwater and spring discharge of the western Great Artesian Basin, Australia: Implications for regional groundwater flow. *Hydrogeology Journal, 28*, 1–11. https://doi.org/10.1007/s10040-019-02107-8

Radke, B. M., Ferguson, J., Cresswell, R. G., Ransley, T. R., & Habermehl, M. A. (2000). *Hydrochemistry and implied hydrodynamics of the Cadna-owie: Hooray Aquifer, Great Artesian Basin, Australia*. Bureau of Rural Sciences, Canberra.

Ransley, T. R., & Smerdon, B. D. (Eds.) (2012). *Hydrostratigraphy, hydrogeology and system conceptualization of the Great Artesian Basin*. A technical report to the Australian Government from the CSIRO Great Artesian Basin Water Resource Assessment. CSIRO Water for a Healthy Country Flagship, Australia. https://publications.csiro.au/rpr/download?pid=csiro:EP132693&dsid=DS5

Ransley, T. R., Radke, B. M., Feitz, A. J., Kellett, J. R., Owens, R., Bell, J., et al. (2015). *Hydrogeological atlas of the Great Artesian Basin*. Geoscience Australia, Canberra. http://dx.doi.org/10.11636/9781925124668

Rousseau-Gueutin, P., Simon, S., Love, A., Post, V., Doublet, C., Simmons, C., et al. (2013). Groundwater flow and hydrodynamics. In A. J. Love et al. (Eds.). *Allocating water and maintaining springs in the Great Artesian Basin, vol. II: Groundwater recharge, hydrodynamics and hydrochemistry of the western Great Artesian Basin*. National Water Commission, Canberra.

Sales, M., Altmann, M., Buick, G., Dowling, C., Bourne, J., & Bennett, A. (2015). Subtle oil fields along the Western flank of the Cooper/Eromanga petroleum system. *The APPEA Journal, 55*(2), 440–440. https://doi.org/10.1071/AJ14075

Sandiford, M., Lawrie, K., & Brodie, R. S. (2020). Hydrogeological implications of active tectonics in the Great Artesian Basin, Australia. *Hydrogeology Journal, 28*, 57–73.

Schenk, C. J., Tennyson, M. E., Mercier, T. J., Klett, T. R., Finn, T. M., Le, P. A., et al. (2016). *Assessment of undiscovered conventional oil and gas resources of the Cooper and Eromanga basins, Australia, 2016: U.S. Geological Survey fact sheet, 2016–3028*. http://dx.doi.org/10.3133/fs20163028

Shand, P., Gotch, T., Love, A., Raven, M., Priestley, S., & Grocke, S. (2016). Extreme environments in the critical zone: Linking acidification hazard of acid sulfate soils in mound spring discharge zones to groundwater evolution and mantle degassing. *Science of the Total Environment, 568*, 1238–1252. https://doi.org/10.1016/j.scitotenv.2016.05.147

Sibenaler, Z. (2010). *Monitoring requirements for water resources in the Arid Lands*. Report for the South Australian Arid Lands Natural Resources Management Board (SAAL NRM), May 2010.

Smerdon, B. D., Ransley, T. R., Radke, B. M., & Kellett, J. R. (2012). *Water resource assessment for the Great Artesian Basin*. A report to the Australian Government from the CSIRO Great Artesian Basin Water Resource Assessment. CSIRO Water for a Healthy Country Flagship, Australia.

Smerdon, B. D., Smith, L. A., Harrington, G. A., Gardener, W. P., Delle Piane, C. & Sarout, J. (2014). Estimating the hydraulic properties of an aquitard from in situ pore pressure measurements. *Hydrogeology Journal, 22*, 1875–1887.

SAAL NRM (2009). *Water allocation plan for the far north prescribed wells area, South Australia*.

Welsh, W. D. (2006). *Great Artesian Basin transient groundwater model*. Bureau of Rural Sciences, Canberra. http://www.southwestnrm.org.au/sites/default/files/uploads/ihub/welsh-wd-2006-great-artesian-basin-transient-groundwater-model.pdf

White, D. C., & Lewis, M. M. (2011). A new approach to monitoring spatial distribution and dynamics of wetlands and associated flows of Australian Great Artesian Basin springs using QuickBird satellite imagery. *Journal of Hydrology, 408*(2011), 140–152.

WMC (1997). Olympic Dam operation, Borefield B Development, Bore Completion Report. Unpublished WMC Report HYD T065.

Wohling, D., Keppel, M., Fulton, S., Costar, A., Sampson, L. & Berens, V. (2013). *Australian Government initiative on coal seam gas and large coal mining: Arckaringa Basin and Pedirka Basin groundwater assessment projects*. DEWNR Technical Report 2013/11. Government of South Australia, Department of Environment, Water and Natural Resources, Adelaide. https://www.waterconnect.sa.gov.au/Content/Publications/DEW/DEWNR-TR-2013-11-Arckaringa-Pedirka-Groundwater-Assessment-project.pdf

Wolaver, B. D., Priestley, S. C., Crossey, L. J., Karlstrom, K., & Love, A. J. (2020). Elucidating sources to arid land Dalhousie Springs in the Great Artesian Basin (Australia) to inform conservation. *Hydrogeology Journal, 28*, 279–296. https://doi.org/10.1007/s10040-019-02072-2

Zeidler, W., & Ponder, W. F. (Eds.) (1989). *Natural history of Dalhousie Springs*. South Australian Museum, Adelaide.

14

Patterns in the Occurrence of Fecal Bacterial Indicators at Public Mineral Springs of Central Victoria, 1986–2013

Andrew Shugg

ABSTRACT

In central Victoria monitoring of the fecal indicators Total Coliform and *Escherichia* Coliform at around 50 mineral spring sites over a period of more than three decades show distinct patterns in contamination. Criteria for examination include both the count levels and occurrence of positive readings. Initially, the springs were unsanitary dug wells, pits, and shallow 3–7 m deep bores. To overcome the recidivism, a replacement program utilizing 30–150 m deep bores was initiated. All the springs had been sited near evidence of subaqueous discharge or exfiltration of mineralized water in the base of highland valleys. Mineral water evolves and circulates in conductive fissure fault and joint systems in the bedrock. Historically a gold-mining region, nearly all sites owe their discovery to alluvial mining activities. Situated in seasonally active mixing zones at the confluence of flow systems, the vulnerability has been reflected in both the development technique and the bacteriological ambience. Catchment use, forest practices, urbanization, subdivision, and the increasing sewerage backlog in settled areas now contribute to a fecal bacterial load in the catchments of the springs and, in particular, the proximal streams and gullies. Within the data set, distinct seasonal patterns appear in the presence, count, ratio between indicators, and/or the absence of fecal bacteria. Spring contamination and bacterial predominance correlate with seasonal catchment hydrodynamics, in spite of event-related episodes. Deep replacement bores reduce or extinguish the occurrence of *Escherichia* Coliform, while some sites may return positive Total Coliform counts, which suggests the need for disinfection during maintenance procedures.

14.1. BACKGROUND AND MONITORING HISTORY

The mineral spring region lies in the low ranges of central Victoria. Spring catchments radiate from a section of the spine of the 3,000 km long Great Dividing Range of eastern Australia. Mineral water occurrence is often only suggested by the appearance of bubbles in pools on streams representing subaqueous discharge in the base of small ephemeral alluviated gullies cut into the deeply weathered Lower Ordovician bedrock. All developed mineral springs owe their discovery and often protection to the activities of alluvial gold miners in the mid-nineteenth century. Ironically, unearthed, buried, or exposed during alluvial mining and sluicing operations this led to preservation in reserves that excluded mining and fossicking.

At undeveloped mineral water sites, bubble formation and ferruginous flocculants in creek beds and pools along rivers remain the only visible evidence of discharge. From the 1850s onward, alluvial and sluicing activities by gold miners in central Victoria sometimes encountered the gassy effusions from the bedrock. Some were dug out or drilled and these became known as the "mineral springs" of central Victoria. In pit development, discharge is characterized by

---

Federation University Australia, Ballarat, Australia

Threats to Springs in a Changing World: Science and Policies for Protection, Geophysical Monograph 275, First Edition.
Edited by Matthew J. Currell and Brian G. Katz.
© 2023 American Geophysical Union. Published 2023 by John Wiley & Sons, Inc.
DOI:10.1002/9781119818625.ch14

small flows of 0.003–0.3 L/s, of 11°C–16°C, high alkalinity spa water with more than 1,500 mg/L HCO_3^-. Public mineral water development in central Victoria follows a stereotypical interception pattern of collecting ascending water from bedrock fracture systems near the base of gullies or beside streams. A consequence of the construction of poorly isolated pits or trenches, collection reservoirs, or shallow bores was that mixing with ambient waters was omnipresent. Increasing catchment clearing, urbanization, and aging of spring infrastructure all have suggested that mineral springs' vulnerability was increasing. Spa water development has focused on tourism-related recreational water consumption, and bath house development is centered in one reserve at Hepburn.

Bacterial pollution of the main public springs has resulted in closure on numerous occasions. In 1964, the main Hepburn Springs were closed by the health department due to fecal contamination (Lawrence, 1965). Soon afterward, the State Development Committee Enquiry into the Mineral Springs 1969–1970 reported on the closure of the public springs (1970) and identified that the capture distance for the fecal bacteria *E. coli* might exceed 200 m. The local authority responded by establishing Spring Protection Buffers in the local planning scheme (Shugg, 1996, 1999).

Major spring dewatering events of 1910–1914 (Dunn et al., 1912 (Note: Unpublished report filed without accompanying drawings Nos. 1239/M/1 and 2); Skeats, 1912 (Note: A. Shugg prepared an annotated version with the mines identified on a plan.); Laing et al., 1997), highlighted that the springs were vulnerable to mine dewatering activities several kilometers away. Underground flow systems had been mapped in the gold mines and bedrock fissure conduits traced through the mines and along the structural corridors in the bedrock (Dunn et al., 1912; Shugg, 2008a). In Victoria, the state parliament reacted by forcing the closure of one of the potentially offending mines (Brady, 1990; Laing et al., 1997). Common to the development of many spa regions are the litany of competing interests that develop, such as the mining, bottling, public health, and tourism industry. The playwright Ibsen (1882) encapsulated some of these themes in the play "An Enemy of the People" depicting the dilemma facing a whistle-blowing doctor in a spa town as spring contamination became evident. In central Victoria, the springs were closed and gradually replaced with deep bores.

Bacterial monitoring of the springs in the state was defined by the Health Act 1958. The local health inspector carried out the monitoring and assessment duty until a change in health regulations in 1989. (The Health Department closed the main springs at Hepburn in 1964 due to fecal coliform pollution, in particular, Pavilion Spring, which provided water to the bath house, bottling plant, and a public outlet in the Pavilion rotunda.) After the change, the Mineral Spring Advisory Committee (MSAC) and the Victorian Mineral Water Committee (VMWC) assumed the responsibility for monitoring, and status reports were presented to the committee (Bannister, 1989, 1992, and minutes of the VMWC). (A continuum in membership and administration existed between the old MSAC and VMWC continued until the end of 2012.)

The monitoring program illustrated that bacterial recidivism was a characteristic of many spring sites. The National Health and Medical Research Council (NHMRC) Australian Drinking Water Guidelines were adopted and the VMWC (2004) established a formal protocol for spring closure. Government funding was directed toward the restoration of springs recording contamination. Monitoring results justified and guided extensive mineral spring restoration works. The mineral springs had been developed either by 5–7 m deep bores drilled in the 1920s or as excavated and lined trenches. Overall, 35 sites were worked on with 27 new 30–150 m deep bores being constructed and several attempts were made excavating domes or collection reservoirs resealing, providing cement grout blankets replacing plumbing (Table 14.1).

Table 14.1 Summary of the average changes in bacterial monitoring results before and after remediation and landscaping activities (Data 1986–2013)

Sites	Apparent increase in average HCO_3^- levels after remediation[a] (mg/L)	Preceding period 1986–2000 bacterial record, % of readings with *E. coli*	Post-2000 or remediation bacterial record, % of readings with *E. coli* to 2013	n
Bores	311	16%	0%	24
Pits	231	15%	18%	8
Pits replaced by bores	210	25%	0%	3

[a] If taken during drilling, a new bore then samples can be influenced by gas sparging effects introduced during DHH drilling. Drilling operations can substantially influence the EC, pH, Fe, HCO_3^-, and minor metals and for variable times. Later appropriately obtained samples may suggest acidity and alkalinity and equilibria changes. The magnitude varies between bore sites, the depth drilled, rapidity of drilling, post drilling relaxation period, and the time and method used for sampling.

14.2. SPRING FLOW SYSTEMS AND OCCURRENCE

The mineral springs occur in localized groundwater discharge zones mainly in the incised valleys of the Campaspe, Lerderderg, Loddon, Moorabool, and Werribee rivers and their upper tributaries that radiate from the crest of the Great Dividing Range. Many mineral water springs or seeps exist in the region; not all have been formally recognized or developed. In the upper portion of the catchments, some were dug up and buried repeatedly by the alluvial gold miners. Within this group, only some have warranted conservation in special mineral spring reserves (Shugg, 1996; Laing et al., 1997). The springs are situated near the valley nadir at the confluence of vadose, phreatic, and surface water flow systems, which provide an opportunity for mixing and bacterial contamination.

The aquifer consists of an open anisotropic and heterogeneous fractured rock continuum. Hydraulic response in the aquifer is well documented in the wealth of early Geological Survey and gold mining literature reports (Newbery, 1867; Dunn et al., 1912; Skeats, 1912; Whitelaw & Baragwanath, 1923). The findings of a detailed hydrogeological assessment between 1974–1976 by the Geological Survey was set out in Laing (1977), and more details of the relationships and hydrogeology of the springs may be found in Shugg (1996, 2004) and in many recent reports by the consultants Nolan ITU and Sinclair Knight Merz (SKM) to the VMWC. The depth of circulation of the meteoric water in the bedrock exceeds the depth of the deepest mines of the goldfields, being 500 m in Daylesford and more than 1,400 m in Bendigo. Near surface rates of movement in the bedrock have been measured using dye tracers and vadose zone transmission rates varied from 0.2 to 1.6 m/hr (SKM, 2003a, 2003b, 2003c, 2007, 2010, 2011, 2012a, 2012b; Shugg, 2004). While the rate of flow in phreatic fissures between the injection site of North Frenchman's Mine and discharge at Wyuna Spring was greater than 0.3 m/hr (7.2 m/day). This value ignores retardation by rehydration of mine workings and modified hydraulic gradients (Dunn et al., 1912; Skeats, 1912; Laing et al., 1997; Shugg, 2004). The transmissive character of the rock and soil in the vicinity contributes to the vulnerability and communication with local and catchment processes.

In the natural state, springs zones are a focus for discharge; when developed they may become an exaggerated vortex in the water table surface inducing and enhancing mixing of ascending carbonated mineral water with reflux, local interfluve recharge, and percolation from streams. Bacterial readings and groundwater surface contouring serve to corroborate these observations. Mixing and open system dynamics are a facet of the springs that can be appreciated after examination of the results of bacterial monitoring.

Individual spring flow is plumbing dependant, with rates falling in the range 0.01–0.3 L/s (Laing et al., 1997; Shugg & Knight, 1994; Shugg, 2004). Spring flow rates exhibit a linear correlation with groundwater hydrographs obtained from adjacent bedrock monitoring bores (Shugg, 2004). Empirically, the mineral water contribution to base flow is much larger than the observed "spring eye" discharge. Flow relationships were quantified in the Hepburn Mineral Spring Reserve using data logger stations and a V notch weir (Shugg, 2004). Pumping tests have been conducted in the spring reserves and produced a wide spectrum of hydraulic characteristics that include response complicated by variable horizontal and vertical anisotropy and sloping or undulating water table. Transmissivity values are usually in single digits (m^2/day) and storativity is indicative of an open system and relating to dilated fracture porosity that will vary and diminish with depth (Stewart, 1979; Shugg, 2004; SKM, 2012a, 2012b).

Recharge to the bedrock terrain is an iterative process changing in depth and composition as the hydrology and hydrogeology of the catchment varies down basin (Mein et al., 1988; Shugg, 2004). Discharge from the bedrock has been examined in substantial detail elsewhere, where it is associated with dry land salinity (Dyson, 1983). Localized reflux to the aquifer has been observed in observation bores below and surrounding the downvalley piped outlet at Lithia Spring (Laing, 1977; Shugg, 2004) and substantiates the open nature of the bedrock system. Static bore hole temperature and electrical conductivity profiles (Laing, 1977; Shugg, 2004) and dynamic profiles logged during compressed air drilling (Shugg, 2004, 2008a, 2008b) illustrate that shallow mixing may be present to a depth of perhaps 12 m below the water table, while in some deep bores the interface has been more than 60 m below the water table. Based on the profiling and outcomes of new bore construction, it is most likely that the zone enriched with soil and fecal bacterial does not usually extend more than 12 m below the water table.

In the spring zones, contributions from elevated ridges and interfluves may have a significant albeit seasonally transient mixing influence. With few exceptions, discharge is valley floor, subaqueous, and buried by lower terrace streambed alluvium. Discharge may not always be apparent or marked only by the rise of bubbles or by iron flocculants coloring quiescent pools. Down the drainage basin interfluve percolates and shallow bedrock groundwater becomes more saline. Notably, chloride concentrations are observed to rise rapidly above the ascendant mineral water concentrations and provide criteria for differentiation. Chloride concentrations matching those in the ascendant mineral water suggest most of the recharge for the low flux mineral water occurs within 2 to 10 km of the catchment crest.

14.3. BACTERIAL MONITORING

The bacterial record examined extends from 1986 to 2013. The frequency of collection, number of sites, and characteristics of individual sites have changed during the period of monitoring. Standard procedures were employed following the Australian NATA Laboratory procedures, and bacterial sampling followed the NHMRC Guidelines. Public health bacterial indicators examined were *E. coli*, Total Coliform, Total Plate Counts, and Streptococci. If positive readings were recorded in a follow-up sample, then the spring was closed to the public until it returned a negative reading. Some early data are prejudiced by attempts to guess the hydrological response to rainfall events, so early data between 1986 and 2000 had a suggested data bias due to collection attempts timed to be 6 days after a rainfall event (Bannister, 1992). Analysis of the data suggests little or no correlation exists ($r^2 < 0.12$) with a 6 day sampling delay after rainfall events. Subsequent shallow vadose zone transmission rates based on the dye tracer tests suggest much shorter delays. Watering spring surrounds, lawn irrigation, or ornamental garden bed fertilizer application are all activities that have produced coliform infection of the springs (Shugg, 2004).

A graphical summary of the monitoring results for the region is presented in Figure 14.1. Bacterial time-series data for all the monitored sites show perennial fecal bacterial contamination. The presence emphatically identifies mixing of waters and a biological contribution to overall water composition. This empirical observation conflicts with the views expressed by the chemists Schaefer and Kecskemeti in 1981 who reflected on the relative constancy of the composition of mineral water. Data logger monitoring in Hepburn Mineral Spring Reserve recorded substantial variation of mineral water composition and physical properties that range from diurnal to seasonal, and in some instances were made more complex due to transient freshwater lensing, interfluve recharge, local reflux, and extraction induced mixing with ambient surface water (Shugg, 2004).

All the creeks in the region carry elevated *E. coli* levels. Ingress of creek water to a spring collection point will result in a positive reading, but the creek is or may not be the only source or pathway. Control samples collected from creeks in spring reserves established ambient levels of *E. coli* in the range of 80 to 400 org per 100 ml. All bacterial indicators used showed a number of peaks during the year. These corresponded in the case of monthly maxima values to an onset of autumn; then the early winter and early summer flushes (Fig. 14.2). With respect to the coliform, speciation or ratios between indicators also changes through the year and may be determined by hydrological catchment processes.

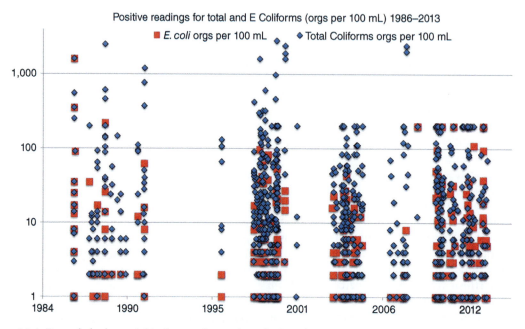

Figure 14.1 Records for bacterial indicators in monitored mineral springs 1986–2013.

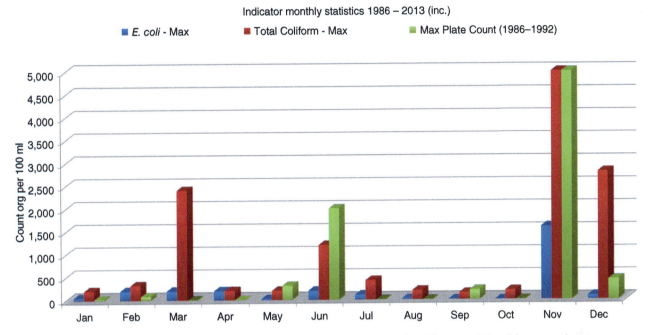

Figure 14.2 Averages for monthly maximum values for Plate Count, Total Coliform, and *E. coli* for mineral springs at Daylesford, Central Victoria.

14.4. CLIMATE AND SPRING HYDROLOGY

The climate changes rapidly in this region of central Victoria with distance from the crest of the Great Dividing Range. Elevations reach little more than 780 m with a rainfall exceeding 600 mm, but usually less than 1,400 mm. The uplands support temperate rain forests. Forty kilometers to the north at Newstead on the Loddon River, near the terminus of some of the flow systems, the elevation drops to 200 m and the median rainfall diminishes to around 450 mm. In the low lands, dry sclerophyll forests are predominate. Flow in the upper catchment streams near many of the spring development sites is ephemeral to intermittent, while perennial flow is only a characteristic of the lower portion of the catchments.

Seasonal variation in bacterial occurrence follows hydrological relationships between spring flow and rainfall, incorporating several hysteresis or lag responses between recharge and discharge. Studies in the Stewarts Creek catchment near the crest of the Dividing Range by Mein et al. (1988) recognized different runoff and recharge relationships, peak and lag factors depending on whether catchments were forested or grassland. Current changes in catchment use with increased urbanization and land clearing may be expected to influence the mixing dynamics at the spring zones also.

Hydrological hysteresis response is also evident in the records of data loggers and observation bore records in the spring reserves (Shugg, 2004). The seasonal hydrological response is presented in the upper panel of (Fig. 14.3).

In this temperate region, groundwater recharge may be proportional to the rainfall or, in particular, the water table elevation follows the average monthly positive rainfall surplus curve (P-Et; Watkins, 1969), and further spring flow is proportional to the groundwater levels (Fig. 14.3). For instance, for the sum of monthly average rainfall surpluses for Daylesford is 192 mm. Surplus is initiated in June and lasts to November. The hydrographs for stream flow and groundwater climb slightly before the rainfall surplus and fall before the end of the rainfall surplus regime. Overall, the hydrographs exhibit strong seasonality with some harmonics probably associated with variations in catchment homogeneity such as the variation between upper catchment grasslands on the basaltic aquifer material and the dry sclerophyll forests on the bedrock.

In examination of the seasonal bacterial response, positive readings and not absolute counts have been presented as this is a statistic more useful to regulation of access to the waters (Fig. 14.3), statistics for absolute, median, and maximum differ slightly. The positive readings for the coliform indicators show a good synergy (Fig. 14.3) with positive readings falling during the winter months, July to October. Implicit is that both Total Coliform and *E. coli* are useful indicators for vulnerability and *E. coli* particularly as an indicator for public health risk. Seasonal variation also exists in the ratio (average Total Coliform/ average *E. coli*) with bimodal maxima occurring in March as 25, and in October as 70. Between the peaks, the ratio falls down to 8. Similar patterns exist for the ratios of the maximum readings, while the peak values of the ratios are

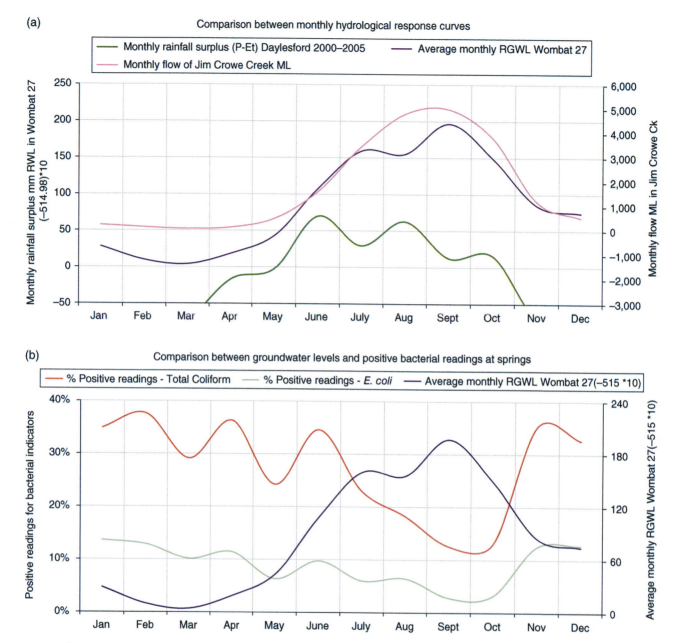

Figure 14.3 Monthly curves for rainfall surplus, stream flow, and groundwater levels (upper panel) and Positive *E. coli* and Total Coliform readings and groundwater levels for Daylesford (lower panel).

smaller. The ratio (Max Total Coliform/Max *E. coli*) falls to a low of 2 in February and a peak in March of 12, and declines until another peak in October of 50, then falls to 3 after October. Within this confusion, it is apparent that Total Coliform remains as a good indicator of public health risk at the spring water supplies.

14.5. OUTCOMES OF REMEDIATION OF SPRING SITES

Another aspect of the spring monitoring data is the impact of the remediation of sites and whether this unduly influences the data set. A summary of the number of sites worked on is found in Table 14.1.

Two types of spring development exist: (1) the dug wells, pits, and trenches and (2) drilled bores. The drilled sites are the most numerous. The bacterial monitoring and patterns of contamination have been useful in justifying the works and in assessing the success of remediation. The remediation process started in antiquity and developed from works by committees of management such as cleaning out bores, resleeving, and basic headworks protection to exhumation of pit springs in the early 1990s. In some instances, the works and disinfection had some small impact, but within a short time, recidivism was reestablished (Fig. 14.1). Exhumation of some sites has revealed many generations of unrecorded alterations or fixes. Preferred materials used include lead or tin pipe,

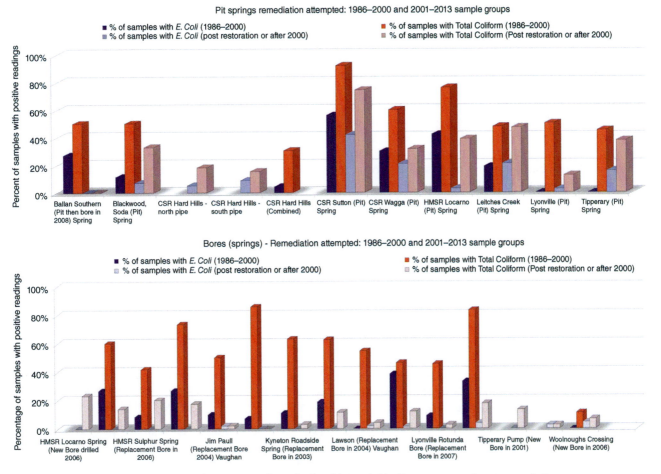

Figure 14.4 The percentage of positive readings for fecal bacterial indicators pre and post remediation.

quartz cobbles, brass fittings, and corrugated galvanized iron collection reservoir covers.

The bacterial indicators Total Coliform and *E. coli* and have been used to access the water quality at sites before and after remediation. To examine the changes at sites pre and post remediation, the percentage of positive readings have been plotted and are presented as Figure 14.4. The bar graphs of the pit and bore development are separated into two panels on Figure 14.4. For the pit or excavated springs, none of the restoration works succeeded. This may have been due to grout failure or that capture efficiency increased and induced leakage from nearby streams that resulted in higher coliform readings (upper panel Fig. 14.4). The lower panel is for sites where shallow 2–7 m deep bores were replaced with bores that logistics depending usually ranged in depth from 30 to 150 m. In new bores, upper casing strings were pressure grouted in place. On the lower panel of Figure 14.4, some positive readings for Total Coliform have been recorded after the new works; this amounts to few actual incidents. Some or most positive readings may be explained as being associated with pump servicing without conducting sterilization. As the bores are located in streamside reserves, some can be inundated by floodwater. Other positive readings may be attributed to careless pump maintenance, foot valve riser withdrawal and replacement, or that screw on bore caps seals were not refitted. Bore purging, disinfection, and bore cap seals when employed usually rectify and reinstate sanitary conditions. These changes in Site performance are embedded in the curves presented in Figure 14.3. Many of the pits are now closed and not monitored, while the new bores dilute the overall statistics.

14.6. CONCLUSIONS

The effervescent nature of the mineral water attracts attention as a geological curiosity, a recreational tasting or quaffing water, and as a resource for spa and bottling development. Disconcertingly, in central Victoria mineral water at spring development sites has exhibited a surprisingly high incidence of positive readings for the fecal indicators Total Coliform and *Escherichia* Coliform (*E. coli*). Over the past three decades, individual sites have recorded positive readings for Total Coliform in 10% to 35% of sampling events and positive readings for *E. coli* in 2% to 14% of sampling events. These values have been tempered by lower or negative readings obtained after restoration works. Also, some recalcitrant sites have been dropped from the monitoring program and closed to the public. Singular contamination events do exist in the data set and some events may correspond with pump or bore servicing and flood inundation. In present terms, the old spring plumbing or development appears poor, inadequate, and vulnerable, but many changes have also taken place in the catchments in the last century, too, and now the old construction techniques are open and no longer suitable. All sites developed using pits and trenches were subject to high levels of bacterial recidivism. To avoid continual spring closure or abandonment, substantial restoration and refurbishment work was undertaken between 1998 and 2013.

The bacterial monitoring data contain evidence of interdependent relationships determined by season and catchment hydrodynamics as well as spring plumbing. Persistent dynamic and seasonally biased flushing and mixing processes may be recognized. These are evident in bacterial levels and in the changes in the ratios between the indicators of Total Coliform and *E. coli*. Hydrologically in this the temperate region, spring flow correlates with water table elevation. In turn, stream flow and water table elevation and catchment flushing relates hydrologically with the rainfall surplus. A recharge and flushing process is initiated as accession to the water table commences just after the autumn equinox in March and continues until after the spring equinox into November. During this period of high stream flow and groundwater accession, the coliform bacteria are less prevalent at the springs. The winter to spring period of June to late October usually has had the lowest occurrence of coliform bacteria at the springs.

Several statistics may be examined within the set of average monthly values, positive readings, maximum values, minimum values, and then the ratio between these statistics. The ratio between monthly average values has been selected. The relative abundance of Total Coliform compared with *E. coli* changes with season. The ratio of the average Total Coliform with the average *E. coli* peaks at 25 in March, then drops through the high hydraulic flux period of the year. A second peak with a ratio of 70 appears in early summer, corresponding to a period of dominance by soil bacteria in the water. The ratio between the maximum Total Coliform and maximum *E. coli* is also multimodal and rises from 2 in February to a peak of 12 by the March equinox then drops during the high winter flow period to rise again between November and December with a value of 50 by the early summer.

Flushing of the catchment is accompanied by a relative reduction of coliform bacteria counts, and may be a dilution effect. Within this period, the soil bacteria indicator Total Coliform appears greatest when the flow systems transition to activity or relax to inactivity. In hydrologically quieter periods, the Total Coliform declines and *E. coli* becomes relatively more significant as the catchment is being flushed of its bacterial load.

ACKNOWLEDGMENTS

The author wishes to acknowledge the work and contributions made by the former members of the Victorian Mineral Water Committees and officers of the Regional Office of the Department of Environment, Water, Land and Planning; and also Ross Bannister late of WaterEcoscience, the staff of Central Highlands Water bacterial laboratories, and the late Philip Clingin of the Victorian Mineral Water Committee.

REFERENCES

Bannister, R. (1992). *Quality of mineral waters of the Daylesford area*. Water, Materials and Science Branch, Rural Water Commission. Report No. 107, February 1992.

Brady, A. (1990). *The mineral springs of the Daylesford and Hepburn: An introductory history and guide to sources*. Minor Thesis MA in Public History, Monash University, February 1990.

Dunn, E. J., Baragwanath, W., Whitelaw, H. S., & Herman, H. (1912). *Hepburn Mineral Springs, Daylesford*. Geological Survey of Victoria Unpublished Report 1912/4. Department of Mines, Victoria.

Dyson, P. R. (1983). Dryland salting and groundwater discharge in the Victorian Uplands. *Proceedings of the Royal Society of Victoria, 95*(3), 113–116.

Laing, A. C. M. (1977). *Daylesford-Hepburn Springs mineral water investigation*. Geological Survey of Victoria Report.

Laing, A. C. M., Shugg, A., & Elder, G. (1997). *Mineral springs of Victoria, hydrogeological archival material*. Department of Natural Resources and Environment Groundwater Report No. 75.

Lawrence, C. R. (1965). *Preliminary report on the Hepburn mineral springs*. Geological Survey of Victoria Unpublished Report 1965/64.

Mein, R. G., Bieniaszewska-Hunter, H., & Papworth, M. (1988). *Land use changes and the hydrologic water balance: Stewarts Creek experimental area*. Hydrology and Water Resources Symposium 1988, ANU Canberra, 1–3 February 1988. Institution of Engineers, Australia National Conference Publication No. 88/1, 129–134.

Newbery, J. C. (1867). Mineral waters of Victoria. *Transactions of the Royal Society of Victoria, 8*, 278–283.

Shugg, A. (1996). *Mineral and spring water resource protection discussion paper*. Department of Natural Resources and Environment, August 1996.

Shugg, A. (1999). *A statewide framework for groundwater contamination assessment and management in Victoria*. Department of Natural Resources and Environment, Victoria, Australia.

Shugg, A. (2004). *Sustainable management of central Victorian mineral waters.* Doctorate of Philosophy Thesis, RMIT University, Melbourne Australia.

Shugg, A. (2008a). Hepburn Spa: Cold carbonated mineral waters of Central Victoria, South Eastern Australia. *Environmental Geology, 58,* 1663–1673. https://doi.org/10.1007/s00254-008-1610-8

Shugg, A. (2008b). *Mixing profiles encountered during drilling of new mineral water bores, Hepburn SE Australia.* Paper presented at the Meeting of the International Association of Hydrogeologists, Proceedings of the Mineral and Thermal Water Meeting, Istanbul Turkey, September 2008.

Shugg, A., & Knight, M. (1994). *Hydrogeology of the mineral springs of central Victoria.* Proceedings of the Water Down Under Conference, 21–25 November 1994. International Association of Hydrogeologists and the Institute of Engineers, Adelaide, Australia.

SKM (2003a). *Dye tracer reports, Locarno Spring.* Report to Victorian Mineral Water Committee. WC02113/L03ajspc.doc.

SKM (2003b). *Dye tracer test report Soda Spring.* Report to Victorian Mineral Water Committee. WC02113/L01ajspc.doc.

SKM (2003c). *Report on the dye testing of Kyneton (Southern) Spring, July 2003.* MS 010 Kyneton 2003 07, WC02519/LR01aspc.doc.

SKM (2007). *Blackwood dye results.* Letter report to the Victorian Mineral Water Committee. June 2007. WC03355, L03 aspc.

SKM (2010). *Hepburn reserve sources: Yield and Pavilion Spring quality.* Hepburn VMWC Grant 2008/01 and 2008/04 Hepburn Springs Investigation. Report prepared for Hepburn Shire Council / VMWC, May 2010.

SKM (2011). *Preliminary report on dye tracing and inspection of the Glen Luce Spring.* Report prepared for Parks Victoria and the Victorian Mineral Water Committee.

SKM (2012a). *Argyle Spring casing condition assessment and dye tracing.* Public Mineral Springs Remediation and Upgrade Program, Report to the VMWC. R01_Argyle Spring Assessment/VW03355.

SKM (2012b). *Replacement of Hepburn Bath House emergency bore, Hepburn mineral springs reserve.* Mineral Spring Remediation and Upgrade Program, Report prepared for the Shire of Hepburn and the Victorian Mineral Water Committee, July 2012.

Skeats, E.W. (1912). Mineral springs at and near Hepburn. *Geological Survey of Victoria, Bulletin 36.*

Stewart, G. (1979). *Hydrogeological investigation of a proposed hazardous waste depot site, Parish of Dargile. Addendum: Further determinations of hydraulic conductivity and water quality.* Victorian Geological Survey Unpublished Report 1979/37.

Watkins, J. R. (1969). The definition of the terms "hydrologically arid" and "humid" for Australia. *Journal of Hydrology, 9,* 167–181.

Whitelaw, H. S., & Baragwanath, W. (1923). The Daylesford Goldfield. *Geological Survey of Victoria, Bulletin 42.*

15

Towards a Collective Effort to Preserve and Protect Springs

Brian G. Katz[1] and Matthew J. Currell[2]

ABSTRACT

Over the past several decades, various anthropogenic activities have resulted in degradation of spring water quality and quantity and their sensitive ecosystems. Numerous studies around the world have documented these detrimental impacts to springs and their associated aquifers due to habitat alteration, groundwater depletion, pollution (from agricultural fertilizers, wastewater disposal, mining, urbanization, and climate change. These impacts to springs have negatively affected human health and ecological systems, and the livelihood of people that depend on springs for their cultural, recreation, and aesthetic values. This volume has provided detailed examples of threatened springs in diverse hydrogeologic settings, relevant information on innovative methods for better understanding the hydrogeology of spring systems, and policy and governance approaches for protecting springs now and for the future. It is our hope that the information in this volume will help to stimulate practitioners, policy makers, scientists, and the public to work together to actively preserve and protect springs and restore impaired spring waters. An effective approach for protecting springs should include continuous monitoring of water quality and quantity; collaboration among scientists, stakeholders, and governmental agencies to develop effective local and regional springs protection strategies, and sufficient funding for current and future scientific research, especially regarding impacts from climate change.

Springs are unique hydrologic systems that represent the transfer of groundwater to the land surface and to other bodies of water. Springs occur in a wide variety of geologic, climatic, and biogeographical settings and support diverse habitats and ecosystems along their spring runs. Spring ecosystems contain rare aquatic and terrestrial plant and animal species that are distinct from other aquatic, wetland, and riparian ecosystems (Stevens et al., 2005; Glazier, 2014). In addition, springs have been important sources of water for human civilization throughout history. Large springs offer outstanding recreational amenities as well as historic and cultural values. Unfortunately, various anthropogenic activities have resulted in degradation of spring water quality and quantity and sensitive ecosystems.

Over the past several decades, there have been hundreds of documented cases around the world of detrimental impacts to springs and their associated aquifers due to habitat alteration, groundwater depletion, and pollution (from agricultural fertilizers, wastewater disposal, mining, urbanization, and climate change) (Glazier, 2014; Cantonati et al., 2020). Numerous published studies have described how these spring maladies have led to substantial degradation of water quality and significant decreases in spring flows (Knight, 2015; Cantonati et al., 2020). For example, degradation of

[1] *Environmental Consultant, Weaverville, North Carolina, USA*
[2] *School of Engineering, Royal Melbourne Institute of Technology, Melbourne, Australia*

Threats to Springs in a Changing World: Science and Policies for Protection, Geophysical Monograph 275, First Edition.
Edited by Matthew J. Currell and Brian G. Katz.
© 2023 American Geophysical Union. Published 2023 by John Wiley & Sons, Inc.
DOI:10.1002/9781119818625.ch15

spring water quality due to elevated nitrate concentrations have been reported in many countries around the world, including Australia, China, Croatia, England, Ethiopia, France, Ireland, Israel, Japan, Jordan, Korea, Slovenia, Spain, Switzerland, Turkey, Uganda, and many locations in the United States. Detrimental impacts to springs have affected not only human health and ecological systems, but also the livelihood of people who depend on springs for their cultural, recreation, and aesthetic values. For example, several national capitals (such as Rome, Vienna, Beirut, Damascus), many small towns and farms rely on springs for potable and agricultural water (Kresic & Stevanovic, 2010).

This volume is the first of its kind to provide detailed examples of threatened springs in diverse hydrogeologic settings, and to combine these examples with relevant information on innovative methods for better understanding the hydrogeology of spring systems, and to present policy and governance approaches and strategies for protecting springs now and for the future.

The chapters in the first part of the book provided up-to-date examples of detrimental impacts to large springs in Florida, Japan, Australia, and Arizona that have been contaminated and/or dewatered due to various anthropogenic activities. These activities include nutrient pollution from agricultural runoff; withdrawal or extraction of large amounts of groundwater from aquifers by mining and industrial operations, agriculture, and by cities for urban water supply; and improper waste disposal from septic tanks and the land application of municipal wastewater. As a result, there have been profound impacts on these spring waters in terms of human health, ecological degradation, recreational activities, and the culture of Indigenous communities. In addition, there are staggering economic costs associated with restoration of impaired springs. For example, over the past two decades the State of Florida in the United States has spent hundreds of millions of dollars for remediation, restoration, and protection of large karst springs that have become contaminated with elevated levels of nitrate (resulting in increased harmful algal abundance). This contamination originates mainly from agricultural activities, the land disposal of municipal wastewaters, and seepage of septic tank effluent into the surficial aquifer and underlying Floridan aquifer (the source of drinking water for millions of residents). In addition, large amounts of monies are being spent on nitrate-reducing capital projects (wastewater, storm water and nonpoint source pollution control projects) and water-quantity projects to protect and restore springs. In addition, the State of Florida has land conservation programs that include acquisition of land in spring recharge zones to prevent further nitrate contamination. Economic impacts from impaired springs in Florida have also included loss of jobs, reduced income from ecotourism, and losses in real estate values (Katz, 2020). In addition, bottled spring waters have important and considerable economic value not only in Florida but also in many other places around the world.

The chapters in the second part of the book have provided several examples of innovative methods and tools to study and better understand processes affecting spring systems and their water quality and quantity. Relevant information was presented on use of environmental tracers (including radioactive isotopes) to assess the origins and timescales of spring water and solute flows; high-resolution climatic and spring monitoring data combined with artificial tracer tests and targeted analysis of micropollutants to assess recharge, flow, and pollution sources. Another chapter showed how direct hydrogeological exploration techniques (including cave mapping) were used to constrain flow patterns and estimation of recharge for springs. The final chapter in this section also showed the value of cave exploration along with direct observations of geology and measurement of water temperatures and flow rates to identify the causes and extent of anthropogenically driven thermal anomalies for a subterranean spring in the Midwest United States.

The chapters in the third section of the book provide examples of different approaches to management of springs, their water quality, and associated values. One study demonstrated the strengths and limitations of Community-Based Water Resources Management (CBWRM) for springs that are vital drinking water sources in rural communities of Timor-Leste. This chapter also discussed the challenges associated with water security issues affected by natural and anthropogenic pressures on these springs. Another study that addressed degradation of spring-water quality in Florida showed the importance of using benthic algal targets as a management strategy for the protection of spring-fed streams. These springs have been experiencing ecological degradation due to increased algal abundance (as mentioned in two chapters in first section). Finally, another chapter showed how scientific evidence-based policies are being used to manage the springs of the southwest Great Artesian Basin in northern South Australia (including those at Kati-Thanda described in the first section). This chapter discussed how the licensing and capping of groundwater extractions can be used to achieve the goal of maintaining artesian pressures near springs. Furthermore, the authors describe how the application of a new springs monitoring, risk assessment, and adaptive management plan can be used to protect the outstanding ecological and cultural values associated with these springs.

Numerous modeling methods have been used to provide critical information for springs protection. For example, various lumped parameter models have been used to estimate the residence time or mean age of the groundwater age distribution discharging from large springs (Maloszewski & Zuber, 1996; Katz, 2004; Jurgens et al., 2012). These models have used various age-dating tracers (e.g., isotopic and other chemical indicators) with known time-varying inputs to the atmosphere. Mahler et al. (2015) linked global and regional climate models with a hydrologic model and a species vulnerability index to evaluate the vulnerability of karst-related flora and fauna in the Edwards aquifer in Texas to climate change during 2011–2050. They projected that spring flow would decrease and the Barton Springs salamander and nine additional species (other salamander species, amphipods, beetles, darter) are vulnerable to climate change. More modeling studies are critically needed to evaluate future impacts from climate change on water quality and ecosystem alterations in other impaired spring systems. Recently, different modeling approaches have been used in other areas to address spring protection and preservation. For example, in northern Italy, Janetti et al. (2021) used a comprehensive framework for springs protection and preservation and a probabilistic risk assessment method associated with the characterization of the groundwater system. Their methodology was applied to a regional-scale hydrogeological setting characterized by a unique system of high-quality natural springs. By incorporating diverse risk pathways in a fault tree model with uncertainty, they were able to evaluate and determine the optimal exploitation that would ensure aquifer system functioning and identify the most vulnerable springs where depletion would first occur. In another study, Javadi et al. (2019) used an index method that analyzes the vulnerability of karst springs to contamination in Iran. Next, they were able to determine spring protection areas more accurately using a second method that integrates the vulnerability index with an analytical model MDHT.

In conclusion, the studies presented herein represent a small sample of the many spring studies around the world that have documented the severe impacts affecting the quality and quantity of springwaters, and the scientific and other methods available to determine the cause(s) of such impacts. This volume highlights land-surface springs; however, other types of springs (underwater springs in rivers, lakes, estuaries, oceans) also need to be protected. While there are numerous strategies for protecting and restoring impaired springs, the chapters in this book have provided relevant examples of current policy and governance approaches for alleviating damage to springs currently and into the future.

It is our hope that the information in this volume will help to stimulate practitioners, policy makers, scientists, and the public to work together to more effectively preserve and protect springs and restore impaired spring waters. To accomplish these goals, there needs to be a four-fold approach on local, regional, and global scales that includes (1) continuous monitoring, collection, and reporting of water quality and quantity data, especially for springs that are impaired and those vulnerable to further decline from human activities; (2) participation and collaboration among scientists, stakeholders, and governmental agencies to develop local and regional strategies for protecting springs from further contamination and dewatering; (3) sufficient funding for the remediation and restoration of impaired springs and contaminated aquifers that feed water to springs; and (4) continued funding for current and future scientific research, especially in light of impacts from climate change.

Future research needs for protecting springs and their valuable resources should include the use of novel and innovative methods for (1) tracking the movement, rate, and timing of various pollution sources to springs and their impacts on springwater quality; (2) investigating the timing and amount of groundwater withdrawals and the impact on springflow declines; (3) accurately delineating springs protection and conservation zones; and (4) determining the impacts of climate change on water quality degradation and water loss on endemic and sensitive ecosystem species that are dependent on springs for their survival. Cantonati et al. (2020) recommended the development of regional and international networks of reference locations with diverse spring types. These networks could be effectively integrated with existing long-term ecological research networks, such as the International Long Term Ecological Research, National Ecological Observatory Network, and monitoring networks for the European Nitrate Directive.

As noted in the introduction and throughout the chapters in the book, the threats to springs will likely intensify due to the continuing anthropogenic pressures created by global climate change and ever-increasing demands to develop water, mineral resources, and land for economic purposes. Climate change would likely result in lower rainfalls, which would result in less recharge to aquifers and lower groundwater levels. This would in turn increase the demand for groundwater with serious consequences, including decreased spring flows. For instance, it is estimated that increasing amounts of groundwater extraction could result in 40%–80% of the world's catchments below environmental flow limits required to maintain ecosystem functioning (De Graf et al., 2019). This in turn will have severe consequences for the well-being and the health of many people and ecological systems and the unique values and features associated with springs. It would be a tragedy if our society allows human activities to continue to degrade springs and we irreparably lose these highly valuable treasures (Cantonati et al., 2020).

REFERENCES

Cantonati, M., Fensham, R. J., Stevens, L. E., Gerecke, R., Glazier, D. S., Goldscheider, N., et al. (2020). Urgent plea for global protection of springs. *Conservation Biology.* https://doi.org/10.1111/cobi.13576

De Graf, I. E. M., Gleeson, T., van Beek, L. P. H., Sutamudjaja, E. H. & Bierkens, M. F. P. (2019). Environmental flow limits to global groundwater pumping. *Nature, 574,* 90–94.

Glazier, D. S. (2014). Springs. In S. A. Elias (Ed.), *Reference module in earth systems and environmental sciences.* Waltham, MA: Elsevier.

Janetti, E. B., Riva, M., & Guadagnini, A. (2021). Natural springs protection and probabilistic risk assessment under uncertain conditions. *Science of the Total Environment, 751,* 14130.

Javadi, S., Moghaddamb, K. M., & Roozbahanib, R. (2019). Determining springs protection areas by combining an analytical model and vulnerability index. *Cadena, 182,* 104167.

Jurgens, B. C., Böhlke, J. K., & Eberts, S. M. (2012). *TracerLPM (version 1): An Excel workbook for interpreting groundwater age distributions from environmental tracer data.* U.S. Geological Survey Techniques and Methods Report 4-F3.

Katz, B. G. (2004). Sources of nitrate contamination and age of water in large karstic springs of Florida. *Environmental Geology, 46,* 689–706.

Katz, B. G. (2020). Nitrate contamination in springs. In *Nitrogen overload: Environmental degradation, ramifications and economic costs.* American Geophysical Union, Geophysical Monograph, 250. John Wiley & Sons, Inc.

Knight, R. L. (2015). *Silenced springs: From tragedy to hope.* Gainesville, Florida: FSI Press.

Kresic, N., & Stevanovic, Z. (2010). *Groundwater hydrology of springs: Engineering, theory, management, and sustainability.* Oxford, United Kingdom: Butterworth Heinemann,

Mahler, B. J., Stamm, J. F., Poteet, M. F., Symstad, A. J., Musgrove, M., Long, A. J., et al. (2015). *Effects of projected climate (2011–50) on karst hydrology and species vulnerability: Edwards aquifer, south-central Texas, and Madison aquifer, western South Dakota.* U.S. Geological Survey Fact Sheet 2014-3046.

Maloszewski, P., & Zuber, A. (1996). Lumped parameter models for the interpretation of environmental tracer data. In *Manual on mathematical models in isotope hydrology.* IAEA-TECDOC 910. Vienna: International Atomic Energy Agency, 9–50.

Stevens, L. E., Stacey, P. B., Jones, A., Duff, D., Gourley, C., & Caitlin, J. C. (2005). A protocol for rapid assessment of southwestern stream riparian ecosystems. In C. I. I. van Riper & D. J. Mattson (Eds.), *Fifth conference on research on the Colorado Plateau* (pp. 397–420). Tucson: University of Arizona Press.

INDEX

Page numbers in *italics* refer to Figures and **bold** refer to Tables.

Acid Mine Drainage (AMD), 151
active tracers, environmental tracers vs., 93–94
ADM. *See* advection dispersion models
advection dispersion models (ADM), 119
AFDW. *See* ash-free dry weight
Aileu district, 159, 163
Ainaro district, 159
Alburni Massif, 3
   caves of, 134–135, 138–139
   dolines in, 133–134, *135*
   endorheic areas, 133–134, *135*
   geological history of, 133
   geological map of, *134*
   hydrogeology of, 132–133, 136, *137*, 139
   karst springs in, 135–136, *139*
   main springs, 133–134, **134**
   materials and methods in study in, 132–133
   recharge in, 136, **137**
   results of study on, 135–138
Alexander, E. C., Jr., 147, 152
Alexander Springs, 175, *176*, 178
algal growth, 16
Alice Springs, 184
Allocating Water and Maintaining Springs in the GAB (AWMSGAB), 188
AMD. *See* Acid Mine Drainage
AMO. *See* Atlantic Multidecadal Oscillation
ANAS. *See* National Water and Sanitation Authority
Anderson, D. E., 38
Anderson-Darling test, 24
Aquagenx, 160
aquitards, 96
Arabana, 70, 82
ArcGIS StoryMaps, 17, 24
Arckaringa basin, 187
ArcMap, 24
ash-free dry weight (AFDW), 173–175, *176*
Assal Spring, 114, 117, 118, 120, 124
   BTC at, *122*
   hydrogeology of, **115**

Atlantic Multidecadal Oscillation (AMO), 20, 29, 30
Atom Trap Trace Analysis (ATTA), 96
Attard, G., 152
Auso Spring, 138–139
Australian Aboriginal peoples, 69. *See also* Indigenous peoples
Austrian Environment Agency, 99, 100
AWMSGAB. *See* Allocating Water and Maintaining Springs in the GAB

bacteriological analysis, 117, 122
Bangkok, 145–146, 151
base-flow recession analysis, karst springs and, 39–40
Basel, 146
Bayer, P., 146
Be'e Timor-Leste E. P. (BTL), 162
Beisner, K. R., 38
Bellino, J. C., 29
Bellucci, F., 139
Benjamini, Y., 24
Bense, V. F., 77
benthic algae
   in FAS, 172–177
   measurement of, 173, 175
   targets, 174–178
Berens, V., 188
Berndt, M. P., 27
Bern University, 99
Biggs, B. J. F., 173
Biggs targets, 174–175, 178
binary mixing models, 117
Black, A. P., 22
Blasch, K. W., 38
Blavoux, 99
Blue Water Audit (BWA), 2–3
   on FAS groundwater, *15*
   on groundwater extraction, *15*
   for nitrogen estimation, 12–13
   nitrogen load estimation by, **14**, *14*
   validation of, 13
bomb pulse, 94

Threats to Springs in a Changing World: Science and Policies for Protection, Geophysical Monograph 275, First Edition.
Edited by Matthew J. Currell and Brian G. Katz.
© 2023 American Geophysical Union. Published 2023 by John Wiley & Sons, Inc.
DOI:10.1002/9781119818625.index

Boniol, D., 24
Bopeechee Spring, 77
Borden Aquifer, 93–94
Boyd, W. E., 71
Brake, L., 71
Bratislava, 152
Brick, Greg, 3, 147
Bryson, J. R., 38
BTC. See tracer breakthrough curve
BTL. See Be'e Timor-Leste E. P.
Bubbler mound spring, *72*
Budd, D. A., 21
Bush, P. W., 16
BWA. See Blue Water Audit

Cadna-Owie aquifer system, 71, **72**
Camp, E. V., 173
Campania, 132, 134
    caves of, 136
Cape York supergroup, 184
carbamazepine (CBZ), 122
Carpentaria basin, 71
Castelcivita Caves, 138
cation exchange capacity (CEC), 52–53
caves. See also specific topics
    of Alburni Massif, 134–135, 138–139
    of Campania, 136
    thermometric survey of, **150**
    with water in karst systems, **136**
CBM. See community-based management
CBOs. See community-based organizations
CBWRM. See community-based water resources management
CBZ. See carbamazepine
CEC. See cation exchange capacity
Cenomanian aquifer, 118
Central City Tunnel System project, 152
CFCs. See chlorofluorocarbons
Chalybeate Springs, 146
Chen, Z., 46, 112
Cherry, J. A., 27
Chiltondiidae, 184
chlorofluorocarbons (CFCs), 97, 98–99
chlorophyll, 173
Chlorophyta, 172, 173, 177, *177*
Clarke, R. A., 15
Clean Water Act, US, 178
climate change, Mediterranean region and, 111–112
Climate Data Guide, 29
Clover Springs, 38, *41*, **42**, **43**, *44*
    event-scale representation, **41**
Coastal Salinity Monitoring Group, 30
Coconino Sandstone, 39

Coldwater Spring, 146
Colorado Plateau, 3
    area of study on, 36–38
    digital elevation model of, *37*
    research methods, 38–39
    results of study on, 39–46
Colorado River Basin, 36
community-based management (CBM), 158, 165
    challenges of, 160
    in Timor-Leste, 159–160
community-based organizations (CBOs), 159
community-based water resources management (CBWRM), 3
    ecological intervention in, 168
    infrastructure, 158–159
    IWRM and, 168–169
    resilience of, 166–167
    social conflicts and, 167–168
    socioecological-technical systems for, 163–164
    technical issues with, 168
    in Timor-Leste, 158, 162–163, 166–169
    willingness to use, 166
convolution integral, 98
Cooper Basin region, *185*, 187
Copeland, Rick, 2, 20, 22, 24
COVID-19, 166
Currell, Matthew J., 3, 73, 79, 80
Cyanobacteria, 172, 173, 177, *177*

Dalhousie Springs, 182
Davis, S., 100
Day, S. J., 168
Decree Law 03/2016, Timor-Leste, 162
Decree Law No. 4/2004, Timor-Leste, 158–159
DFM. See distributed process-based flow model
DGAS. See General Directorate of Water and Sanitation
Digital Elevation Models, 132
distributed process-based flow model (DFM), 124
DNSA. See National Directoriate for Water Services
dolines, 136
    in Alburni Massif, 133–134, *135*
Donovan, Keegan, 3, 38
Dornberg, D., 147, *148*
Doummar, Joanna, 3
Drinking Water Database, 16
dry weight (DW), 173–175
DW. See dry weight

Egypt, 52
EIS. See environmental impact statement
El Gran Sasso Springs, 112
Eller, K., 12
El Niño, 36, 160
El Niño southern oscillation (ENSO), 20, 29

encroachment
    additional monitoring needs, 30–31
    in FAS, 22, 28–29
    passive, 28–30
    unresolved issues, 30
endorheic areas, 136
    Alburni Massif, 133–134, *135*
    in karst springs, 133
ENSO. *See* El Niño southern oscillation
enterococci, 122
environmental impact statement (EIS), 73
Environmental Protection Agency, 30
environmental tracers, 87–88, 112–113
    active tracers vs., 93–94
    applications of, 90
    classification of, 90
    dating, 97
    exclusion principle, 97
    in Fischa-Dagnitz spring, 99–102
    geophysics of, 89–90
    for groundwater, 91–92
    groundwater flow and, *90*
    groundwater levels and, 88
    hydrochemistry of, 89–90
    hydrograph analysis and, 94
    ideal, 95
    inclusion principle, 97–98
    infiltration conditions and, 91–92
    input functions of, *94*
    in karst springs, 88
    least certain, 96–97
    in Lebanon study, 117, *121*
    noble gases as, 95–96
    pre-application, 88–90
    propagation of pressure vs. transport of water and, 88
    properties of, 94–95
    radioactive decay and, 91
    snapshot vs. time series, 94
    solute origin and, 92
    in springs vs. wells, 88
    time information from, 97–99
    timescales for, 88
    transport properties of, 90–91
    types of, 90–99
    water molecule components as, 95
    for water movement timescale, 92–99
Environment Protection and Biodiversity Conservation Act (EPBC), 184, 190
EnvStats, 24
EPBC. *See* Environment Protection and Biodiversity Conservation Act
Epting, J. M., 145–146
Eromanga basin, 71, 187

Escherichia coli, 160, 202, 203, *204*, 205–206
    in Timor-Leste, *161*
evapotranspiration, 46
excess air, 91
exclusion principle, 97

Fairfax, R. J., 70
FAO. *See* Food and Agriculture Organization
Far North Numerical Groundwater Model, 195
Far North Prescribed Wells Area (FNPWA), 184, 186
    monitoring program, *191*
FAS. *See* Floridan aquifer system
FDEP. *See* Florida Department of Environmental Protection
fecal coliform, 122
Fensham, R. J., 70
Ferguson, G., 146
fertilizer. *See* nitrogen fertilizer
Fetter, C. W., 27, 28, 30
field fluorometers, 117
Fiorillo, Francesco, 3, 132
Fischa-Dagnitz spring
    age of water at, 100–101, *103*
    environmental tracers in, 99–102
    flow system, *99*
    hydrogeology of, 99–100
    measured concentrations at, *102*
Florida Administrative Code, 22
Florida Climate Center, 23
Florida Department of Environmental Protection (FDEP), 12, 16, *17*, 177
Floridan aquifer system (FAS), 20–21
    benthic algae in, 172–177
    BWA on groundwater in, *15*
    conceptual model of study on, 27–28
    discharge points, 10–11
    encroachment in, 22, 28–29
    footprints, 13
    karst springs in, 171
    key findings of study on, 31
    monitoring sites, **23**
    passive encroachment in, 28–30
    regional occurrence and magnitude, 9–10
    results of study on, 25–26
    salinity in, 27–28
    statistical methods for study on, 24–25
    study area, 22–24
    upper and lower divisions, 21–22
Florida Platform, 9
Florida Springs Institute, 11, 16
Florida Springs Region, *10*
    ecological impairments in, 11–12
    groundwater chemistry in, 172
    groundwater consumption in, 13–14

Florida Springs Region (*cont'd*)
   groundwater withdrawal estimation in, 13
   nitrates in, *12*
   nitrogen loading in, 13
   nutrient mass balances for protection and recovery of, 15–16
   public informing in, 16–17
   water quality changes in, 10–11
   water quality in, 172
Florida Springs Restoration Areas, *10*
Florida Statewide Agricultural Irrigation Demand (FSAID), 13
Florida Water Resources Monitoring Council, 30
FNPWA. *See* Far North Prescribed Wells Area
FNPWA Water Allocation Plan (FNWAP), 186, 190–192
Food and Agriculture Organization (FAO), 52
fossil fuels, 2
Freeze, R. A., 27
freshwater, stresses to, 111–112
FSAID. *See* Florida Statewide Agricultural Irrigation Demand
Fu, G., 188

GAB. *See* Great Artesian Basin
GAB Consultive Committee (GABCC), 189
Gabriel, G., 89
GAB Springs Adaptive Management Plan and Template (GABSAMP), 193–194, **194**
GAB Strategic Management Plan (GABSMP), 189–190
GAB Sustainability Initiative (GABSI), 190
GAB Water Resources Assessment, 188
gas industry, 2
GDEs. *See* groundwater-dependent ecosystems
General Directorate of Water and Sanitation (DGAS), 163
geographic information system (GIS), 12, 16–17
Geoparks, 138
geophysics, of environmental tracers, 89–90
Gerber, Christoph, 3
Ghyben-Herzberg relationship, 27, 29
GIS. *See* geographic information system
Global Network of Isotopes in Precipitation (GNIP), 96
Global Water Partnership, Technical Advisory Committee (GWP-TAC), 161
GMF. *See grupu maneja fasilidade*
GNIP. *See* Global Network of Isotopes in Precipitation
Goldscheider, N., 36
Gotch, T., 182, 184, 189
Government Resolution 43/2020, Timor-Leste, 158
Grava del Fumo, 139
Great Artesian Basin (GAB), 3, 4
   discharge and recharge of, 188
   economic importance of, 186–187
   geography of, 187
   geology and hydrogeology of, 71
   as groundwater resource, 181
   Indigenous Australians supported by springs of, 184
   licensed volumes sourced from, *186*
   monitoring, 192–193
   national regulatory response to threats to, 189–190
   pastoral use of, 70
   predicted vs. actual impacts on, *78*
   size of, 69–70
   spring protection, 80–82
   spring supergroups, *182*, *183*
   state management of, 80–81
   subbasins, 71
   threats to, 187–189
   Wellfield A, 73–82, *74*, *76*, 80, **81**
   Wellfield B, 73, *74*, *76*, 77
   wellfields of, and Olympic Dam, 72–73
Great Dividing Range, 199
Greece, 112
Green, G., 188–189
Griffiths, R. E., 38
Grotta del Falco, 138–139
groundwater chemistry
   in Florida Springs Region, 172
   tea leaf production and changes in, 63–65
groundwater-dependent ecosystems (GDEs), 188
groundwater flow
   conceptualization of, *90*
   environmental tracers and, *90*
   monitoring, 115–116
groundwater lens, *24*
groundwater levels
   BWA on, **15**
   environmental tracers and, 88
   spring flows correlated with, 78–80
   Wellfield A and, 73–80
groundwater pollution, from nitrogen fertilizer, 52–53
groundwater resources
   in Florida Springs Region, 13–14
   GAB as, 181
groundwater temperatures. *See also* subsurface urban heat island
   anthropogenic changes in, 146–147
   in Minneapolis, 146
   in Platteville Limestone, 152
   in Schieks Cave, 151–152
   urban development and, 145–146
grupu maneja fasilidade (GMF), 159, 160, 165, 168
Guan, 173, 177–178
Gum Slough, 175
GWP-TAC. *See* Global Water Partnership, Technical Advisory Committee

Habermehl, M. A., 71
Hamel, S., 159
Hartmann, A., 112
Hawke, A., 192
Heburn Springs, 200

helium, 95
Hemmerle, H., 146
Hermit Hill Springs, 77, 79
Hermit Shale, 40
Hochberg, Y., 24
Hooray aquifer system, 71, **72**, 187
Horner, R. R., 173
Hoxworth Spring, 38, 39, 40, 42, **42**, *43*, *44*
Human Research Ethics Committee, 166
Hungary, 138
hydrochemistry, of environmental tracers, 89–90
hydrogen, 95
hydrogeology
   of Alburni Massif, 132–133, 136, *137*, 139
   of Assal Spring, **115**
   of Fischa-Dagnitz spring, 99–100
   of GAB, 71
   of karst springs, 38–39, 131–132
   of Laban Spring, **115**
   of Qachqouch Spring, **115**
   of Timor-Leste, 160
Hydrogeology Journal, 70
hydrograph analysis
   environmental tracers and, 94
   of karst springs, 39, 40–41
   in Lebanon study, *118*

IAEA. *See* International Atomic Energy Agency
Ichetucknee River, 175, 178
IGABDR. *See* Improving GAB Drought Resilience
Ii, Hiroyuki, 3
ILG. *See* Irrigated Lands Geodatabase
Illstrup, Carl J., 147
Impermeable Fault Zone, *78*
Improving GAB Drought Resilience (IGABDR), 190
inclusion principle, 97–98
Indigenous peoples, 69, 189
   GAB springs supporting, 184
infiltration conditions, environmental tracers and, 91–92
integrated water resource management (IWRM), 158
   CBWRM and, 168–169
   in Timor-Leste, 161–162
International Atomic Energy Agency (IAEA), 99–100
International Conference on Water (1977), 161
Irrigated Lands Geodatabase (ILG), 13
Irrwanyere, *183*
Italian Hydrogeography Service, 132
Italy, 112
   karst springs in, 136
IWRM. *See* integrated water resource management

Jagalingou People, 2
Japan, 51–52, 54, 145–146
   results of study in, 64–65

Japanese Environmental Standard, 59, 65
Jinan Springs, 112
Johnston, R. H., 16
Juukan Gorge, 70

Kaibab Formation, 39
Kaibab plateau, 36, 38
Kanivetsky, R., *147*
Karst Disturbance Index, 138
karst springs, 35–36
   in Alburni Massif, 135–136, *139*
   anthropogenic activity impacting, 138
   area of study on, 36–38
   base-flow recession analysis and, 39–40
   caves in karst systems, **136**
   classification of, **38**
   drainage properties, 44–45
   endorheic areas in, 133
   environmental tracers in, 88
   in FAS, 171
   hydrogeology of, 38–39, 131–132
   hydrograph analysis of, 39, 40–41
   in Italy, 136
   in Mediterranean region, 112
   recharge in, 133
   regression modeling and, 39, 41
   research methods, 38–39
   results of study on, 39–46
   seasonal water storage, 45–46
   snowpack and, 46
   stable isotope analysis and, 39
Kasahara, S., 146
Kati Thanda, 3, 4, *183*
   locations of bores and springs, *75*
   naming of, 70
Kati Thanda Springs, 70
Katz, B. G., 12, 28, 210
Keane, D., 77
Kenya, 54
Keppel, Mark, 3–4, 71, 182, 189
Kessler, H., 138
Kiku, *55*, 66
   tea leaf production in, 59–62
   water chemistry at, 59–62
Klein, H., 21
Knight, Robert L., 2, 15
Konikow, L. F., 73
Koohbor, 124
Kralik, Martin, 100
Krause, R. E., 22, 27
Krauthausen site, 93–94
Krcmar, D., 152
Kress, A., 147
Krothe, N. C., 45

krypton, 91
K-S, 71
Kumar, C., 168
Kuniansky, E. L., 21

Laban Spring, 114, 117, 118, 120
   BTC at, *122*
   hydrogeology of, **115**
Lake Eyre, *183*. *See also* Kati Thanda
Lake Eyre Basin Water Knowledge Project (LEBWKP), 188
Lake Frome, 182
Landscapes South Australia Act, 186
La Niña, 36, 160
Lautem district, 159
Lawton, J. E., 147
Leake, S. A., 73
Lebanon, 112. *See also specific topics*
   conceptual model construction, **116**
   data collection on, 115–117
   environmental tracers in study on, 117, *121*
   field site in study of, 113–115
   high-resolution data in study on, 118
   hydrograph analysis in study on, *118*
   investigation methods in study of, 115–118
   MP analysis in, 117–118
   numerical models in study in, 122–124
   results of study on, 118–124
   spring intrinsic vulnerability in, 119–120
   spring specific vulnerability in, 120–122
   wastewater treatment plants in, 114–115
LEBWKP. *See* Lake Eyre Basin Water Knowledge Project
Lee, E. S., 45
Leibundgut, C., 93
linear regression models, *42*
Lithia Spring, 201
Little Minnehaha Falls (LMF), 147, *148*, 149, 151
Local Meteoric Water Line (LMWL), 45–46
Longnecker, J., 147
Lowess curve, 23, *24*

Maillet's equation, 39
Makinohara Plateau, 59
MAL. *See* maximum admissible limits
Maloszewski, P., 97, 98
Mann-Kendall (MK) test, 24
Marella, Richard, 13
Markolf, S. A., 163, 164
Mattson, Robert A., 3, 173
maximum admissible limits (MAL), 117
Mazor, Emmanuel, 91
McLachlan spring, 77
mean residence time (MRT), 98, 101

Mediterranean region
   climate change and, 111–112
   karst springs in, 112
Meeks, Angeline, 2
Meinzer, O. E., 171
Menberg, K., 146
micropollutants (MPs), 112–113
   analysis of, in Lebanon, 117–118
   in Nahr El Kalb River, *114*, **123**
   in Qachqouch Spring, 122
MIKE SHE, 125
Miller, J. A., 21
Mineral Spring Advisory Committee (MSAC), 200
Mineral Spring region, central Victoria
   bacterial monitoring of, 202, **202**
   changes after remediation, **200**
   climate and spring hydrology, 203–204
   monitoring history, 199–200
   remediation outcomes, 204–205, *205*
   spring flow systems and occurrence, 201–202
mining, 2
Minneapolis, 145–146
   background on study of, 147–149
   geological cross section of, at Schieks Cave, *150*
   groundwater temperatures in, 146
   SUHIs in, 152
Minnesota Pollution Control Agency, 146
Minnesota Well Index (MWI), 149
Miocene, 184
MK test. *See* Mann-Kendall test
M. L. & T. Co., 149
MODFLOW, 152
Mogollon Rim, 36–37
monsoon recession events, **41**
Montecollina Bore, *185*
Moore, P. J., 27
Morita, A., 53, 62
Mound Springs, 3
MPs. *See* micropollutants
MRT. *See* mean residence time
MSAC. *See* Mineral Spring Advisory Committee
Mudd, Gavin M., 3, 77
Mulligan Spring complex, *183*
multiparameter probes, 116
Municipal Water, Sanitation and Environment Services (SMASA), 159
MWI. *See* Minnesota Well Index

N_2O emissions, reduction of, 54
NADP. *See* National Atmospheric Deposition Program
Nahr El Kalb River, 113
   MPs in, *114*, **123**
Nakasone, H., 53

National Atmospheric Deposition Program (NADP), 13
National Directoriate for Water Services (DNSA), 159
National Health and Medical Research Council (NHMRC), 200
National Oceanic and Atmospheric Administration (NOAA), 29
National Suco Development Program (PNDS), 159
National Water and Sanitation Authority (ANAS), 162
National Water Resources Management Policy, 161–162
Native Titles, 70
Nerantzaki, S. D., 112
Neuendorf, K. K. E., 22
NGOs. *See* nongovernment organizations
NHMRC. *See* National Health and Medical Research Council
Nicollet, J. N., 146
Nikolaidis, N. P., 112
nitrates, 11–12, 16
   in Florida Springs Region, *12*
   origin of, 92
nitrite, 52
nitrogen, 172
   absorption of, in tea leaf production, 53–54
   Blue Water Audit tool for estimation of, 12–13
   combination land use and contamination by, 65
nitrogen fertilizer
   acidic groundwater pollution from, 52–53
   reduction of, 65
   in tea leaf production, 51–53
   toxicity, 52
nitrogen loading
   BWA estimation of, **14**, *14*
   in Florida Springs Region, 13
Nitrogen Source Inventory Loading Tool (NSILT), 12–13
NOAA. *See* National Oceanic and Atmospheric Administration
noble gases, 91
   as environmental tracers, 95–96
nongovernment organizations (NGOs), in Timor-Leste, 158
North Canyon Springs, 39, 40, *41*, 42, **42**, *43*, 44
Northwest Florida Water Management District (NWFWMD), 26, 30
Norwest Fault Zone, 73, 77, 80
NSILT. *See* Nitrogen Source Inventory Loading Tool
NWFWMD. *See* Northwest Florida Water Management District

OAW. *See* Old Artesian Well
Odum, H. T., 175
oil industry, 2
Old Artesian Well (OAW), 149, *149*, 151
Old Finniss Spring Vents, 77
Olympic Dam, 70, 82, 184
   GAB wellfields, 72–73
   location of, *74*
Oodnadatta Track, 184

Ordovician Platteville-Glenwood Formation (OPG), *147*
Osaka, 145–146, 151
Overland Telegraph, 184
oxygen, 95
ozone layer, 97

Paris, 146
Paris Observatory, 152
passive encroachment
   drivers of, 29–30
   in FAS, 28–30
   rainfall and, 29
   sea-level rise and, 29–30
Pedirka basin, 187
periphyton, 172–173. *See also* benthic algae
Permatil, 158, 162–163, 165
Pertosa Cave, 138
phosphorus, 59–62
Pirdalinha, 182, *183*
piston flow model (PM), 98
Plan International, 164
Platteville Limestone, 146, 149, 151
   groundwater temperatures in, 152
Pleistocene Epoch, 28
PM. *See* piston flow model
PNDS. *See* National Suco Development Program
Pointon, R. K., 195
Policy resolution No. 38/2020, Timor-Leste, 162
Ponce, V. M., 29
Potamogeton crispus, 174
Power, M. E., 173–174
Prinos, S. T., 22
PRISM, 38
Proceedings of the Royal Society of Queensland, 70

Qachqouch Spring, 113, 122
   flood waters of, 118
   hydrogeology of, **115**
   MPs in, 122
   spring intrinsic vulnerability, 119–120
Quinlan, E. L., 175

radioactive decay, environmental tracers and, 91
radiocarbon, 93–94, 96–97
radon, 95–96
rainfall
   passive encroachment and, 29
   recharge and, 29
Randolph, R. B., 22, 27
Ransley, T. R., 188
Raper, S. C. D., 30
Rapid Periphyton Survey (RPS), 177
recharge coefficient, 133, 137–138

Red Cross, 164
Redwall-Muav aquifer, 38
reference conditions, 172–173
Regional Inventory of Caves of Campania, 136
regional-Kendall (RK) test, 24
    results of, **26**, **27**
regression modeling, karst springs and, 39, 41
resource extraction, 2
retention basins, *163*
RK test. *See* regional-Kendall test
Robber's Roost, 39, 40, *41*, 42, **42**, *43*, *44*, 45
Rock Springs Run, 175
Rossini, R. A., 70, 195
Roxby Downs, 73
RPS. *See* Rapid Periphyton Survey
Rutherford, J., 89

SAAL Landscape Region. *See* socioecological-technical systems; South Australia Arid Lands Landscape Region
Safe Water Drinking Act, 16
Sagittaria kurziana, 173
St. Johns River Water Management District (SJRWMD), 22, 26, 30, 174
St. Paul, 145
St. Peter Sandstone, *148*, 149, 152
salinity. *See also* encroachment
    in FAS, 27–28
SA Mining Act, 192
Sangun Formation, 65
sanitation, 2
Sauer, 39
SAV. *See* submerged aquatic vegetation
Schieks Cave, 146
    aquifer thickness at, *151*
    background on, 147
    concrete chamber in, *148*
    geological cross section of Minneapolis at, *150*
    geology of, 149
    groundwater temperatures in, 151–152
    location of, *147*
    maps of, *147*, *148*
    methods of study on, 149
    Old Artesian Well, 149, *149*
    results of study in, 149–151
    thermal anomaly in, 147–149, 151
    thermometric survey of, **150**
    Washington Avenue tunnel, *147*, *148*
Schwarza River, *99*, 101
Scott, T. M., 21, 22, 28
SDG. *See* Sustainable Development Goals
sea-level rise, passive encroachment and, 29–30
SEGS. *See* Southeastern Geological Society
Seltzer, A. M., 91

Semipermeable Fault Zone, *78*
Seoul, 151
SETS. *See* socioecological-technical systems
Shand, P., 189
Shimizu, *54*, *55*, *56*, *57*, 62–63
Shizuoka Prefecture, 54, 66
    water chemistry at, 59–62
Shugg, Andrew, 4, 201
Silver Glen Run, 175, 178
Silver Springs, 11–12, 175
SJRFWMD. *See* St. Johns River Water Management District
S. Maria karst system, 139
SMASA. *See* Municipal Water, Sanitation and Environment Services
Smerdon, B. D., 188
snapshots, environmental tracer, 94
snowpack, karst springs and, 46
socioecological-technical systems (SETS), 163, 164
    water system resilience and, **167**
Solomon, Kip, 98, 100
solute origin, environmental tracers and, 92
South Australia Arid Lands (SAAL) Landscape Region, 184, 192
Southeastern Geological Society (SEGS), 21
Southern Vienna Basin (SVB), 99–100
Southwest Florida Water Management District (SWFWMD), 22, 26, 30
Spechler, R. M., 22, 27
Special Water Licenses for Wellfield A, 73
spring flow, Wellfield A and, 73–80
spring flows, groundwater levels correlated with, 78–80
spring intrinsic vulnerability
    in Lebanon, 119–120
        Qachqouch Spring, 119–120
        upper Catchment Springs, 119–120
spring monitoring sites, *20*, **23**
    statistical methods, 24–25
spring specific vulnerability, in Lebanon, 120–122
stable isotope analysis, *43*, *44*
    collection for, 116–117
    karst springs and, 39, 42
Stefan, H. G., 145, 146, 151
Stewarts Creek, 203
Stolp, B., 100
Strategic Management Plan 2020, 80
strontium, 92
submerged aquatic vegetation (SAV), 172
subsurface urban heat island (SUHI), 146
    in Minneapolis, 152
Suckow, Axel, 3, 98
SUHI. *See* subsurface urban heat island
Supai Formation, 40
Surat basin, 71
Sustainable Development Goals (SDG), 115, 158

Sustainable Water in Timor-Leste Municipalities (SWiM), 159, 160, 164, 167
   survey data, **165**
Suwannee River Water Management District (SRWMD), 22, 28
SVB. *See* Southern Vienna Basin
SWFWMD. *See* Southwest Florida Water Management District
SWiM. *See* Sustainable Water in Timor-Leste Municipalities
SWRMD. *See* Suwannee River Water Management District
Szafraniec, M. L., 173

Taiwan, 52, 53–54
Tanagro River, 138
Taniguchi, M., 145–146, 151
Tanno Pond, 59
Tanno River, 59
tara bandu, 168
Taylor, C. A., 145, 146, 151
TAZ. *See* thermally affected zones
tea leaf production, *55*
   in catchments, 62–63
   groundwater chemistry changes and, 63–65
   in Kiku, 59–62
   methods for study on, 54
   nitrogen absorption in, 53–54
   nitrogen fertilizer in, 51–53
   soil water chemistry and, 54–59
   on volcanic loam, 54–59
ternary mixing models, 117
Thailand, 145–146
Theis, C. V., 29, 73
thermally affected zones (TAZ), 152
Thirrka, 182
timescales
   available, 92–93
   for environmental tracers, 88
   of water movement and environmental tracers, 92–99
time series
   environmental tracer, 94
   tritium, *101*
Timor-Leste, 3
   CBM in, 159–160
   CBWRM in, 158, 162–163, 166–169
   community member interviews, *165*
   Decree Law 03/2016, 162
   Decree Law No. 4/2004, 158–159
   Escherichia coli in, *161*
   flooding in, 166
   government of, 158, 162
   Government Resolution 43/2020, 158
   hydrogeology of, 160
   IWRM in, 161–162
   methodology of study in, 164–166
   NGOs in, 158
   Policy resolution No. 38/2020, 162
   Strategic Development Plan 2011-2030, 158
   water infrastructure in, 158–159, *159*
   water quality in, 160–161
   water quantity in, 160
   water security in, 157, 159–161
Tissen, C., 151
Tokyo, 151
Toroweap Formation, 39
tracer breakthrough curve (BTC), 119–120, *120*
   at Assal Spring, *122*
   at Laban Spring, *122*
Trichoderma viride, 54
tritium, 94, 95–96
   time series, *101*
Twin Cities, 145
Two-Region Nonequilibrium model (2NREM), 119

UNESCO, 138
UN-Water, 157
Upchurch, S. B., 27, 28
Upper Catchment Springs, spring intrinsic vulnerability, 119–120
urban development, groundwater temperatures and, 145–146

Vacher, H. L., 21
Vallone Lontrano-Petina, 139
variable saturated flow (VSF) model, 124
Verdi, R. J., 29
Victorian Mineral Water Committee (VMWC), 200
Vogel, J. C., 98
volcanic loam, tea leaf production on, 54–59
VSF model. *See* variable saturated flow model
Vulnerability Assessment of Climate Change Impacts on Groundwater Resources in Timor-Leste, 160

Wabma Kadaru National Park, 182, *183*
Wacissa River, 178
Wallace, 160
Walton, 29–30
Wangan People, 2
Waralanha, *183*
Washington Avenue tunnel, Schieks Cave, 147, *148*
waste management, 2
wastewater treatment plants, in Lebanon, 114–115
water, age of, 92
   at Fischa-Dagnitz spring, 100–101, *103*
WaterAid, 160
Water Information Network (WIN), 22
water infrastructure, in Timor-Leste, 158–159, *159*
water management districts (WMDs), 13, 19. *See also specific topics*

water quality, 157
   in Florida Springs Region, 10–11, 172
   in Timor-Leste, 160–161
water quantity, in Timor-Leste, 160
Water Resources Act, Australian, 190
water scarcity, 112
water security
   challenges, 159–161
   defining, 157
   socioecological-technical systems for, 163–164
   in Timor-Leste, 157, 159–161
water sources, springs as, 1
Wellfield A, GAB
   groundwater levels and, 73–80
   hydrological impacts due to, **81**
   impacts of, on springs and bores, 73–82
   location of, *74*
   Special Water Licenses for, 73
   spring flow, 73–80
   water extraction by, *76*
Wellfield B, GAB, 73
   development of, 77
   location of, *74*
   water extraction by, *76*
Werner, A. S., 28
Whitford, L. A., 175

WHO. *See* World Health Organization
Wigley, T. M. L., 30
Williams, L. J., 21
WIN. *See* Water Information Network
Winnipeg, 146
WMDs. *See* water management districts
Woeber, Andy, 20, 22, 24
Wood, A. J., 38
Woodbury, A. D., 146
World Health Organization (WHO), 53, 161
World Vision, 164
Wyuna Spring, 201

Xanthophyta, 172
xenon, 91
Xu, Z., 22

Yalcin, T., 146
Yamamoto, T., 53
Yamazaki, S., 53
Yame district, *54*, 63
Yetemen, O., 146

Zalusky, 147, 149
Zhu, K., 146
Zuber, A., 98